CMake Best Practices

Discover proven techniques for creating and maintaining programming projects with CMake

Dominik Berner

Mustafa Kemal Gilor

BIRMINGHAM—MUMBAI

CMake Best Practices

Associate Group Product Manager: Richa Tripathi
Publishing Product Manager: Gebin George
Senior Editor: Kinnari Chohan
Technical Editor: Maran Fernandes
Copy Editor: Safis Editing
Project Coordinator: Manisha Singh
Proofreader: Safis Editing
Indexer: Rekha Nair
Production Designer: Aparna Bhagat
Marketing Coordinator: Sonakshi Bubbar

First published: May 2022

Production reference: 1110522

Published by Packt Publishing Ltd.
Livery Place
35 Livery Street
Birmingham
B3 2PB, UK.
ISBN 978-1-80323-972-9

www.packt.com

To all the extraordinarily talented people who helped me in pursuing the mastery of the craft of software development.

– Dominik Berner

To my loving wife, to my family, and my friends. Except for the math teacher who yelled at me in fourth grade, and the people who thought investing time in software engineering was not going to be worth it. Well, guess there was no objection in saying that; that thought has not aged very well.

– Mustafa Kemal Gilor

Contributors

About the authors

Dominik Berner is a software engineer, blogger, and conference speaker with 20 years of professional software development under his belt. He codes mainly in C++ and has worked on many software projects, from writing bleeding-edge software for surgical simulators in a startup and maintaining large legacy platforms for large corporations in the MedTech industry to creating IoT solutions for companies in between. He believes that well-designed and maintained build environments are one of the key elements to enabling teams to write software efficiently and excel at creating quality software. When he is not writing code, he occasionally writes articles for his blog or speaks at conferences about software development.

"I would like to thank the people who contributed to this book: my co author, Mustafa Gilor, for complementing my expertise, Kinnari Chohan for being a superb editor, Gebin George for kickstarting this book, and the whole team at Packt. Many thanks go to our technical reviewers, Richard von Lehe, Horváth V., and Toni Solarin-Sodara. And finally, thanks to my family, Brigitte, Alice, and Theo, for having the patience and giving me all their support when writing this book!"

Mustafa Kemal Gilor is an experienced professional working in performance-critical software development for the telecommunication and defense industries and open source software. His expertise is in high-performance and scalable software design, network technologies, DevOps, and software architecture. His interest in computers emerged very early on in his childhood. He learned programming to hack MMORPG games at around the age of 12, and he has been writing software ever since. His favorite programming language is C++, and he enjoys doing framework design and system programming. He is also a strong advocate of CMake; he has maintained many code bases and ported many legacy projects to CMake throughout his career.

"First, I'd like to thank my co author, Dominik Berner, Packt Publishing, and all the technical reviewers who made this book possible with their valuable work. I also would like to thank the kindest and most understanding person in my life – my wife, Büşra – for providing me with the support and encouragement I needed during the writing phase. Lastly, I'd like to give credit to my family and also my friends Gürcan Pehlevan, Ramazan Cömert, Mustafa Hergül, Habip İşler, and Ahmet Aksoy for believing in me and backing me up throughout the process."

About the reviewers

Richard Von Lehe lives in Minnesota in the Twin Cities area. He has spent considerable time using CMake in the past several years on software projects, including orthodontic modeling, building control, drone collision avoidance, and specialty printers. In his free time, he enjoys relaxing with his family and their pet rabbit, Gus. He also enjoys biking and playing guitar.

Toni Solarin-Solada is a software engineer specializing in the design of cross-platform programming libraries that abstract low-level operating system services.

Table of Contents

3

Creating a CMake Project

Part 2: Practical CMake – Getting Your Hands Dirty with CMake

4

Packaging, Deploying, and Installing a CMake Project

5

Integrating Third-Party Libraries and Dependency Management

6

Automatically Generating Documentation with CMake

7

Seamlessly Integrating Code Quality Tools with CMake

8

Executing Custom Tasks with CMake

9

Creating Reproducible Build Environments

10

Handling Big Projects and Distributed Repositories in a Superbuild

11
Automated Fuzzing with CMake

Part 3: Mastering the Details

12
Cross-Platform Compiling and Custom Toolchains

13
Reusing CMake Code

14
Optimizing and Maintaining CMake Projects

15
Migrating to CMake

16
Contributing to CMake and Further Reading Material

Preface

The software world and the tooling we use to create software are evolving every single day. CMake is no exception here, as after over 20 years of constantly evolving, it can now be considered something of an industry standard when it comes to building C++ applications. But while CMake is very feature-rich and its documentation is very comprehensive, real-world examples and guidelines on how to use the features together are rare. This is where *CMake Best Practices* jumps in.

Instead of explaining every last detail and feature of CMake, this book contains examples to illustrate how CMake is best for various tasks when building software without covering every single edge case. There are other books for that. The aim of this book is to keep things as simple as possible while covering the recommended best practices for getting things done. The rationale behind this approach is that you don't need to know about all of CMake's capabilities to achieve your everyday tasks.

We will try to explain a concept first and then illustrate it with concrete examples. This way, you will learn by practice and be able to apply that knowledge to your daily work with CMake. Since the audience of this book will be mostly engineers, we have tailored the book's content accordingly. While writing this book, we are software engineers first and then authors. As a result, the content of the book is more practical than theoretical. It is a compendium of carefully selected, proven techniques that you can use in your everyday CMake workflow.

From engineers to engineers, we hope that you enjoy this book.

Who this book is for

This book is for software engineers and build-system maintainers working with C or C++ on a regular basis and trying to use CMake to improve their everyday tasks. Basic C++ and general programming knowledge will help you to better understand the examples covered in the book.

What this book covers

Chapter 1, Kickstarting CMake, explains what CMake is in a nutshell and then jumps right into installing CMake and building something with CMake. You will learn how to install the latest stable version manually, even if it's not provided by your package manager. You'll also learn about the basic concepts behind CMake and why it is a build system generator and not a build system itself. You will learn how it fits into modern software development with C++ (and C).

Chapter 2, Accessing CMake in the Best Ways, shows how to best use CMake from the command line with a GUI and how CMake integrates with some common IDEs and editors.

Chapter 3, Creating a CMake Project, takes you through setting up a project to build an executable and a library and linking the two together.

Chapter 4, Packaging, Deploying, and Installing a CMake Project, shows you how to create a distributable version of your software project. You will learn how to add installation instructions and package the project using CMake and CPack (CMake's packaging program).

Chapter 5, Integrating Third-Party Libraries and Dependency Management, explains how to integrate existing third-party libraries into your project. It also shows you how to add libraries already installed on your system, external CMake projects, and non-CMake projects.

Chapter 6, Automatically Generating Documentation, explores how to generate documentation from your code with CMake as part of the build process with doxygen, dot (graphviz), and plantuml.

Chapter 7, Seamlessly Integrating Code-Quality Tools with CMake, shows you how to integrate unit testing, code sanitizers, static code analysis, and code coverage tools into your project. It will show you how CMake can help to discover and execute tests.

Chapter 8, Executing Custom Tasks with CMake, explains how you can integrate almost any tool into your build process. You will learn how to wrap external programs into custom targets or hook them into the build process to execute them. We will cover how custom tasks can be used to generate files and how they can consume files produced by other targets. You will also learn how to execute system commands during the configuration of the CMake build and how to create platform-agnostic commands using the CMake script mode.

Chapter 9, Creating Reproducible Build Environments, shows how you can build an environment portable between various machines including CI/CD pipelines, and how to work with Docker, sysroots, and CMake presets to make your build work "out of the box" everywhere.

Chapter 10, Handling Big Projects and Distributed Repositories in a Superbuild, simplifies managing projects that are distributed across multiple git repositories with CMake. You will learn how to create a super-build that allows you to build specific versions as well as the latest nightly builds. You will explore what prerequisites it needs and how to combine them.

Chapter 11, Automated Fuzzing with CMake, shows how you can integrate and use fuzzing tools with CMake.

Chapter 12, Cross-Platform Compiling and Custom Toolchains, demonstrates how you can use cross-platform toolchains. You will also learn how to write your own toolchain definitions and conveniently use different toolchains with CMake.

Chapter 13, Reusing CMake Code, explains CMake modules and how you can generalize your CMake files. You will learn how to write broadly used modules, which you can ship individually from your project.

Chapter 14, Optimizing and Maintaining CMake Projects, suggests how to get faster build times and provides tips and tricks for keeping a CMake project neat and tidy over a long period of time.

Chapter 15, Migrating to CMake, explains a high-level strategy on how to migrate a large existing codebase to CMake without the need to completely stop your development.

Chapter 16, Contributing to CMake and Further Reading Material, suggests where to go if you want to contribute, what is looked for, and basic contributing guidelines. It will also guide you on where to find additional in-depth information or more specific literature.

To get the most out of this book

You will need CMake version 3.21 or newer and a modern C++ compiler that understands at least C++14 installed on your computer to run the examples. Some examples may require additional software to run, which will be mentioned in the relevant chapters. All software used for the examples are open source and available for free.

Software/hardware covered in the book	Operating system requirements
CMake 3.21 or newer	Linux, Windows, or macOS
GCC, Clang, or MSVC	Linux, Windows, or macOS
git	Linux, Windows, or macOS

If you are using the digital version of this book, we advise you to type the code yourself or access the code from the book's GitHub repository (a link is available in the next section). Doing so will help you avoid any potential errors related to the copying and pasting of code.

Download the example code files

You can download the example code files for this book from GitHub at `https://github.com/PacktPublishing/CMake-Best-Practices`. If there's an update to the code, it will be updated in the GitHub repository.

We also have other code bundles from our rich catalog of books and videos available at `https://github.com/PacktPublishing/`. Check them out!

Download the color images

We also provide a PDF file that has color images of the screenshots and diagrams used in this book. You can download it here: `https://static.packt-cdn.com/downloads/9781803239729_ColorImages.pdf`

Conventions used

There are a number of text conventions used throughout this book.

`Code in text`: Indicates code words in text, database table names, folder names, filenames, file extensions, pathnames, dummy URLs, user input, and Twitter handles. Here is an example: "`FILES_MATCHING` cannot be used after `PATTERN` or `REGEX` but it can be done vice versa."

A block of code is set as follows:

```
include(GNUInstallDirs)
install(DIRECTORY dir1 DESTINATION ${CMAKE_INSTALL_
  LOCALSTATEDIR} FILES_MATCHING PATTERN "*.x")
install(DIRECTORY dir2 DESTINATION ${CMAKE_INSTALL_
  LOCALSTATEDIR} FILES_MATCHING PATTERN "*.hpp"
    EXCLUDE PATTERN "*")
install(DIRECTORY dir3 DESTINATION ${CMAKE_INSTALL_
  LOCALSTATEDIR} PATTERN "bin" EXCLUDE)
```

When we wish to draw your attention to a particular part of a code block, the relevant lines or items are set in bold:

```
# ...
-- Installing: /tmp/install-test/qbin/ch4_ex01_executable
```

Any command-line input or output is written as follows:

```
install(DIRECTORY dir1 dir2 dir3 TYPE LOCALSTATE)
```

Bold: Indicates a new term, an important word, or words that you see onscreen. For instance, words in menus or dialog boxes appear in **bold**. Here is an example: "To start configuring a project, select the project's root directory by clicking the **Browse Source...** button."

> **Tips or Important Notes**
> Appear like this.

Get in touch

Feedback from our readers is always welcome.

General feedback: If you have questions about any aspect of this book, email us at `customercare@packtpub.com` and mention the book title in the subject of your message.

Errata: Although we have taken every care to ensure the accuracy of our content, mistakes do happen. If you have found a mistake in this book, we would be grateful if you would report this to us. Please visit `www.packtpub.com/support/errata` and fill in the form.

Piracy: If you come across any illegal copies of our works in any form on the internet, we would be grateful if you would provide us with the location address or website name. Please contact us at `copyright@packt.com` with a link to the material.

If you are interested in becoming an author: If there is a topic that you have expertise in and you are interested in either writing or contributing to a book, please visit `authors.packtpub.com`.

Share Your Thoughts

Once you've read *CMake Best Practices*, we'd love to hear your thoughts! Scan the QR code below to go straight to the Amazon review page for this book and share your feedback.

https://packt.link/r/1-803-23972-7

Your review is important to us and the tech community and will help us make sure we're delivering excellent quality content.

Part 1: The Basics

In the first chapter, you will learn how to invoke CMake and get a high-level overview of its basic concepts, as well as a brief introduction to the CMake language.

The second chapter will be about using CMake from the command line, as a GUI, or in various IDEs and editors. It will illustrate how to change various configuration options and select different compilers.

In *the third chapter*, we will cover creating a simple CMake project to build executables and libraries.

This part contains the following chapters:

- *Chapter 1, Kickstarting CMake*
- *Chapter 2, Accessing CMake in the Best Ways*
- *Chapter 3, Creating a CMake Project*

1
Kickstarting CMake

If you're developing software using C++ or C, you have probably heard about CMake before. Over the last 20 years, CMake has evolved into something that's an industry standard when it comes to building C++ applications. But CMake is more than just a build system – it is a build system generator, which means it produces instructions for other build systems such as Makefile, Ninja, Visual Studio, Qt Creator, Android Studio, and Xcode. And it does not stop at building software – CMake also includes features that support installing, packaging, and testing software.

As a de facto industry standard, CMake is a must-know technology for any C++ programmer.

In this chapter, you will get a high-level overview of what CMake is and learn about the necessary basics to build your first program. We will have a look at CMake's build process and provide an overview of how to use the CMake language to configure build processes.

In this chapter, we will cover the following topics:

- CMake in a nutshell
- Installing CMake
- The CMake build process
- Writing CMake files
- Different toolchains and build configurations

Let's begin!

Technical requirements

To run the examples in this chapter, you will need a recent C++ compiler that understands C++17. Although the examples are not complex enough to require the functionality of the new standard, the examples have been set up accordingly.

We recommend using any of the compilers listed here to run the examples:

- **Linux**: GCC 9 or newer, Clang 10 or newer
- **Windows**: MSVC 19 or newer or MinGW 9.0.0 or newer
- **macOS**: Apple Clang 10 or newer

> Note
>
> To try out any examples in this book, we have provided a ready-made Docker container that contains all the requirements.
>
> You can find it at `https://github.com/PacktPublishing/CMake-Best-Practices`.

CMake in a nutshell

CMake is open source and available on many platforms. It is also compiler-independent, making it a very strong tool when it comes to building and distributing cross-platform software. All these features make it a valuable tool for building software in a modern way – that is, by relying heavily on build automation and built-in quality gates.

CMake consists of three command-line tools:

- `cmake`: CMake itself, which is used to generate build instructions
- `ctest`: CMake's test utility, which is used to detect and run tests
- `cpack`: CMake's packaging tool, which is used to pack software into convenient installers, such as deb, RPM, and self-extracting installers

There are also two interactive tools:

- `cmake-gui`: A GUI frontend to help with configuring projects
- `ccmake`: An interactive terminal UI for configuring CMake

`cmake-gui` can be used to conveniently configure a CMake build and select the compiler to be used:

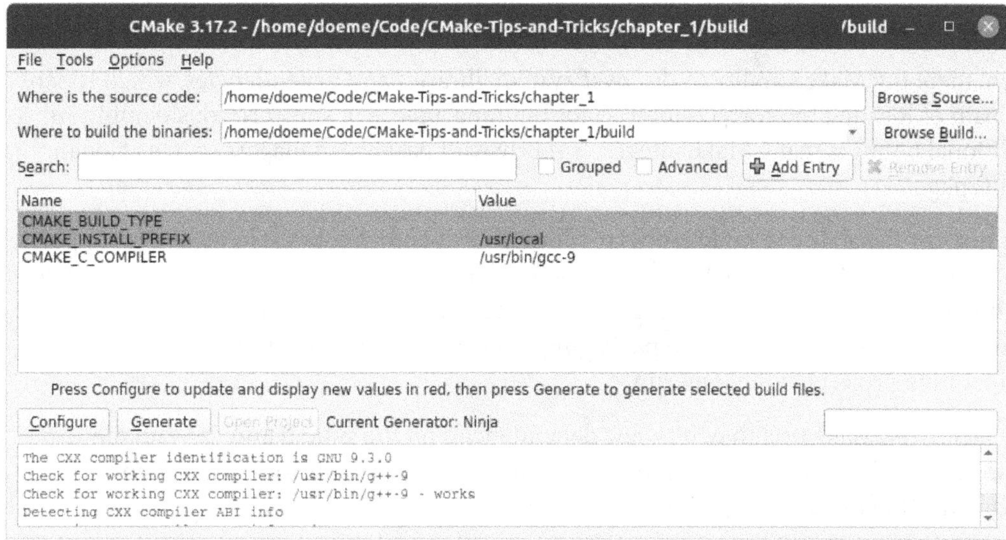

Figure 1.1 – cmake-gui after configuring a project

If you're working on the console but still want to have an interactive configuration of CMake, then `ccmake` is the right tool. While not as convenient as `cmake-gui`, it offers the same functionality. This is especially useful when you must configure CMake remotely over an `ssh` shell or similar:

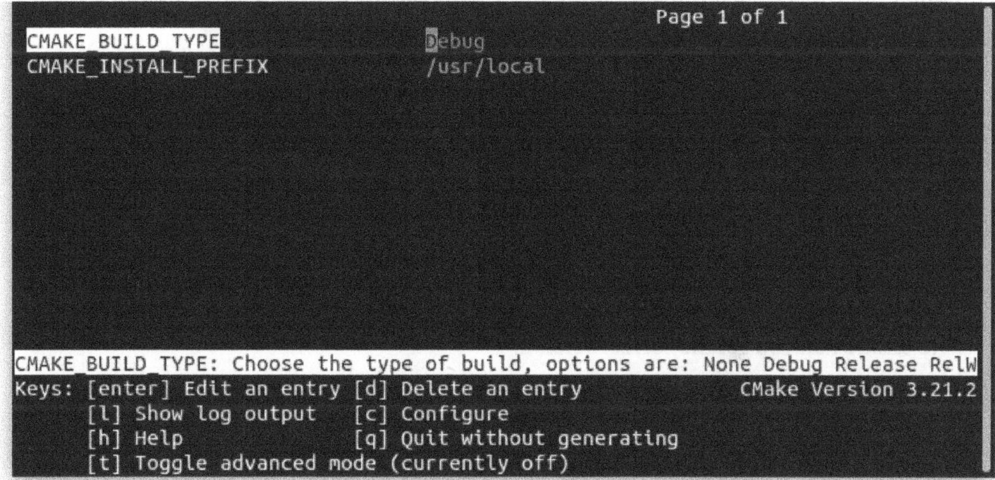

Figure 1.2 – Configuring a project using ccmake

The advantage of CMake over a regular build system is manyfold. First, there is the cross-platform aspect. With CMake, it is much easier to create build instructions for a variety of compilers and platforms without the need to know the specifics of the respective build system in depth.

Then, there is CMake's ability to discover system libraries and dependencies, which lessens the pain of locating the correct libraries for building a piece of software considerably. An additional bonus is that CMake integrates nicely with package managers such as Conan and vcpkg.

It is not just the ability to build software for multiple platforms, but also its native support for testing, installing, and packaging software that makes CMake a much better candidate for building software than just a single build system. Being able to define everything from building and over-testing to packaging at a single point helps tremendously with maintaining projects in the long run.

The fact that CMake itself has very few dependencies on the system and can run on the command line without user interaction makes it very suitable for build system automatization in CI/CD pipelines.

Now that we've covered briefly what CMake can do, let's learn how to install CMake.

Installing CMake

CMake is freely available to download from `https://cmake.org/download/`. It is available as either a precompiled binary or as source code. For most use cases, the precompiled binary is fully sufficient, but since CMake itself has very few dependencies, building a version is also possible.

Any major Linux distribution offers CMake over its package repositories. Although the pre-packaged versions of CMake are not usually the latest releases, these installations are often sufficient to use if the system is regularly updated.

> **Note**
>
> The minimum version of CMake to use with the examples in this book is 3.21. We recommend that you download the appropriate version of CMake manually to ensure that you get the correct version.

Building CMake from source

CMake is written in C++ and uses Make to build itself. Building CMake from scratch is possible, but for most use cases, using the binary downloads will do just fine.

After downloading the source package from `https://cmake.org/download/`, extract it to a folder and run the following command:

```
./configure make
```

If you want to build `cmake-gui` as well, configure it with the `--qt-gui` option. This requires Qt to be installed. Configuring will take a while, but once it's succeeded, CMake can be installed using the following command:

```
make install
```

To test whether the installation was successful, you can execute the following command:

```
cmake --version
```

This will print out the version of CMake, like this:

```
cmake version 3.21.2
CMake suite maintained and supported by Kitware (kitware.com/
cmake).
```

Building your first project

Now, it's time to get your hands dirty and see whether your installation worked. We have provided an example of a simple `hello world` project that you can download and build right away. Open a console, type in the following, and you'll be ready to go:

```
git clone https://github.com/PacktPublishing/CMake-Best-
Practices.git
cd CMake-Best-Practices/chapter_1
mkdir build
cd build
cmake ..
cmake --build .
```

This will result in an executable called `Chapter_1` that prints out `Welcome to CMake Best Practices` on the console.

Let's have a detailed look at what happened here:

1. First, the example repository is checked out using Git and then the build folder is created. The file structure of the example CMake project will look like this before the build:

   ```
   .
   ├── CMakeLists.txt
   └── build
   └── src
       └── main.cpp
   ```

   Apart from the folder containing the source code, there is a file called CMakeLists.txt. This file contains the instructions for CMake on how to create build instructions for the project and how to build it. Every CMake project has a CMakeLists.txt file at the root of the project, but there might be many files with that name in various subfolders.

2. After cloning the repository, the build process is started with cmake. CMake's build process is a two-stage process. The first step, which is usually called *configuration*, reads the CMakeLists.txt file and generates an instruction for the native build toolchain of the system. In the second step, these build instructions are executed and the executables or libraries are built.

   During the configuration step, the build requirements are checked, the dependencies are resolved, and the build instructions are generated.

3. Configuring a project also creates a file called CMakeCache.txt that contains all the information that's needed to create the build instructions. The next call to cmake --build . executes the build by internally calling CMake; if you are on Windows, it does so by invoking the Visual Studio compiler. This is the actual step for compiling the binaries. If everything went well, there should be an executable named Chapter1 in the build folder.

For brevity, we cd'd into the build directory in the previous examples and used relative paths to find the source folders. This is often convenient, but if you want to call CMake from somewhere else, you can use the --S option to select the source file and the --B option to select the build folder:

```
cmake -S /path/to/source -B /path/to/build
cmake -build /path/to/build
```

Explicitly passing the build and source directories often comes in handy when using CMake in a continuous integration environment since being explicit helps with maintainability. It is also helpful if you want to create different build directories for different configurations, such as when you're building cross-platform software.

A minimal CMakeLists.txt file

For a very simple `hello world` example, the `CMakeLists.txt` file only consists of a few lines of instructions:

```
cmake_minimum_required(VERSION 3.21)

project(
  "chapter1"
  VERSION 1.0
  DESCRIPTION "A simple project to demonstrate basic CMake
    usage" LANGUAGES CXX)

add_executable(Chapter1)
target_sources(Chapter1 PRIVATE src/main.cpp)
```

Let's understand these instructions in a bit more detail:

- The first line defines the minimum version of CMake that's required to build this project. Every `CMakeLists.txt` file starts with this directive. This is used to warn the user if the project uses features of CMake that are only available from a certain version upward. Generally, we recommend setting the version to the lowest version that supports the features that are used in the project.
- The next directive is the name, version, and description of the project to be built, followed by the programming languages that are used in the project. Here, we use `CXX` to mark this as a C++ project.
- The `add_executable` directive tells CMake that we want to build an executable (as opposed to a library or a custom artifact, which we will cover later in this book).
- The `target_sources` statement tells CMake where to look for the sources for the executable called `Chapter1` and that the visibility of the sources is limited to the executable. We will go into the specifics of the single commands later in this book.

Congratulations – you are now able to create software programs with CMake. But to understand what is going on behind the commands, let's look at the CMake build process in detail.

Understanding the CMake build process

CMake's build process works in two steps, as shown in the following diagram. First, if it's invoked without any special flags, CMake scans the system for any usable toolchains during the configuration process and then decides what its output should be. The second step, which is when `cmake --build` is invoked, is the actual compilation and building process:

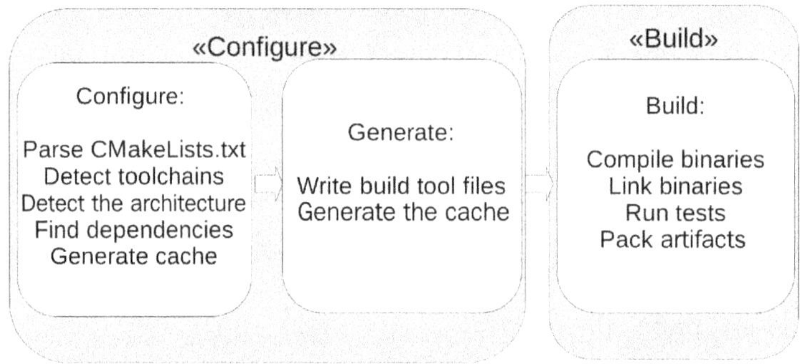

Figure 1.3 – CMake's two-stage build process

The standard output is Unix Makefiles unless the only detected compiler is Microsoft Visual Studio, in which case a Visual Studio solution (`.sln`) will be created.

To change the generator, you can pass the `-G` option to CMake, like this:

```
cmake .. -G Ninja
```

This will generate files to be used with Ninja (`https://ninja-build.org/`), an alternative build generator. Many generators are available for CMake. A list of the ones that are supported natively can be found on CMake's website: `https://cmake.org/cmake/help/latest/manual/cmake-generators.7.html`.

There are two main types of generators – the ones where there are many Makefile flavors and Ninja generators, which are generally used from the command line, and the ones that create build files for an IDE such as Visual Studio or Xcode.

CMake addition differentiates between *single-configuration generators* and *multi-configuration generators*. For single-configuration generators, the build files have to be rewritten each time the configuration is changed; multi-configuration build systems can manage different configurations without the need to regenerate. Although the examples in this book use single-configuration generators, they would also work on multi-configuration generators. For most of the examples, the chosen generator is irrelevant; otherwise, it will be mentioned:

Generator	Multi-Configuration
Makefiles (all flavors)	No
Ninja	No
Ninja Multi-Config	Yes
Xcode	Yes
Visual Studio	Yes

In addition, there are extra generators that use a normal generator but also produce project information for an editor or IDE, such as Sublime Text 2, Kate Editor, Code::Blocks, and Eclipse. For each, you can select whether the editor should use Make or Ninja to internally build the application.

After the call, CMake will create a lot of files in the `build` folder, with the most notable being the `CMakeCache.txt` file. This is where all the detected configurations are stored. Note that when you're using `cmake-gui`, the first step is split into configuring the project and generating the build file. However, when it's run from the command line, the steps are merged into one. Once configured, all the build commands are executed from the `build` folder.

Source folders and build folders

In CMake, two logical folders exist. One is the `source` folder, which contains a hierarchical set of projects, while the other is a `build` folder, which contains the build instructions, cache, and all the generated binaries and artifacts.

The root of the `source` folder is wherever the top `CMakeLists.txt` file is located. The `build` folder can be placed inside the `source` folder, but some people prefer to have it in another location. Both are fine; note that for the examples in this book, we decided to keep the `build` folder inside the `source` folder. The `build` folder is often called just `build`, but it can take any name, including prefixes and suffixes for different platforms. When using a `build` folder inside the source tree, it is a good idea to add it to `.gitignore` so that it does not get checked in accidentally.

When configuring a CMake project, the project and folder structure of the source folder is recreated inside the build folder so that all the build artifacts are in the same position. In each mapped folder, there is a subfolder called CMakeFiles that contains all the information that's generated by CMake's configuration step.

The following code shows an example structure for a CMake project:

```
├── chapter_1
│   ├── CMakeLists.txt
│   └── src
│       └── main.cpp
├── CMakeLists.txt
```

When you execute the CMake configuration, the file structure of the CMake project is mapped into the build folder. Each folder containing a CMakeLists.txt file will be mapped and a subfolder called CMakeFiles will be created, which contains the information that's used by CMake for building:

```
├── build
│   ├── chapter_1
│   │   └── CMakeFiles
│   └── CMakeFiles
```

So far, we have used existing projects to learn about the CMake build process. We learned about the configuration and the build step, as well as generators, and that we need CMakeLists.txt files to pass the necessary information to CMake. So, let's go a step further and see what the CMakeLists.txt files look like and how the CMake language works.

Writing CMake files

When you're writing CMake files, there are a few core concepts and language features that you need to know about. We won't cover every detail of the language here as CMake's documentation does a pretty good job at this – especially when it comes to being comprehensive. In the following sections, we will provide an overview of the core concepts and language features. Further chapters will dive into the details of different aspects.

The full documentation for the language can be found at https://cmake.org/cmake/help/latest/manual/cmake-language.7.html.

The CMake language – a 10,000-feet overview

CMake uses configuration files called `CMakeLists.txt` files to determine build specifications. These files are written in a scripting language, often called CMake as well. The language itself is simple and supports variables, string functions, macros, function definitions, and importing other CMake files.

Apart from lists, there is no support for data structures such as structs or classes. But it is this relative simplicity that makes the CMake project inherently maintainable if done properly.

The syntax is based on keywords and whitespace-separated arguments. For example, the following command tells CMake which files are to be added to a library:

```
target_sources(MyLibrary
                PUBLIC include/api.h
                PRIVATE src/internals.cpp src/foo.cpp)
```

The `PUBLIC` and `PRIVATE` keywords denote the visibility of the files when they're linked against this library and serve as delimiters between the lists of files.

Additionally, the CMake language supports so-called "generator expressions," which are evaluated during build system generation. These are commonly used to specify special information for each build configuration. They will be covered extensively in *Chapter 3*, *Creating a CMake Project*.

Projects

CMake organizes the various build artifacts such as libraries, executables, tests, and documentation into projects. There is always exactly one root project, although projects can be encapsulated into each other. As a rule, there should only be one project per `CMakeLists.txt` file, which means that each project has to have a separate folder in the source directory.

Projects are described like this:

```
 project(
"chapter1"
VERSION 1.0
DESCRIPTION "A simple C++ project to demonstrate basic CMake
  usage" LANGUAGES CXX
)
```

The current project that's being parsed is stored in the PROJECT_NAME variable. For the root project, this is also stored in CMAKE_PROJECT_NAME, which is useful for determining whether a project is standalone or encapsulated in another. Since version 3.21, there's also a PROJECT_IS_TOP_LEVEL variable to directly determine whether the current project is the top-level project. Additionally, with <PROJECT-NAME>_IS_TOP_LEVEL, you can detect whether a specific project is a top-level project.

The following are some additional variables regarding the projects. All of them can be prefixed with CMAKE_ to the value for the root project. If they're not defined in the project() directive, the strings are empty:

- PROJECT_DESCRIPTION: The description string of the project
- PROJECT_HOMEPAGE_URL: The URL string for the project
- PROJECT_VERSION: The full version that's given to the project
- PROJECT_VERSION_MAJOR: The first number of the version string
- PROJECT_VERSION_MINOR: The second number of the version string
- PROJECT_VERSION_PATCH: The third number of the version string
- PROJECT_VERSION_TWEAK: The fourth number of the version string

Each project has a source and binary directory, and they may be encapsulated in each other. Let's assume that each of the CMakeFiles.txt files in the following example defines a project:

```
.
├── CMakeLists.txt #defines project("CMakeBestPractices"...)
├── chapter_1
│   ├── CMakeLists.txt # defines project("Chapter 1"...)
```

When parsing the CMakeLists.txt file in the root folder, PROJECT_NAME and CMAKE_PROJECT_NAME will both be CMakeBestPractices. When you're parsing chapter_1/CMakeLists.txt, the PROJECT_NAME variable will change to "Chapter_1" but CMAKE_PROJECT_NAME will stay as CMakeBestPractices, as set in the file in the root folder.

Although projects can be nested, it is good practice to write them in a way that they can work standalone. While they may depend on other projects that are lower in the file hierarchy, there should be no need for a project to live as a child of another. It is possible to put multiple calls to `project()` in the same `CMakeLists.txt` file, but we discourage this practice as it tends to make projects confusing and hard to maintain. In general, it is better to create a `CMakeLists.txt` file for each project and organize the structure with subfolders.

This book's GitHub repository, which contains the examples in this book, is organized in a hierarchical way, where each chapter is a separate project that may contain even more projects for different sections and examples.

While each example can be built on its own, you can also build this whole book from the root of the repository.

Variables

Variables are a core part of the CMake language. Variables can be set using the `set` command and deleted using `unset`. Variable names are case-sensitive. The following example shows how to set a variable named `MYVAR` and assign a value of `1234` to it:

```
set(MYVAR "1234")
```

To delete the `MYVAR` variable, we can use `unset`:

```
unset(MYVAR)
```

The general code convention is to write variables in all caps. Internally, variables are always represented as strings.

You can access the value of a variable with the `$` sign and curly brackets:

```
message(STATUS "The content of MYVAR are ${MYVAR}")
```

Variable references can even be nested and are evaluated inside out:

```
${outer_${inner_variable}_variable}
```

Variables might be scoped in the following way:

- **Function scope**: Variables that are set inside a function are only visible inside the function.
- **Directory scope**: Each of the subdirectories in a source tree binds variables and includes any variable bindings from the parent directory.
- **Persistent cache**: Cached variables can be either system- or user-defined. These persist their values over multiple runs.

Passing the PARENT_SCOPE option to set() makes the variable visible in the parent scope.

CMake comes with a wide variety of predefined variables. These are prefixed with CMAKE_. A full list is available at https://cmake.org/cmake/help/latest/manual/cmake-variables.7.html.

Lists

Even though CMake stores variables as strings internally, it is possible to work with lists in CMake by splitting values with a semicolon. Lists can be created by either passing multiple *unquoted* variables to set() or directly as a semicolon-separated string:

```
set(MYLIST abc def ghi)
set(MYLIST "abc;def;ghi")
```

Manipulating lists by modifying their contents, reordering, or finding things can be done using the list command. The following code will query MYLIST for the index of the abc value and then retrieve the value and store it in the variable called ABC:

```
list(FIND MYLIST abc ABC_INDEX)
list(GET MYLIST ${ABC_INDEX} ABC)
```

To append a value to a list, we can use the APPEND keyword. Here, the xyz value is appended to MYLIST:

```
list(APPEND MYLIST "xyz")
```

Cached variables and options

CMake caches some variables so that they run faster in subsequent builds. The variables are stored in CMakeCache.txt files. Usually, you don't have to edit them manually, but they are great for debugging builds that do not behave as expected.

All the variables that are used to configure the build are cached. To cache a custom variable called ch1_MYVAR with the foo value, you can use the set command, like this:

```
set(ch1_MYVAR foo CACHE STRING "Variable foo that configures
   bar")
```

Note that cached variables must have a type and a documentation string that provides a quick summary of them.

Most of the cached variables that are automatically generated are marked as advanced, which means they are hidden from the user in `cmake-gui` and `ccmake` by default. To make them visible, they have to be toggled explicitly. If additional cache variables are generated by a `CMakeLists.txt` file, they can also be hidden by calling the `mark_as_advanced(MYVAR)` command:

Figure 1.4 – Left – cmake-gui does not show variables marked as advanced. Right – Marking the "Advanced" checkbox displays all the variables marked as advanced

As a rule of thumb, any option or variable that the user should change should be marked as advanced. This should happen rarely.

For simple Boolean cache variables, CMake also provides the `option` keyword. `option` has a default value of `OFF` unless specified otherwise. They can also depend on each other via the `CMakeDependentOption` module:

```
option(CHAPTER1_PRINT_LANGUAGE_EXAMPLES "Print examples for
  each language" OFF)
include(CMakeDependentOption)
cmake_dependent_option(CHAPTER1_PRINT_HELLO_WORLD "print a
  greeting from chapter1 " ON CHAPTER1_PRINT_LANGUAGE_EXAMPLES
    ON)
```

Options are often a convenient way to specify simple project configuration. They are cache variables of the `bool` type. If a variable with the same name as the option already exists, a call to `option` does nothing.

Properties

Properties in CMake are values that are attached to a specific object or scope of CMake, such as a file, target, directory, or test case. Properties can be set or changed by using the `set_property` function. To read the value of a property, you can use the `get_property` function, which follows a similar pattern. By default, `set_property` overwrites the values that are already stored inside a property. Values can be added to the current value by passing `APPEND` or `APPEND_STRING` to `set_property`.

The full signature is as follows:

```
set_property(<Scope> <EntityName>
             [APPEND] [APPEND_STRING]
             PROPERTY <propertyName> [<values>])
```

The scope specifier may have the following values:

- `GLOBAL`: Global properties that affect the whole build process.
- `DIRECTORY <dir>`: Properties that are bound to the current directory or the directories specified in `<dir>`. These can also be set directly using the `set_directory_properties` command.
- `TARGET <targets>`: Properties of specific targets. They can also be set using the `set_target_properties` function.
- `SOURCE <files>`: Applies a property to a list of source files. They can also be set directly using `set_source_files_properties`. Additionally, there are the `SOURCE DIRECTORY` and `SOURCE TARGET_DIRECTORY` extended options:
    - `DIRECTORY <dirs>`: This sets the property for the source files in the directory's scope. The directory must already be parsed by CMake by either being the current directory or by being added with `add_subdirectory`.
    - `TARGET_DIRECTORY <targets>`: This sets the property to the directory where the specified targets are created. Again, the targets must already exist at the point where the property is set.
- `INSTALL <files>`: This sets the properties for installed files. These can be used to control the behavior of `cpack`.
- `TEST <tests>`: This sets the properties for tests. They can also be set directly using `set_test_properties`.
- `CACHE <entry>`: This sets the properties for cached variables. The most common ones include setting variables as advanced or adding documentation strings to them.

The full list of supported properties, sorted by their different entities, can be found at `https://cmake.org/cmake/help/latest/manual/cmake-properties.7.html`.

It is good practice to use direct functions such as `set_target_properties` and `set_test_properties` when modifying properties instead of the more general `set_property` command. Using explicit commands avoids making mistakes and confusion between the property names and is generally more readable. There's also the `define_property` function, which creates a property without setting the value. We advise that you don't use this as properties should always have a sane default value.

Loops and conditions

Like any programming language, CMake supports conditional and loop blocks. Conditional blocks are in-between `if()`, `elseif()`, `else()`, and `endif()` statements. Conditions are expressed using various keywords.

Unary keywords are prefixed before the value, as shown here:

```
if(DEFINED MY_VAR)
```

The unary keywords to be used in conditions are as follows:

- `COMMAND`: True if the supplied value is a command
- `EXISTS`: Checks whether a file or a path exists
- `DEFINED`: True if the value is a defined variable

Additionally, there are unary filesystem conditions:

- `EXISTS`: True if the passed file or directory exits
- `IS_DIRECTORY`: Checks whether the supplied path is a directory
- `IS_SYMLINK`: True if the supplied path is a symbolic link
- `IS_ABSOULTE`: Checks whether a supplied path is an absolute path

Binary tests compare two values and are placed between the values to be compared, like this:

```
if(MYVAR STREQUAL "FOO")
```

The binary operators are as follows:

- `LESS`, `GREATER`, `EQUAL`, `LESS_EQUAL`, and `GREATER_EQUAL`: These compare numeric values.
- `STRLESS`, `STREQUAL`, `STRGREATER`, `STRLESS_EQUAL`, and `STRGREATER_EQUAL`: These lexicographically compare strings.
- `VERSION_LESS`, `VERSION_EQUAL`, `VERSION_GREATER`, `VERSION_LESS_EQUAL`, and `VERSION_GREATER_EQUAL`: These compare version strings.
- `MATCHES`: This compares against a regular expression.
- `IS_NEWER_THAN`: Checks which of the two files that passed has been modified recently.
- `IS_NEWER_THAN`: Unfortunately, this is not very precise because if both files have the same timestamp, it also returns `true`. There is also more confusion because if either of the files is missing, the result is also `true`.

Finally, there's the Boolean `OR`, `AND`, and `NOT` operators.

Loops are either achieved by `while()` and `endwhile()` or `foreach()` and `endforeach()`. Loops can be terminated using `break()`; `continue()` aborts the current iteration and starts the next one immediately.

`while` loops take the same conditions as an `if` statement. The following example loops as long as `MYVAR` is less than `5`. Note that to increase the variable, we are using the `math()` function:

```
set(MYVAR 0)
while(MYVAR LESS "5")
  message(STATUS "Chapter1: MYVAR is '${MYVAR}'")
  math(EXPR MYVAR "${MYVAR}+1")
endwhile()
```

In addition to `while` loops, CMake also knows loops for iterating over lists or ranges:

```
foreach(ITEM IN LISTS MYLIST)
# do something with ${ITEM}
endforeach()
```

`for` loops over a specific range can be created by using the `RANGE` keyword:

```
foreach(ITEM RANGE 0 10)
# do something with ${ITEM}
endforeach()
```

Although the `RANGE` version of `foreach()` could work with only a `stop` variable, it is good practice to always specify both the start and end values.

Functions

Functions are defined by `function()/endfunction()`. Functions open a new scope for variables, so all the variables that are defined inside are not accessible from the outside unless the `PARENT_SCOPE` option is passed to `set()`.

Functions are case-insensitive and are invoked by calling `function`, followed by parentheses:

```
function(foo ARG1)
# do something
endfunction()
# invoke foo with parameter bar
foo("bar")
```

Functions are a great way to make parts of your CMake reusable and often come in handy when you're working on larger projects.

Macros

CMake macros are defined using the `macro()/endmacro()` commands. They are a bit like functions, with the difference that in functions, the arguments are true variables, whereas in macros, they are string replacements. This means that all the arguments of a macro must be accessed using curly brackets.

Another difference is that by calling a function, control is transferred to the functions. Macros are executed as if the body of the macro had been pasted into the place of the calling state. This means that macros are not creating scopes regarding variables and control flow. Consequently, it is highly recommended to avoid calling `return()` in macros as this would stop the scope from executing where the macro is called.

Targets

The build system of CMake is organized as a set of logical targets that correspond to an executable, library, or custom command or artifact, such as documentation or similar.

There are three major ways to create a target in CMake – add_executable, add_library, and add_custom_target. The first two are used to create executables and static or shared libraries, while the third can contain almost any custom command to be executed.

Targets can be made dependent on each other so that one target has to be built before another.

It is good practice to work with targets instead of global variables when you're setting properties for build configurations or compiler options. Some of the target properties have visibility modifiers such as PRIVATE, PUBLIC, or INTERFACE to denote which requirements are transitive – that is, which properties have to be "inherited" by a dependent target.

Generator expressions

Generator expressions are small statements that are evaluated during the configuration phase of the build. Most functions allow generator expressions to be used, with a few exceptions. They take the form of $<OPERATOR:VALUE>, where OPERATOR is applied or compared to VALUE. You can think of generator expressions as small inline if-statements.

In the following example, a generator expression is being used to enable the -Wall compiler flag for my_target if the compiler is either GCC, Clang, or Apple Clang. Note that GCC is identified as COMPILER_ID "GNU":

```
target_compile_options(my_target PRIVATE
  "$<$<CXX_COMPILER_ID:GNU,Clang,AppleClang>:-Wall>")
```

This example tells CMake to evaluate the CXX_COMPILER_ID variable to the comma-separated GNU, Clang, AppleClang list and that if it matches either, append the -Wall option to the target – that is, my_target. Generator expressions come in very handy for writing platform- and compiler-independent CMake files.

In addition to querying values, generator expressions can be used to transform strings and lists:

```
$<LOWER_CASE:CMake>
```

This will output `cmake`.

You can learn more about generator expressions at `https://cmake.org/cmake/help/latest/manual/cmake-generator-expressions.7.html`.

Since CMake supports a variety of build systems, compilers, and linkers, it is often used to build software for different platforms. In the next section, we will learn how CMake can be told which toolchain to use and how to configure the different build types, such as debug or release.

CMake policies

For the top-level `CMakeLists.txt` file, `cmake_minimum_required` must be called before any call to the project as it also sets which internal policies for CMake are used to build the project.

Policies are used to maintain backward compatibility across multiple CMake releases. They can be configured to use the `OLD` behavior, which means that `cmake` behaves backward compatible, or as `NEW`, which means the new policy is in effect. As each new version will introduce new rules and features, policies will be used to warn you of backward-compatibility issues. Policies can be disabled or enabled using the `cmake_policy` call.

In the following example, the `CMP0121` policy has been set to a backward-compatible value. `CMP0121` was introduced in CMake version 3.21 and checks whether index variables for the `list()` commands are in a valid format – that is, whether they are integers:

```
cmake_minimum_required(VERSION 3.21)
cmake_policy(SET CMP0121 OLD)

list(APPEND MYLIST "abc;def;ghi")
list(GET MYLIST "any" OUT_VAR)
```

By setting `cmake_policy(SET CMP0121 OLD)`, backward compatibility is enabled and the preceding code will not produce a warning, despite the access to `MYLIST` with the `"any"` index, which is not an integer.

Setting the policy to `NEW` will throw an error – `[build] list index: any is not a valid index` – during the configuration step of CMake.

> **Avoid Setting Single Policies Except When You're Including Legacy Projects**
>
> Generally, policies should be controlled by setting the `cmake_minimum_required` command and not by changing individual policies. The most common use case for changing single policies is when you're including legacy projects as subfolders.

So far, we have covered the basic concepts behind the CMake language, which is used to configure build systems. CMake is used to generate build instructions for different kinds of builds and languages. In the next section, we will learn how to specify the compiler to use and how builds can be configured.

Different toolchains and build types

The power of CMake comes from the fact that you can use the same build specification – that is, `CMakeLists.txt` – for various compiler toolchains without the need to rewrite anything. A toolchain typically consists of a series of programs that can compile and link binaries, as well as creating archives and similar.

CMake supports a variety of languages that the toolchains can be configured for. In this book, we will focus on C++. Configuring the toolchain for different programming languages is done by replacing the CXX part of the following variables with the respective language tag:

- C
- CXX – C++
- CUDA
- OBJC – Objective C
- OBJCXX – Objective C++
- Fortran
- HIP – HIP C++ runtime API for NVIDIA and AMD GPUs
- ISPC – C-based SPMD programming language
- ASM – Assembler

If a project does not specify its language, it's assumed that C and CXX are being used.

CMake will automatically detect the toolchain to use by inspecting the system, but if needed, this can be configured by environment variables or, in the case of cross-compiling, by providing a toolchain file. This toolchain is stored in the cache, so if the toolchain changes, the cache must be deleted and rebuilt. If multiple compilers are installed, you can specify a non-default compiler by either setting the environment variables as `CC` for C or `CXX` for a C++ compiler before calling CMake. Here, we're using the CXX environment variable to overwrite the default compiler to be used in CMake:

```
CXX=g++-7 cmake /path/to/the/source
```

Alternatively, you can overwrite the C++ compiler to use by passing the respective `cmake` variable using `-D`, as shown here:

```
cmake -D CMAKE_CXX_COMPILER=g++-7 /path/to/source
```

Both methods ensure that CMake is using GCC version 7 to build instead of whatever default compiler is available in the system. Avoid setting the compiler toolchain inside the `CMakeLists.txt` files as this clashes with the paradigm that states that CMake files should be platform- and compiler-agnostic.

By default, the linker is automatically selected by the chosen compiler, but it is possible to select a different one by passing the path to the linker executable with the `CMAKE_CXX_LINKER` variable.

Build types

When you're building C++ applications, it is quite common to have various build types, such as a debug build that contains all debug symbols and release builds that are optimized.

CMake natively provides four build types:

- `Debug`: This is non-optimized and contains all the debug symbols. Here, all the asserts are enabled. This is the same as setting `-O0 -g` for GCC and Clang.
- `Release`: This is optimized for speed without debugging symbols and asserts disabled. Usually, this is the build type that is shipped. This is the same as `-O3 -DNDEBUG`.
- `RelWithDebInfo`: This provides optimized code and includes debug symbols but disabled asserts, which is the same as `-O2 -g -DNDEBUG`.
- `MinSizeRel`: This is the same as `Release` but optimized for a small binary size instead of speed, which would be `-Os -DNDEBUG`. Note that this configuration is not supported for all generators on all platforms.

Note that the build types must be passed during the configuration state and are only relevant for single-target generators such as CMake or Ninja. For multi-target generators such as MSVC, they are not used, as the build-system itself can build all build types. It is possible to create custom build types, but since they do not work for every generator, this is usually not encouraged.

Since CMake supports such a wide variety of toolchains, generators, and languages, a frequent question is how to find and maintain working combinations of these options. Here, presets can help.

Maintaining good build configurations with presets

A common problem when building software with CMake is how to share good or working configurations to build a project. Often, people and teams have a preferred way of where the build artifacts should go, which generator to use on which platform, or just the desire that the CI environment should use the same settings to build as it does locally. Since CMake 3.19 came out in December 2020, this information can be stored in `CMakePresets.json` files, which are placed in the root directory of a project. Additionally, each user can superimpose their configuration with a `CMakeUserPresets.json` file. The basic presets are usually placed under version control, but the user presets are not checked into the version system. Both files follow the same JSON format, with the top-level outline being as follows:

```
{
"version": 3,
"cmakeMinimumRequired": {
"major": 3,
"minor": 21,
"patch": 0
},
"configurePresets": [...],
"buildPresets": [...],
"testPresets": [...]
}
```

1. The first line, `"version": 3`, denotes the schema version of the JSON file. CMake 3.21 supports up to version 3, but it is expected that new releases will bring new versions of the schema.

2. Next, `cmakeMinimumRequired{ ... }` specifies which version of CMake to use. Although this is optional, it is good practice to put this in here and match the version with the one specified in the `CMakeLists.txt` file.
3. After that, the various presets for the different build stages can be added with `configurePresets`, `buildPresets`, and `testPresets`. As the name suggests, `configurePresets` applies to the configure stage of CMake's build process, while the other two are used for the build and test stages. The build and test presets may inherit one or more configure presets. If no inheritance is specified, they apply to all the previous steps.

To see what presets have been configured in a project, run `cmake --list-presets` to see a list of available presets. To build using a preset, run `cmake --build --preset name`.

To see the full specification of the JSON schema, go to `https://cmake.org/cmake/help/v3.21/manual/cmake-presets.7.html`.

Presets are a good way to share knowledge about how to build a project in a very explicit way. At the time of writing, more and more IDEs and editors are adding support for CMake presets natively, especially for handling cross-compilation with toolchains. Here, we're only giving you the briefest overview of CMake presets; they will be covered in more depth in *Chapter 12, Cross-Platform Compiling and Custom Toolchains*.

Summary

In this chapter, you were provided with a brief overview of CMake. First, you learned how to install and run a simple build. Then, you learned about the two-stage build process of CMake before touching on the most important language features for writing CMake files.

By now, you should be able to build the examples provided in this book's GitHub repository: `https://github.com/PacktPublishing/CMake-Best-Practices`. You learned about the core features of the CMake language such as variables, targets, and policies. We briefly covered functions and macros, as well as conditional statements and loops for flow control. As you continue reading this book, you will use what you have learned so far to discover further good practices and techniques to move from simple one-target projects to complex software projects that keep being maintainable through a good CMake setup.

In the next chapter, we will learn how some of the most common tasks in CMake can be performed and how CMake works together with various IDEs.

Further reading

To learn more about the topics that were covered in this chapter, take a look at the following resources:

- The official CMake documentation: `https://cmake.org/cmake/help/latest/`
- The official CMake tutorial: `https://cmake.org/cmake/help/latest/guide/tutorial/index.html`

Questions

Answer the following questions to test your knowledge of this chapter:

1. How do you start the configure step of CMake?
2. How do you start the build step of CMake?
3. Which executable from CMake can be used to run tests?
4. Which executable from CMake is used for packaging?
5. What are targets in CMake?
6. What is the difference between properties and variables?
7. What are CMake presets used for?

2
Accessing CMake in Best Ways

In the previous chapter, we got acquainted with CMake and learned about CMake's fundamental concepts. Now, we are going to learn how to interact with CMake. We shall do so by learning to configure, build, and install the existing projects. This will enable you to interact with CMake projects.

This chapter will take a look into what CMake has to offer as an interface and inspect some of the popular IDE and editor integrations. This chapter will cover the following:

- Using CMake via a command-line interface
- Using `cmake-gui` and `ccmake` interfaces
- IDE and editor integrations (Visual Studio, Visual Studio Code, and Qt Creator)

Since we have a lot to cover, let's not waste any time and get started with the technical requirements.

Technical requirements

Before going further into detail, there are some requirements that need to be satisfied to follow the examples:

- **The CMake-Best-Practices Git repository**: This is the main repository that contains all the exemplary content for the book. It is accessible online at `https://github.com/PacktPublishing/CMake-Best-Practices`.

Using CMake via a command-line interface

The most common way of using CMake is via the **command-line interface** (**CLI**). CLIs exist in virtually every environment, thus, learning to use CMake in a CLI is essential. In this section, we are going to learn how to perform the most basic CMake operations using the CLI.

Interactions with the CMake CLI can be done by issuing the `cmake` command in your operating system's terminal, assuming that CMake is installed and the `cmake` executable is included in your system's `PATH` variable (or equivalent). You can verify that by issuing `cmake` in your terminal without any parameters, as shown in the following figure:

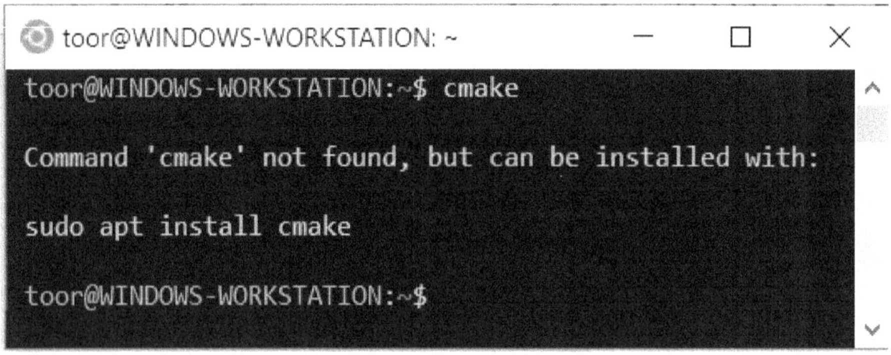

```
toor@WINDOWS-WORKSTATION: ~
toor@WINDOWS-WORKSTATION:~$ cmake

Command 'cmake' not found, but can be installed with:

sudo apt install cmake

toor@WINDOWS-WORKSTATION:~$
```

Figure 2.1 – Invoking the cmake command

If your terminal is complaining about a missing command, then you should either install CMake (explained in *Chapter 1*, *Kickstarting CMake*) or make it discoverable by adding it to your system's `PATH` variable. Consult your operating system's guide about how to add a path to the system's `PATH` variable.

After installing CMake and adding it to the PATH variable (if required), you should test whether CMake is usable. The most basic command you can execute in the command line is cmake --version, which allows you to check CMake's version.

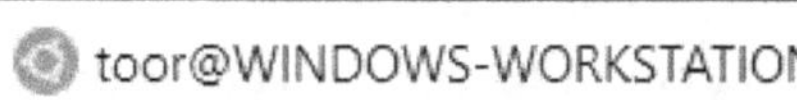


```
toor@WINDOWS-WORKSTATION:~$ cmake --version
cmake version 3.21.2

CMake suite maintained and supported by Kitware (kitware.com/cmake).
```

Figure 2.2 – Checking CMake's version in the terminal

CMake will output a version string in the form of cmake version <maj.min.rev>. You should see an output that contains the version number of CMake you've installed on your system.

> **Note**
>
> If the version does not match with the installed one, then you probably have multiple installations of CMake on your system. Since this book contains examples written for CMake version 3.21 and above, it is recommended to have that issue fixed before going any further.

After installing CMake, you should install your build system and compiler as well. For Debian-like operating systems (for example, Debian and Ubuntu), this can be easily done by issuing the sudo apt install build-essential command. This package essentially contains gcc, g++, and make.

The CLI usage will be illustrated in the Ubuntu 20.04 environment. Apart from the minor edge cases, the usage is the same in other environments as well. Those edge cases will be mentioned as we go on.

Learning the basics of the CMake CLI

The three basic things you should learn about using the CMake CLI are listed here:

- Configuring a CMake project
- Building a CMake project
- Installing a CMake project

After learning the basics, you will be able to build and install any CMake project of your choice. Let's get started with configuring.

Configuring a project via the CLI

To configure a CMake project via the command line, you can use the cmake -G "Unix Makefiles" -S <project_root> -B <output_directory> construct. The -S argument is used for specifying the CMake project to be configured, whereas -B specifies the configure output directory. Lastly, the -G argument allows us to specify the generator that will be used for the build system generation. The result of the configuration process will be written to <output_directory>.

As an illustration, let's configure our book's example project in the project root build directory:

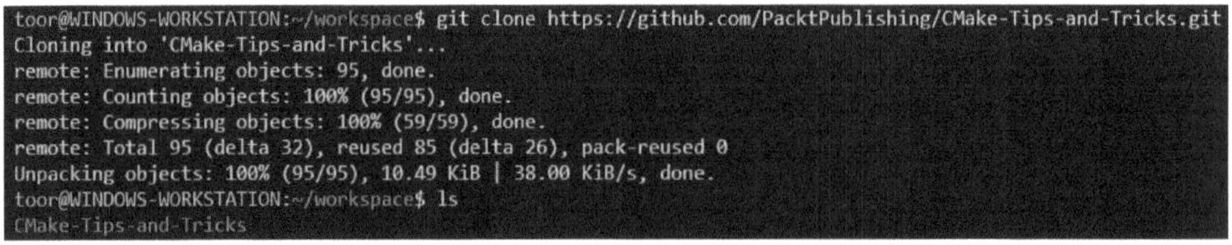

```
toor@WINDOWS-WORKSTATION:~/workspace$ git clone https://github.com/PacktPublishing/CMake-Tips-and-Tricks.git
Cloning into 'CMake-Tips-and-Tricks'...
remote: Enumerating objects: 95, done.
remote: Counting objects: 100% (95/95), done.
remote: Compressing objects: 100% (59/59), done.
remote: Total 95 (delta 32), reused 85 (delta 26), pack-reused 0
Unpacking objects: 100% (95/95), 10.49 KiB | 38.00 KiB/s, done.
toor@WINDOWS-WORKSTATION:~/workspace$ ls
CMake-Tips-and-Tricks
```

Figure 2.3 – Cloning the example code repository

> **Important Note**
>
> The project must be already present in your environment. If not, clone it via Git by issuing git clone https://github.com/PacktPublishing/CMake-Best-Practices.git in your terminal.

Now go into the CMake-Best-Practices directory and issue cmake -G "Unix Makefiles" -S . -B ./build, as shown in the following figure:

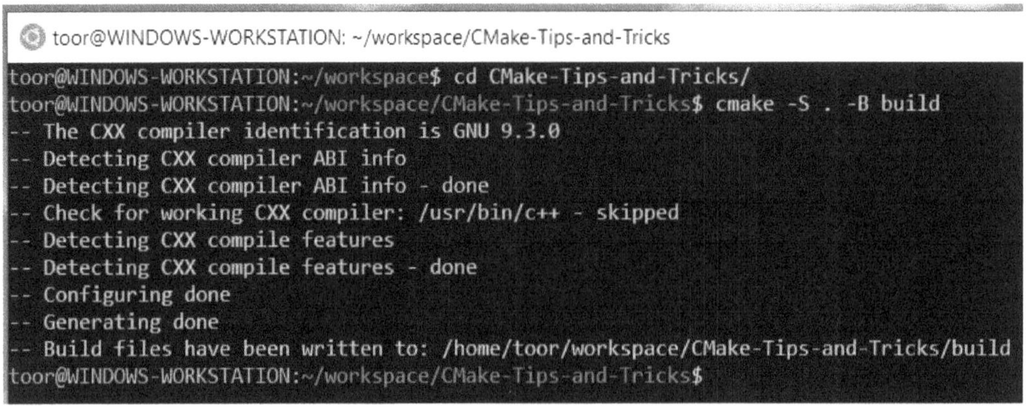

```
toor@WINDOWS-WORKSTATION: ~/workspace/CMake-Tips-and-Tricks
toor@WINDOWS-WORKSTATION:~/workspace$ cd CMake-Tips-and-Tricks/
toor@WINDOWS-WORKSTATION:~/workspace/CMake-Tips-and-Tricks$ cmake -S . -B build
-- The CXX compiler identification is GNU 9.3.0
-- Detecting CXX compiler ABI info
-- Detecting CXX compiler ABI info - done
-- Check for working CXX compiler: /usr/bin/c++ - skipped
-- Detecting CXX compile features
-- Detecting CXX compile features - done
-- Configuring done
-- Generating done
-- Build files have been written to: /home/toor/workspace/CMake-Tips-and-Tricks/build
toor@WINDOWS-WORKSTATION:~/workspace/CMake-Tips-and-Tricks$
```

Figure 2.4 – Configuring the example code with CMake

This command is like saying to CMake, *Use the "Unix Makefiles" (-G "Unix Makefiles") generator to generate a build system for the CMake project in the current directory (-S .) to build (-B ./build) directory.*

CMake will configure the project located in the current folder in the `build` folder. As we omitted the build type, CMake used the `Debug` build type (the default `CMAKE_BUILD_TYPE` value for the project).

In subsequent sections, we are going to learn about the fundamental settings that are used in the configure step.

Changing the build type

CMake does not assume any build type by default. In order to set the build type, an additional variable named `CMAKE_BUILD_TYPE` must be supplied to the `configure` command. To supply additional variables, the variable must be prefixed with `-D`.

To get the `Release` build instead of `Debug`, add the `CMAKE_BUILD_TYPE` variable in the `configure` command, as mentioned previously: `cmake -G "Unix Makefiles" -DCMAKE_BUILD_TYPE:STRING=Release -S . -B ./build`.

> **Note**
>
> The `CMAKE_BUILD_TYPE` variable only makes sense for single-configuration generators, such as Unix Makefiles and Ninja. In multiple-configuration generators, such as Visual Studio, the build type is a build-time parameter instead of a configuration-time parameter, thus, it cannot be configured by using the `CMAKE_BUILD_TYPE` parameter.

Changing the generator type

Depending on the environment, CMake attempts to select an appropriate generator by default. To specify a generator explicitly, the `-G` argument must be supplied with a valid generator name. For example, if you want to use Ninja as a build system instead of make, you can change it as follows:

```
cmake -G "Ninja" -DCMAKE_BUILD_TYPE:STRING=Debug -S . -B ./
build
```

The output should be similar to the command output shown in the following figure:

```
toor@WINDOWS-WORKSTATIC
toor@WINDOWS-WORKSTATION:~/workspace/CMake-Tips-and-Tricks$ cmake -G "Ninja" -DCMAKE_BUILD_TYPE:STRI
NG=Debug -S . -B build
-- The CXX compiler identification is GNU 9.3.0
-- Detecting CXX compiler ABI info
-- Detecting CXX compiler ABI info - done
-- Check for working CXX compiler: /usr/bin/c++ - skipped
-- Detecting CXX compile features
-- Detecting CXX compile features - done
Current build type Debug
-- Configuring done
-- Generating done
-- Build files have been written to: /home/toor/workspace/CMake-Tips-and-Tricks/build
toor@WINDOWS-WORKSTATION:~/workspace/CMake-Tips-and-Tricks$ cd build/
toor@WINDOWS-WORKSTATION:~/workspace/CMake-Tips-and-Tricks/build$ ls
CMakeCache.txt  CMakeFiles  build.ninja  chapter_1  chapter_2  cmake_install.cmake
```

Figure 2.5 – Checking the CMake's Ninja generator output

This will cause CMake to generate Ninja build files instead of makefiles.

In order to see all available generator types for your environment, issue the `cmake --help` command. Available generators will be listed at the end of the *Help text generators* section, as shown here:

```
toor@WINDOWS-WORKSTATIC

The following generators are available on this platform (* marks default):
  Green Hills MULTI            = Generates Green Hills MULTI files
                                 (experimental, work-in-progress).
* Unix Makefiles               = Generates standard UNIX makefiles.
  Ninja                        = Generates build.ninja files.
  Ninja Multi-Config           = Generates build-<Config>.ninja files.
  Watcom WMake                 = Generates Watcom WMake makefiles.
  CodeBlocks - Ninja           = Generates CodeBlocks project files.
  CodeBlocks - Unix Makefiles  = Generates CodeBlocks project files.
  CodeLite - Ninja             = Generates CodeLite project files.
  CodeLite - Unix Makefiles    = Generates CodeLite project files.
  Eclipse CDT4 - Ninja         = Generates Eclipse CDT 4.0 project files.
  Eclipse CDT4 - Unix Makefiles= Generates Eclipse CDT 4.0 project files.
  Kate - Ninja                 = Generates Kate project files.
  Kate - Unix Makefiles        = Generates Kate project files.
  Sublime Text 2 - Ninja       = Generates Sublime Text 2 project files.
  Sublime Text 2 - Unix Makefiles
                               = Generates Sublime Text 2 project files.
```

Figure 2.6 – List of available generators in help

The generator with one asterisk next to it is the default for the environment you're currently in.

Changing the compiler

In CMake, the compilers to be used are specified on a per-language basis via the CMAKE_<LANG>_COMPILER variables. In order to change the compiler for a language, CMAKE_<LANG>_COMPILER must be supplied to the Configure command. For a C/C++ project, the variables usually overridden are CMAKE_C_COMPILER (C compiler) and CMAKE_CXX_COMPILER (C++ compiler). Compiler flags are similarly controlled by the CMAKE_<LANG>_FLAGS variable. This variable can be used for holding configuration-independent compiler flags.

As an example, let's try to use g++-10 as a C++ compiler in an environment where it is not the default compiler:

```
cmake -G "Unix Makefiles" -DCMAKE_CXX_COMPILER=/usr/bin/g++-10
-S . -B ./build
```

Here, we can see g++-10 is used instead of the system's default compiler, g++-9:

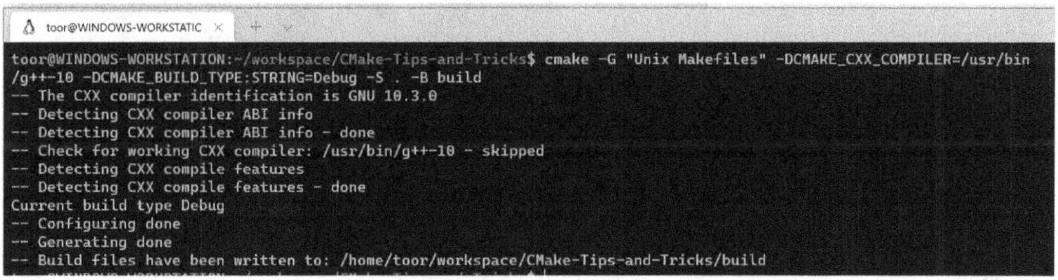

```
toor@WINDOWS-WORKSTATION:~/workspace/CMake-Tips-and-Tricks$ cmake -G "Unix Makefiles" -DCMAKE_CXX_COMPILER=/usr/bin
/g++-10 -DCMAKE_BUILD_TYPE:STRING=Debug -S . -B build
-- The CXX compiler identification is GNU 10.3.0
-- Detecting CXX compiler ABI info
-- Detecting CXX compiler ABI info - done
-- Check for working CXX compiler: /usr/bin/g++-10 - skipped
-- Detecting CXX compile features
-- Detecting CXX compile features - done
Current build type Debug
-- Configuring done
-- Generating done
-- Build files have been written to: /home/toor/workspace/CMake-Tips-and-Tricks/build
```

Figure 2.7 – Configuring the project using a different compiler (g++-10)

Without the compiler specification, CMake prefers to use g++-9 in this environment:

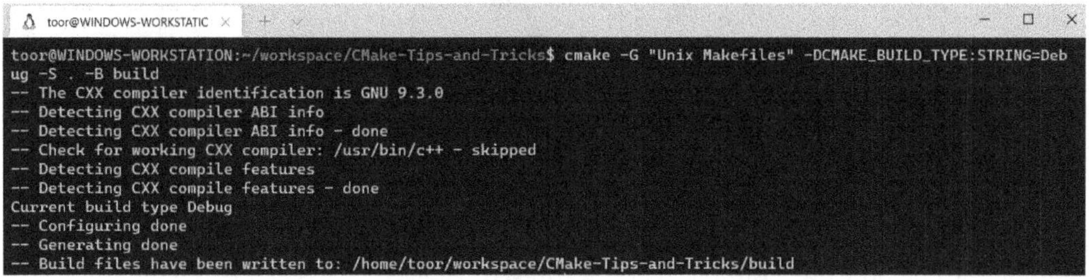

```
toor@WINDOWS-WORKSTATION:~/workspace/CMake-Tips-and-Tricks$ cmake -G "Unix Makefiles" -DCMAKE_BUILD_TYPE:STRING=Deb
ug -S . -B build
-- The CXX compiler identification is GNU 9.3.0
-- Detecting CXX compiler ABI info
-- Detecting CXX compiler ABI info - done
-- Check for working CXX compiler: /usr/bin/c++ - skipped
-- Detecting CXX compile features
-- Detecting CXX compile features - done
Current build type Debug
-- Configuring done
-- Generating done
-- Build files have been written to: /home/toor/workspace/CMake-Tips-and-Tricks/build
```

Figure 2.8 – Configuring behavior without a compiler preference

Passing flags to the compiler

To illustrate how to specify compiler flags, suppose that you want to enable all warnings and treat them as an error. These behaviors are controlled with `-Wall` and `-Werror` compiler flags, respectively, in the `gcc` toolchain; thus, we need to pass these flags to the C++ compiler. The following code specifies how to do it:

```
cmake -G "Unix Makefiles" -DCMAKE_CXX_FLAGS:STRING="-Wall
-Werror" - S  .  B  ./build S . -B ./build
```

We can see the flags specified in the command (`-Wall` and `-Werror`) are passed into the compiler in the following example:

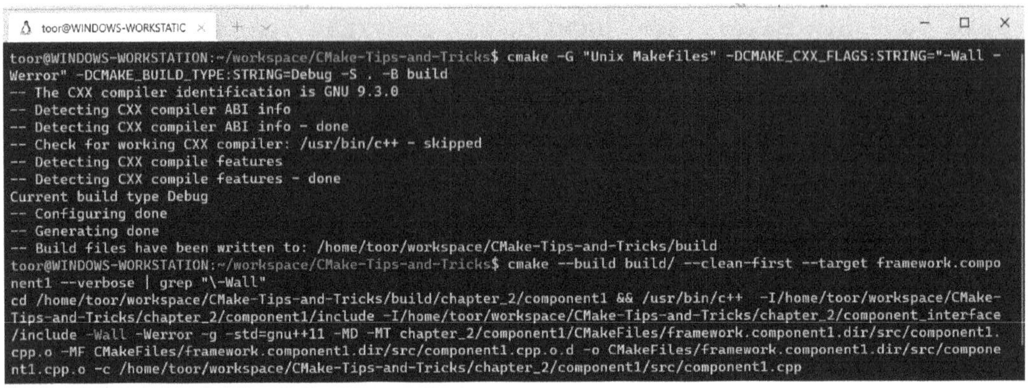

```
toor@WINDOWS-WORKSTATIC
toor@WINDOWS-WORKSTATION:~/workspace/CMake-Tips-and-Tricks$ cmake -G "Unix Makefiles" -DCMAKE_CXX_FLAGS:STRING="-Wall -
Werror" -DCMAKE_BUILD_TYPE:STRING=Debug -S . -B build
-- The CXX compiler identification is GNU 9.3.0
-- Detecting CXX compiler ABI info
-- Detecting CXX compiler ABI info - done
-- Check for working CXX compiler: /usr/bin/c++ - skipped
-- Detecting CXX compile features
-- Detecting CXX compile features - done
Current build type Debug
-- Configuring done
-- Generating done
-- Build files have been written to: /home/toor/workspace/CMake-Tips-and-Tricks/build
toor@WINDOWS-WORKSTATION:~/workspace/CMake-Tips-and-Tricks$ cmake --build build/ --clean-first --target framework.compo
nent1 --verbose | grep "\-Wall"
cd /home/toor/workspace/CMake-Tips-and-Tricks/build/chapter_2/component1 && /usr/bin/c++  -I/home/toor/workspace/CMake-
Tips-and-Tricks/chapter_2/component1/include -I/home/toor/workspace/CMake-Tips-and-Tricks/chapter_2/component_interface
/include -Wall -Werror -g -std=gnu++11 -MD -MT chapter_2/component1/CMakeFiles/framework.component1.dir/src/component1.
cpp.o -MF CMakeFiles/framework.component1.dir/src/component1.cpp.o.d -o CMakeFiles/framework.component1.dir/src/compone
nt1.cpp.o -c /home/toor/workspace/CMake-Tips-and-Tricks/chapter_2/component1/src/component1.cpp
```

Figure 2.9 – Passing flags to the C++ compiler

Build flags can be customized for a per-build type by suffixing them with capitalized build type string. There are four variables for four different build types, as listed next. They are useful for specifying build types depending on compiler flags. Flags specified in such variables are only valid when the configuration build type matches:

- `CMAKE_<LANG>_FLAGS_DEBUG`
- `CMAKE_<LANG>_FLAGS_RELEASE`
- `CMAKE_<LANG>_FLAGS_RELWITHDEBINFO`
- `CMAKE_<LANG>_FLAGS_MINSIZEREL`

In addition to the previous example, if you want to treat warnings as errors only in the `Release` builds, build type-specific compiler flags allow you to do so.

Here is an example that illustrates the usage of the build type-specific compiler flags:

```
cmake -G "Unix Makefiles" -DCMAKE_CXX_FLAGS:STRING="-Wall
-Werror" -DCMAKE_CXX_FLAGS_RELEASE:STRING="-O3" -DCMAKE_BUILD_
TYPE:STRING=Debug -S . -B ./build
```

Notice that an additional `CMAKE_CXX_FLAGS_RELEASE` parameter is present in the preceding command. The contents in this variable will only be passed to the compiler when the build type is `Release`. Since the build type is specified as `Debug`, we can see the `-O3` flag is not present in the flags passed to the compiler, as shown in the following figure:

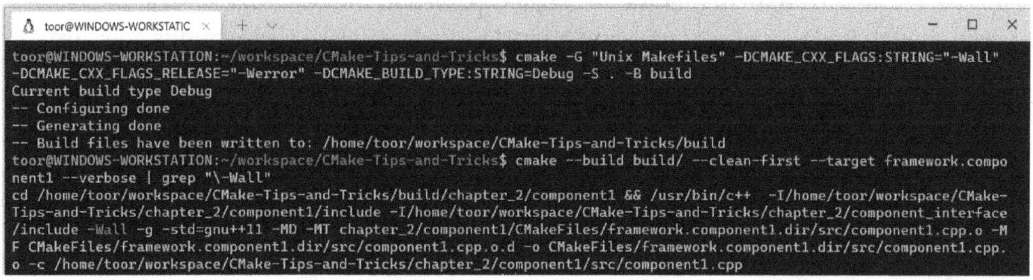

```
toor@WINDOWS-WORKSTATIC
toor@WINDOWS-WORKSTATION:~/workspace/CMake-Tips-and-Tricks$ cmake -G "Unix Makefiles" -DCMAKE_CXX_FLAGS:STRING="-Wall"
-DCMAKE_CXX_FLAGS_RELEASE="-Werror" -DCMAKE_BUILD_TYPE:STRING=Debug -S . -B build
Current build type Debug
-- Configuring done
-- Generating done
-- Build files have been written to: /home/toor/workspace/CMake-Tips-and-Tricks/build
toor@WINDOWS-WORKSTATION:~/workspace/CMake-Tips-and-Tricks$ cmake --build build/ --clean-first --target framework.compo
nent1 --verbose | grep "\-Wall"
cd /home/toor/workspace/CMake-Tips-and-Tricks/build/chapter_2/component1 && /usr/bin/c++  -I/home/toor/workspace/CMake-
Tips-and-Tricks/chapter_2/component1/include -I/home/toor/workspace/CMake-Tips-and-Tricks/chapter_2/component_interface
/include -Wall -g -std=gnu++11 -MD -MT chapter_2/component1/CMakeFiles/framework.component1.dir/src/component1.cpp.o -M
F CMakeFiles/framework.component1.dir/src/component1.cpp.o.d -o CMakeFiles/framework.component1.dir/src/component1.cpp.
o -c /home/toor/workspace/CMake-Tips-and-Tricks/chapter_2/component1/src/component1.cpp
```

Figure 2.10 – Specifying flags based on the build type; the –O3 flag is missing in the Debug build

In *Figure 2.10*, notice that CMake is issuing a warning about a specified but unused variable, `CMAKE_CXX_FLAGS_RELEASE`. This confirms that the `CMAKE_CXX_FLAGS_RELEASE` variable is not used in the `Debug` build type. When the build type is specified as `Release`, we can see that the `-O3` flag is present:

```
cmake -G "Unix Makefiles" -DCMAKE_CXX_FLAGS:STRING="-Wall
-Werror" -DCMAKE_CXX_FLAGS_RELEASE:STRING="-O3"
-DCMAKE_BUILD_TYPE:STRING= "Release" -S . -B ./build
```

In this line, you are saying to CMake, *Configure the CMake project located at the current directory to build/folder using the "Unix Makefiles" generator. For all build types, pass the -Wall flag to the compiler unconditionally. If the build type is Release, pass the -O3 flag as well.*

Here is the output of the command when the build type is set to `Release`:

```
toor@WINDOWS-WORKSTATION:~/workspace/CMake-Tips-and-Tricks$ cmake -G "Unix Makefiles" -DCMAKE_CXX_FLAGS:STRING="-Wall"
-DCMAKE_CXX_FLAGS_RELEASE="-Werror" -DCMAKE_BUILD_TYPE:STRING=Release -S . -B build
Current build type Release
-- Configuring done
-- Generating done
-- Build files have been written to: /home/toor/workspace/CMake-Tips-and-Tricks/build
toor@WINDOWS-WORKSTATION:~/workspace/CMake-Tips-and-Tricks$ cmake --build build/ --clean-first --target framework.compo
nent1 --verbose | grep "\-Wall"
cd /home/toor/workspace/CMake-Tips-and-Tricks/build/chapter_2/component1 && /usr/bin/c++  -I/home/toor/workspace/CMake-
Tips-and-Tricks/chapter_2/component1/include -I/home/toor/workspace/CMake-Tips-and-Tricks/chapter_2/component_interface
/include -Wall -Werror -std=gnu++11 -MD -MT chapter_2/component1/CMakeFiles/framework.component1.dir/src/component1.cpp
.o -MF CMakeFiles/framework.component1.dir/src/component1.cpp.o.d -o CMakeFiles/framework.component1.dir/src/component1
.cpp.o -c /home/toor/workspace/CMake-Tips-and-Tricks/chapter_2/component1/src/component1.cpp
```

Figure 2.11 – Specifying flags based on build type; the -O3 flag is present in the Release build

In *Figure 2.11*, we can confirm that the `-O3` flag is passed to the compiler as well. Be aware that even though `RelWithDebInfo` and `MinSizeRel` are also release builds, they are separate from the `Release` build type, and so flags specified in the `CMAKE_<LANG>_FLAGS_RELEASE` variable will not apply to them.

Listing cached variables

You can list all the cached variables by issuing the `cmake -L ./build/` command (see *Figure 2.12*). This, by default, does not show the advanced variables and help strings associated with each variable. To show them as well, use the `cmake -LAH ./build/` command instead.

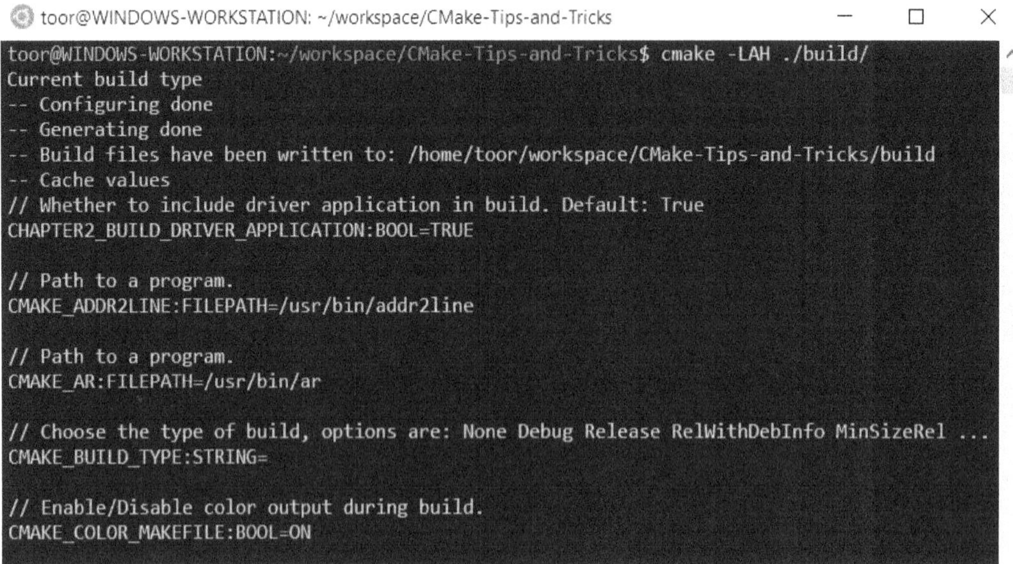

toor@WINDOWS-WORKSTATION: ~/workspace/CMake-Tips-and-Tricks
toor@WINDOWS-WORKSTATION:~/workspace/CMake-Tips-and-Tricks$ cmake -LAH ./build/
Current build type
-- Configuring done
-- Generating done
-- Build files have been written to: /home/toor/workspace/CMake-Tips-and-Tricks/build
-- Cache values
// Whether to include driver application in build. Default: True
CHAPTER2_BUILD_DRIVER_APPLICATION:BOOL=TRUE
// Path to a program.
CMAKE_ADDR2LINE:FILEPATH=/usr/bin/addr2line
// Path to a program.
CMAKE_AR:FILEPATH=/usr/bin/ar
// Choose the type of build, options are: None Debug Release RelWithDebInfo MinSizeRel ...
CMAKE_BUILD_TYPE:STRING=
// Enable/Disable color output during build.
CMAKE_COLOR_MAKEFILE:BOOL=ON

Figure 2.12 – List of cached variables dumped by CMake

Building a configured project via CLI

To build the configured project, issue the `cmake --build ./build` command.

This command tells DCMake to *Build the CMake project already configured in the build folder.*

You can also equivalently issue `cd build && make`. The benefit of using `cmake --build` is that it saves you from invoking build system-specific commands. It is especially helpful when building CI pipelines or build scripts. In this way, you can change your build system generator without changing your build command.

You can see an example output for the `cmake --build ./build` command in the following example:

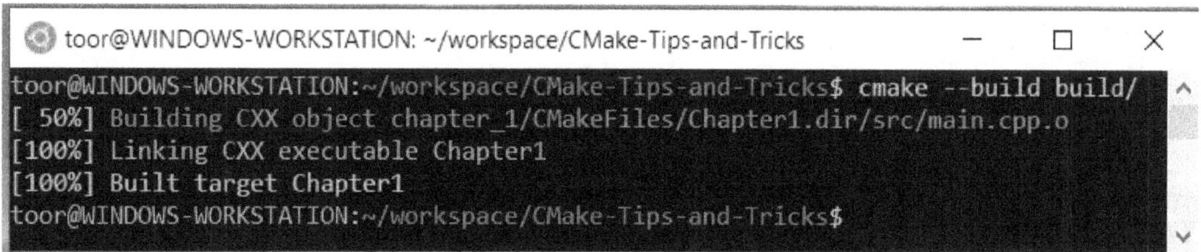

Figure 2.13 – Building a configured project

Building in parallel

You can also customize build time details while issuing the build command. The most prominent build time configuration is the number of jobs that will be used to build the project. To specify job count, append `--parallel <job_count>` to your `cmake --build` command.

To build in parallel, issue `cmake --build ./build --parallel 2`, where the number `2` specifies the job count. The recommended number of jobs for a build system is, at most, *one job per hardware thread*. In multi-core systems, it is also recommended to use at least one less than the available hardware thread count to not affect the system's responsivity during the build process.

> **Note**
>
> You can usually use more than one job per hardware thread and get faster build times since the build process is mostly I/O bound, but your mileage may vary. Experiment and observe.
>
> Also, some of the build systems, such as Ninja, will try to utilize as many hardware threads as are available in the system, so it is redundant to specify the job count for such build systems if your target is to use all hardware threads in your system. You can retrieve the hardware thread count by issuing the `nproc` command in Linux environments.

It is a good practice not to use fixed values for environment-dependent variables in the commands that are expected to be invoked in different environments, such as CI/CD scripts and build scripts. Here is an example `build` command that utilizes `nproc` to determine the number of parallel jobs dynamically:

```
cmake --build ./build/ --parallel $(($(nproc)-1))
```

Let's observe how different job counts affect the build time. We will use the `time` tool to measure how long each command invocation is. The environment details are as follows:

- **OS**: Ubuntu 20.04.3 LTS (Focal Fossa)
- **CPU**: AMD Ryzen Threadripper 1950X 16-Core Processor (32 hardware threads)
- **RAM**: 32 GB

With one job (`--parallel 1`), the build time result would be as follows:

```
toor@WINDOWS-WORKSTATIC
toor@WINDOWS-WORKSTATION:~/workspace/CMake-Tips-and-Tricks$ /usr/bin/time --format='%C took
 %e seconds' cmake --build build --clean-first --parallel 1 1> /dev/null
cmake --build build --clean-first --parallel 1 took 1.26 seconds
```

Figure 2.14 – Parallelized build time result with one job

The build time result with two jobs (`--parallel 2`) would be as follows:

```
toor@WINDOWS-WORKSTATIC
toor@WINDOWS-WORKSTATION:~/workspace/CMake-Tips-and-Tricks$ /usr/bin/time --format='%C took
 %e seconds' cmake --build build --clean-first --parallel 2 1> /dev/null
cmake --build build --clean-first --parallel 2 took 0.96 seconds
```

Figure 2.15 – Parallelized build time result with two jobs

The build time result with three jobs (`--parallel 3`) would be as follows:

```
toor@WINDOWS-WORKSTATIC
toor@WINDOWS-WORKSTATION:~/workspace/CMake-Tips-and-Tricks$ /usr/bin/time --format='%C took
 %e seconds' cmake --build build --clean-first --parallel 3 1> /dev/null
cmake --build build --clean-first --parallel 3 took 0.76 seconds
```

Figure 2.16 – Parallelized build time result with three jobs

This is the build time result with four jobs (`--parallel 4`):

```
toor@WINDOWS-WORKSTATIC
toor@WINDOWS-WORKSTATION:~/workspace/CMake-Tips-and-Tricks$ /usr/bin/time --format='%C took
 %e seconds' cmake --build build --clean-first --parallel 4 1> /dev/null
cmake --build build --clean-first --parallel 4 took 0.75 seconds
```

Figure 2.17 – Parallelized build time result with four jobs

Even though invoked on a very simple project, we can clearly see how extra jobs help get faster build times. Going from one job to two jobs reduces the build time by 0.3 seconds, whereas going from two jobs to three jobs gives us an additional 0.2 seconds. But, going from three jobs to four jobs makes only a 0.01-second difference, which means we've reached the limit of build parallelism for this project. From this point on, throwing more jobs will not achieve a notable difference in build time.

Building specific target(s) only

By default, CMake will build all available targets that are configured. Since building all the targets is not always desirable, CMake allows building a subset of targets via the `--target` sub-option. This sub-option may be specified multiple times, as shown here:

```
cmake --build ./build/ --target "ch2_framework_component1"
--target "ch2_framework_component2"
```

This command will limit the build scope to just the `ch2_framework_component1` and `ch2_framework_component2` targets. If these targets also depend on other targets, they will be built as well.

Removing previous build artifacts before the build

If you want to run a clean build, you may want to remove the artifacts from the previous run first. To do that, the `--clean-first` sub-option can be used. This sub-option will invoke a special target that cleans all the artifacts generated by the build process (for example, invokes make clean).

Here is an example of how you can do it for a build folder named `build`:

```
cmake --build ./build/ --clean-first
```

Debugging your build process

As we did in the *Passing flags to the compiler* section previously, you may want to inspect which commands are invoked with which arguments in the build process. The `--verbose` sub-command instructs CMake to invoke all build commands with verbose mode given that verbose mode is supported by the command. This enables us to investigate nasty compilation and linkage errors with ease.

To build a folder named `build` in verbose mode, invoke `--build`, as shown in the following example:

```
cmake --build ./build/ --verbose
```

Passing command-line arguments to the build tool

If you ever need to pass arguments to the underlying build tool, you can append `--` at the end of the command and write the arguments that will be given:

```
cmake --build ./build/ -- --trace
```

In the preceding case, `--trace` will be directly forwarded to the build tool, which is `make` in our case. This will cause `make` to print tracing information for each recipe built.

Installing a project via the CLI

CMake natively allows installation of artifacts in the environment, if desired. In order to do that, CMake code must be already using CMake `install()` instructions to specify what to install when `cmake --install` (or the build system equivalent) is invoked. The content of `chapter_2` is already configured in such a way for illustrating the command.

We'll learn how to make CMake targets installable later in *Chapter 4, Packaging, Deploying, and Installing a CMake Project.*

The `cmake --install` command requires an already configured and built project. Configure and build the CMake project if you haven't done it yet. Afterward, issue the `cmake --install <project_binary_dir>` command to install the CMake project. Since, in our examples, `build` is used as a project binary directory, `<project_binary_dir>` will be replaced with `build`.

The following figure shows an example of the `install` command:

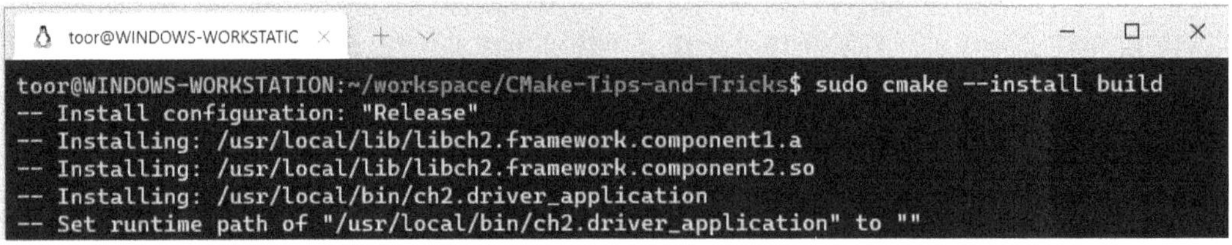

```
toor@WINDOWS-WORKSTATION:~/workspace/CMake-Tips-and-Tricks$ sudo cmake --install build
-- Install configuration: "Release"
-- Installing: /usr/local/lib/libch2.framework.component1.a
-- Installing: /usr/local/lib/libch2.framework.component2.so
-- Installing: /usr/local/bin/ch2.driver_application
-- Set runtime path of "/usr/local/bin/ch2.driver_application" to ""
```

Figure 2.18 – Installing a project

The default installation directory varies between environments. For Unix-like environments, it defaults to `/usr/local`, whereas in a Windows environment, it defaults to `C:/Program Files`.

> **Tip**
>
> Keep in mind that the project must be already built before trying to install the project.
>
> In order to be able to install the project successfully, you must have the appropriate rights/permissions to write to the installation target directory.

Changing the default installation path

To change the default installation directory, you may specify the additional `--prefix` parameter, as shown here, to change the installation directory:

```
cmake --install build --prefix /tmp/example
```

The following figure shows the contents of the `/tmp/example` folder after invoking `cmake --install` with the `/tmp/example` prefix:

```
toor@WINDOWS-WORKSTATION:~/workspace/CMake-Tips-and-Tricks$ cmake --install build --prefix /tmp
/example
-- Install configuration: "Release"
-- Installing: /tmp/example/lib/libch2.framework.component1.a
-- Installing: /tmp/example/lib/libch2.framework.component2.so
-- Installing: /tmp/example/bin/ch2.driver_application
-- Set runtime path of "/tmp/example/bin/ch2.driver_application" to ""
toor@WINDOWS-WORKSTATION:~/workspace/CMake-Tips-and-Tricks$ ls -lRah /tmp/example/
/tmp/example/:
total 16K
drwxr-xr-x 4 toor toor 4.0K Sep 19 18:56 .
drwxrwxrwt 4 root root 4.0K Sep 19 18:56 ..
drwxr-xr-x 2 toor toor 4.0K Sep 19 18:56 bin
drwxr-xr-x 2 toor toor 4.0K Sep 19 18:56 lib

/tmp/example/bin:
total 32K
drwxr-xr-x 2 toor toor 4.0K Sep 19 18:56 .
drwxr-xr-x 4 toor toor 4.0K Sep 19 18:56 ..
-rwxr-xr-x 1 toor toor  24K Sep 19 18:48 ch2.driver_application

/tmp/example/lib:
total 40K
drwxr-xr-x 2 toor toor 4.0K Sep 19 18:56 .
drwxr-xr-x 4 toor toor 4.0K Sep 19 18:56 ..
-rw-r--r-- 1 toor toor 9.0K Sep 19 18:48 libch2.framework.component1.a
-rw-r--r-- 1 toor toor  18K Sep 19 18:48 libch2.framework.component2.so
```

Figure 2.19 – Installing a project to a different path

As can be seen here, the installation root is successfully changed to `/tmp/example`.

Stripping binaries while installing

In the software world, build artifacts are usually bundled with some extra information, for example, a symbol table required for debugging. This information may not be necessary for executing the end product and may drastically increase binary sizes. If you're looking to reduce your end product's storage footprint, stripping binaries may be a good option. One additional benefit of stripping is that it makes it harder to reverse engineer binaries since essential symbol information is stripped away from the binaries.

CMake's `--install` command allows the stripping of binaries while installing the operation. It can be enabled by specifying an additional `--strip` option in the `--install` command, as shown next:

```
cmake --install build --strip
```

In the following example, you can observe the size difference between unstripped and stripped binaries. Note that stripping static libraries has its own limitations and CMake does not perform it by default. You can see the size of the unstripped binaries in this figure:

```
toor@WINDOWS-WORKSTATION:~/workspace/CMake-Tips-and-Tricks$ ls -lrah /usr/local/lib | grep libc
h
-rw-r--r--  1 root root   18K Sep 19 18:48 libch2.framework.component2.so
-rw-r--r--  1 root root  9.0K Sep 19 18:48 libch2.framework.component1.a
```

Figure 2.20 – Artifact size (unstripped)

With a stripped (`cmake -install build --strip`) binary, the size difference looks as shown in the following figure:

```
toor@WINDOWS-WORKSTATION:~/workspace/CMake-Tips-and-Tricks$ ls -lrah /usr/local/lib | grep libc
h
-rw-r--r--  1 root root   15K Sep 19 20:14 libch2.framework.component2.so
-rw-r--r--  1 root root  9.0K Sep 19 18:48 libch2.framework.component1.a
```

Figure 2.21 – Artifact size (stripped)

Installing specific components only (component-based install)

If the project is using CMake's `COMPONENT` feature in the `install()` commands, you may install specific components by specifying their component names. The `COMPONENT` feature allows the installation to be separated into sub-parts. For illustrating this functionality, the `chapter_2` example is structured into two components named `libraries` and `executables`.

In order to install a specific component, an additional `--component` argument is needed along with the `cmake --install` command:

```
cmake --install build --component executables
```

Here is an example invocation:

```
toor@WINDOWS-WORKSTATION:~/workspace/CMake-Tips-and-Tricks$ sudo cmake --install build --compon
ent executables
-- Install configuration: ""
-- Installing: /usr/local/bin/ch2.driver_application
-- Set runtime path of "/usr/local/bin/ch2.driver_application" to ""
```

Figure 2.22 – Installing a specific component only

Installing a specific configuration (for multiple-configuration generators only)

Some of the generators support multiple configurations for the same build configuration (for example, Visual Studio). For that kind of generator, the `--install` option provides an additional `--config` argument to specify which configuration of binaries is intended to be installed.

Here's an example:

```
cmake --install build --config Debug
```

> **Note**
>
> As you may have noticed, the command parameters used in examples are pretty long and explicit. This is intentional. Explicitly specifying arguments allows us to get consistent results in each run, no matter which environment we're running our commands in. For example, without the `-G` argument, CMake will default to the environment's preferred build system generator, which may not be our intention. Our motto here is, *Being explicit is almost always better than being implicit*. The former makes our intention clearer and naturally enables more future-proof and maintainable CMake code in CI systems/scripts as well.

We have covered the fundamentals of CMake command-line usage. Let's continue to learn about the other available interface form – the graphical interface of the CMake.

Advanced configuration using CMake-GUI and ccmake

Albeit having different looks, most interfaces tend to do the same thing; thus, most of the things we have already covered in the previous section are also valid here. Remember, we are going to change our form of the interaction, not the tool we're actually interacting with.

> **Note**
>
> Before going any further, check whether the `ccmake` command is available in your terminal. If not, verify your `PATH` variable is set correctly and check your installation as well.

Learning how to use ccmake (CMake curses GUI)

`ccmake` is a terminal-based **graphical user interface** (**GUI**) for CMake, which allows users to edit cached CMake variables. Instead of calling it a GUI, the term **terminal user interface** (**TUI**) may suit better since there are no traditional shell UI elements such as windows and buttons. These elements are rendered in the terminal using a text-based interface framework named `ncurses`.

Since `ccmake` is a part of the default CMake installation, no extra installation is needed besides CMake. Using `ccmake` is exactly the same as using CMake in a CLI, except it lacks the ability to invoke build and install steps. The main difference is that `ccmake` will show a terminal-based graphical interface for editing cached CMake variables interactively. This is a handy tool when you are experimenting with the settings. The status bar of `ccmake` will display a description for each setting and its possible values.

To start using `ccmake`, use `ccmake` instead of `cmake` in the project configure step. In our example, we will exactly replicate the CLI example we did previously in the *Configuring a project via the CLI* section:

```
ccmake -G "Unix Makefiles" -S . -B ./build
```

The following shows an example output for the preceding command:

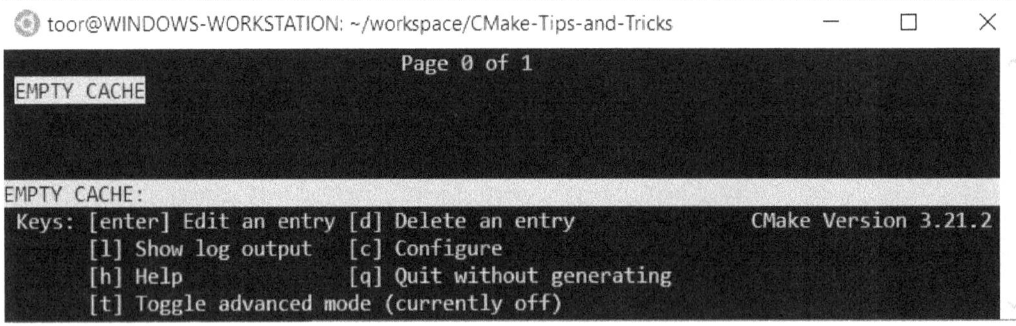

Figure 2.23 – ccmake main screen

After running the command, a terminal-based UI will appear. The initial page is the main page where CMake variables can be edited. `EMPTY CACHE` means no prior configuration has been made and the CMake cache file (`CMakeCache.txt`) is currently empty. In order to start editing variables, the project must be configured first. To configure, press the *C* key on the keyboard, as indicated in the `Keys:` section.

After pressing the *C* key, the CMake configure step will be executed and the log output screen will be displayed with configuration output:

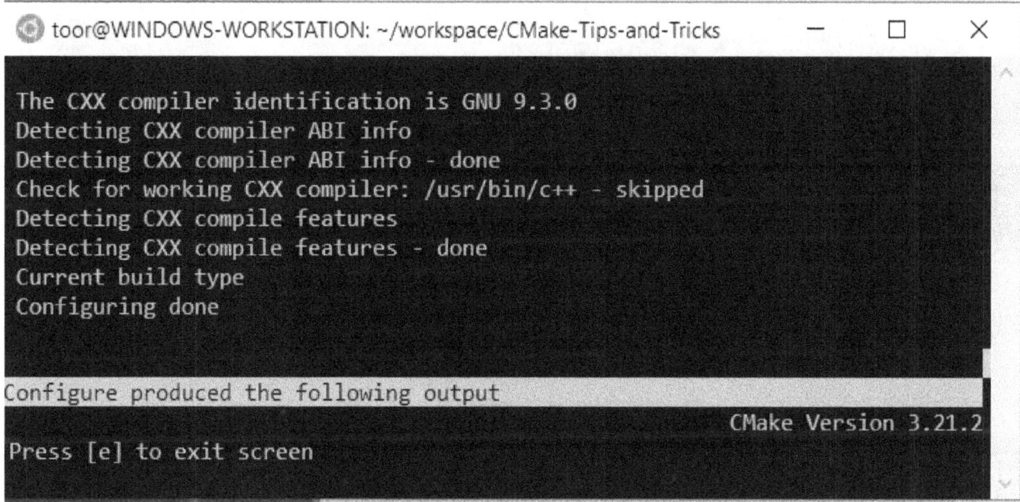

Figure 2.24 – ccmake log screen after configuration

To close the log output screen and return to the main screen, press *E*. Upon return, you will notice that `EMPTY CACHE` is replaced with variable names in the `CMakeCache.txt` file. To select a variable, use the up and down arrow keys on your keyboard. The currently selected variable will be highlighted in white, as seen in the next figure:

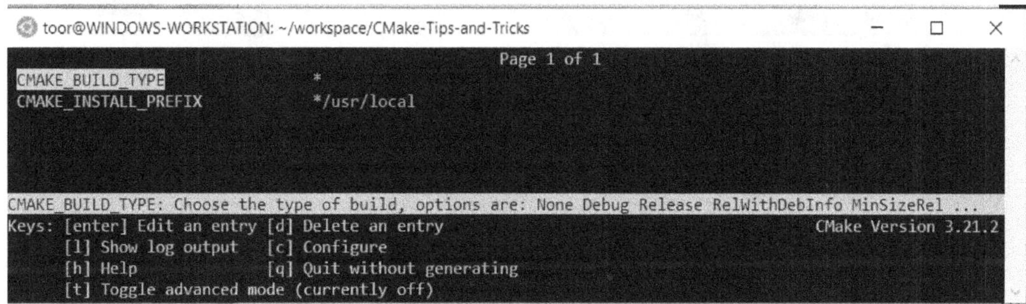

Figure 2.25 – ccmake main screen after configuration

In the preceding screenshot, the `CMAKE_BUILD_TYPE` variable is selected. On the right-hand side, the current value of the CMake variable is displayed. For `CMAKE_BUILD_TYPE`, it is empty right now. An asterisk next to the value of a variable means that the variable's value has just changed with the prior configuration. You can either edit it by pressing the *Enter* key or delete it by pressing the *D* key on the keyboard. The following figure shows what the `ccmake` main screen looks like after changing the variable:

```
toor@WINDOWS-WORKSTATION: ~/workspace/CMake-Tips-and-Tricks
                                             Page 1 of 1
CMAKE_BUILD_TYPE                 Release
CMAKE_INSTALL_PREFIX             /usr/local

CMAKE_BUILD_TYPE: Choose the type of build, options are: None Debug Release RelWithDebInfo MinSizeRel ...

Editing option, press [enter] to confirm
                press [esc] to cancel
```

Figure 2.26 – ccmake main screen following a variable change

Let's set `CMAKE_BUILD_TYPE` to `Release` and configure again:

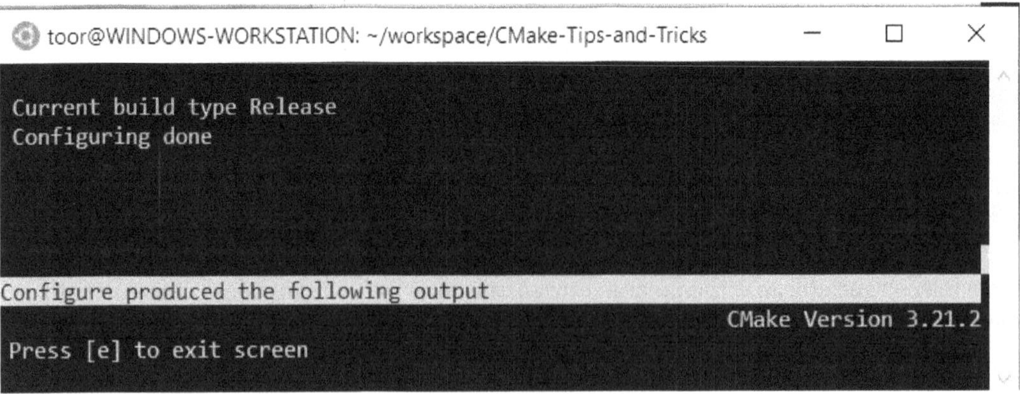

Figure 2.27 – ccmake configuration output (Release)

We can observe that the build type is now set to `Release`. Return to the previous screen and save the changes by pressing the `g` (generate) button. The changes can be discarded by pressing the `q` (quit without generating) button.

To edit other variables, such as `CMAKE_CXX_COMPILER` and `CMAKE_CXX_FLAGS`, the advanced mode should be turned on. These variables are, by default, marked as advanced flags by calling the `mark_as_advanced()` CMake function; thus, they are hidden on graphical interfaces by default. On the main screen, press `t` to toggle to advanced mode.

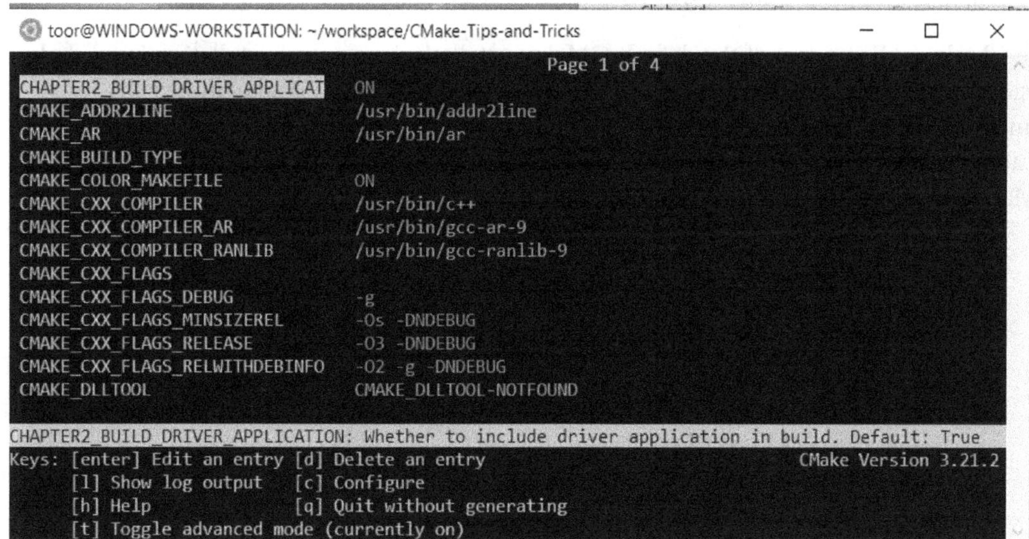

Figure 2.28 – ccmake in advanced mode

After activating the advanced mode, a whole new set of options become visible. You can observe and alter their values, just like normal variables. You may have noticed that a previously hidden variable named `CHAPTER2_BUILD_DRIVER_APPLICATION` is now present. This is a user-defined CMake variable. This variable is defined as follows:

```
# Option to exclude driver application from build.
set(CHAPTER2_BUILD_DRIVER_APPLICATION TRUE CACHE BOOL "Whether
to include driver application in build. Default: True")
# Hide this option from GUI's by default.
mark_as_advanced(CHAPTER2_BUILD_DRIVER_APPLICATION)
```

The `CHAPTER2_BUILD_DRIVER_APPLICATION` variable is defined as a cache variable with a Boolean type, having a default value of `true`. It is marked as advanced, which is why it was not present in the non-advanced mode.

Using CMake via cmake-gui

If you are the type of person who finds CLIs counter-intuitive, or you prefer GUI over CLI, CMake has a cross-platform GUI, too. In contrast to `ccmake`, `cmake-gui` has more to offer, such as `Environment Editor` and `Regular Expression Explorer`.

The CMake GUI is a part of the default CMake installation; no extra installation is needed besides CMake. Its main purpose is to allow a user to configure a CMake project. To launch `cmake-gui`, issue the `cmake-gui` command in your terminal. For Windows, it can be also located from the start menu. If none of these methods work, go into your CMake installation path and it should be present in the `bin\` directory.

> **Note**
>
> If you are launching `cmake-gui` in a Windows environment and you intend to use the toolchain provided by Visual Studio, launch `cmake-gui` from the appropriate Native Tools Command Prompt of your IDE. If you have multiple versions of IDEs, ensure that you are using the correct Native Tools Command Prompt. Otherwise, CMake may fail to discover the required tools, such as a compiler, or may find incorrect ones. Refer to `https://docs.microsoft.com/en-us/visualstudio/ide/reference/command-prompt-powershell?view=vs-2019` for further information.

Here is the main window of the CMake GUI:

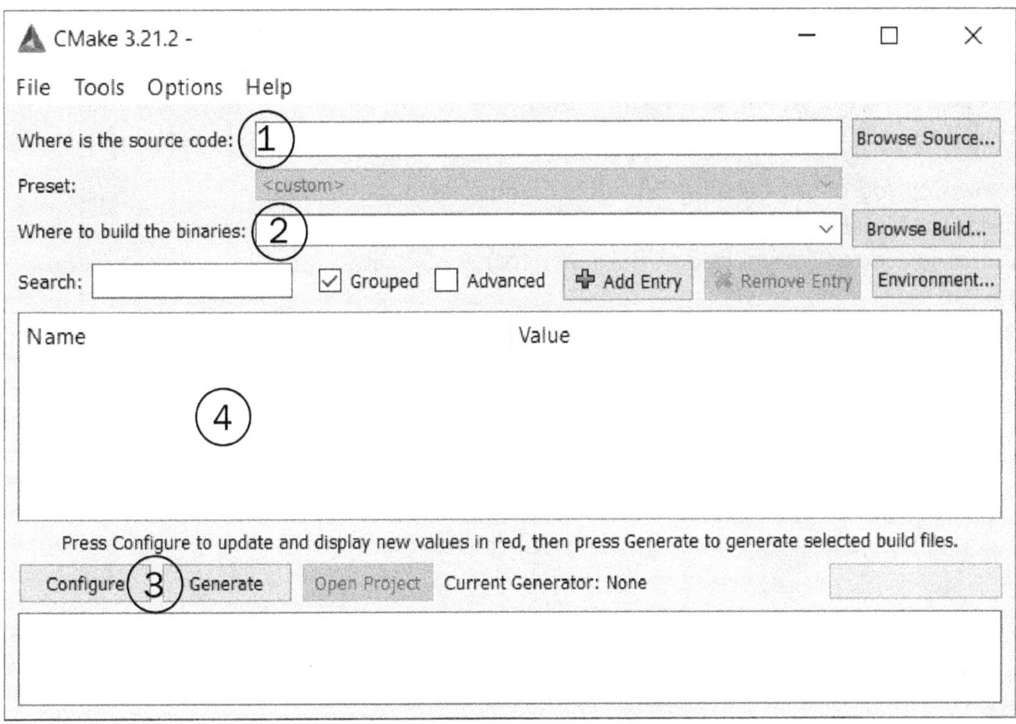

Figure 2.29 – CMake GUI main window

The main screen of the CMake GUI essentially contains the following:

- Source code path field
- Output path field
- Configure and generate buttons
- Cache variable list

These are the four essential things we are going to interact with. To start configuring a project, select the project's root directory by clicking the **Browse Source...** button. Consequently, select an output directory for the project by clicking the **Browse Build...** button. This path will be the path for the generated output files by the selected generator.

After setting source and output paths, click **Configure** to start configuring the selected project. The CMake GUI will let you choose details such as the generator to be used, platform selection (if supported by the generator), toolset, and compiler, as shown in the following figure:

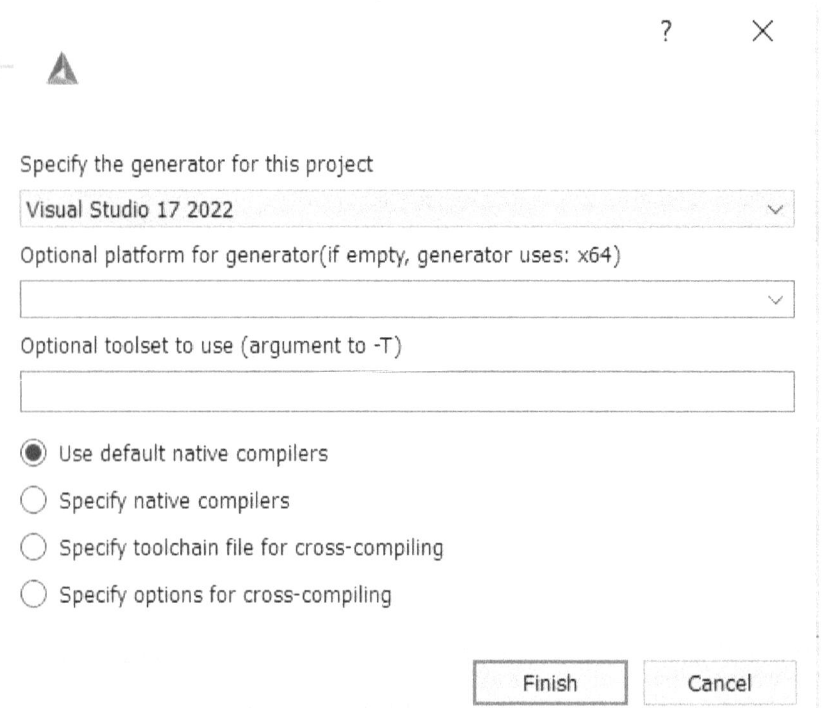

Figure 2.30 – CMake GUI generator selection screen

After filling in these details according to your environment, click **Finish** to continue. The CMake GUI will start configuring your project with the given details and report the output in the log section. Upon successful configuration, you should also see the cache variables in the cache variable list section as well:

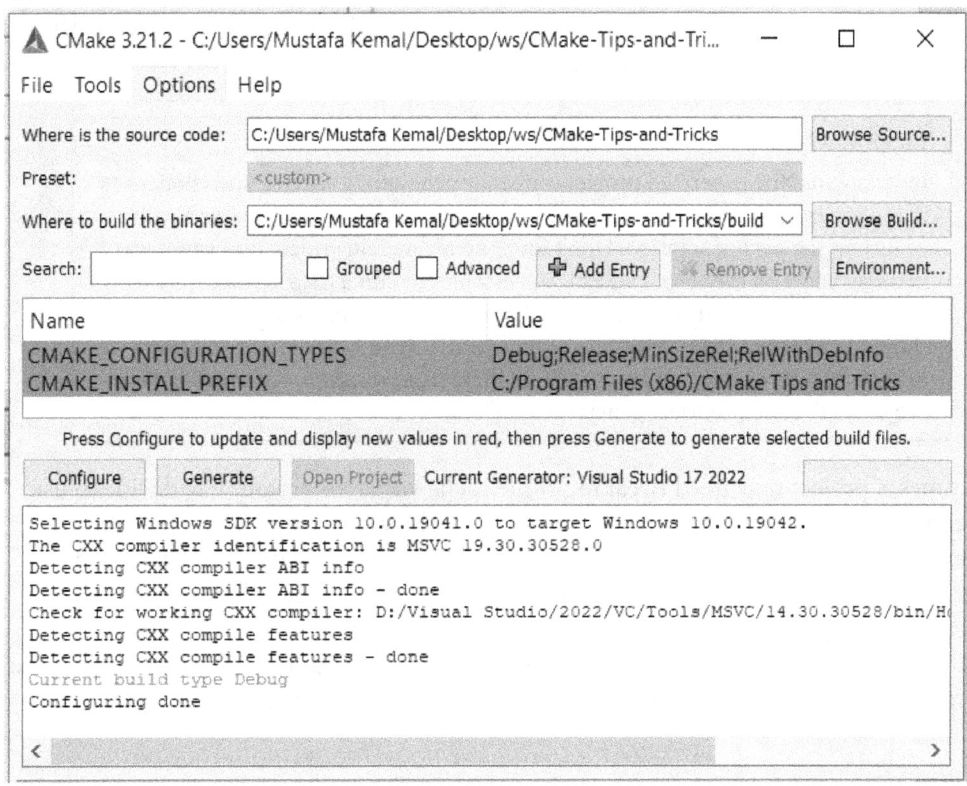

Figure 2.31 – CMake GUI after configuration

If everything seems to be in order, press the **Generate** button to generate the build files that are needed by the build system of your choice. For Visual Studio generators, the generated files are `.sln` and `.cxxproj` along with other stuff. After generating the project, the **Open Project** button will become enabled. It will cause the generated project to be opened with the appropriate editor or IDE. If the build system is not associated with any IDE (for example, `makefiles`), then generated files will be displayed instead. After that, you can use your IDE to build the project.

> **Important Note**
>
> Be aware that the generated project is just the generator's artifact, and changes to the generated project files (`.sln`, `.cxxproj`) will not be saved and will be lost on the next generation. Don't forget to regenerate project files when you make a change to the `CMakeLists.txt` files or edit a `CMakeCache.txt` file (either directly or indirectly). For the version-control aspect, you should treat generated project files as build artifacts and should not add them to version control. You can always obtain them from scratch by generating the project with an appropriate generator via CMake.

Sometimes, a project may need tweaking some cache variables or you may decide to use a different build type, for example. To change any cache variables, click on the value of the desired cache variable; it should become editable. Depending on the variable type, a checkbox may be displayed instead of a string. If the desired variable is not visible on the list, it may be an *advanced* variable, which can only be visible when the **Advanced** checkbox on the window is checked. You can also use the search box to locate the variable with more ease. Here, you can see `cmake-gui` in `Advanced Mode`:

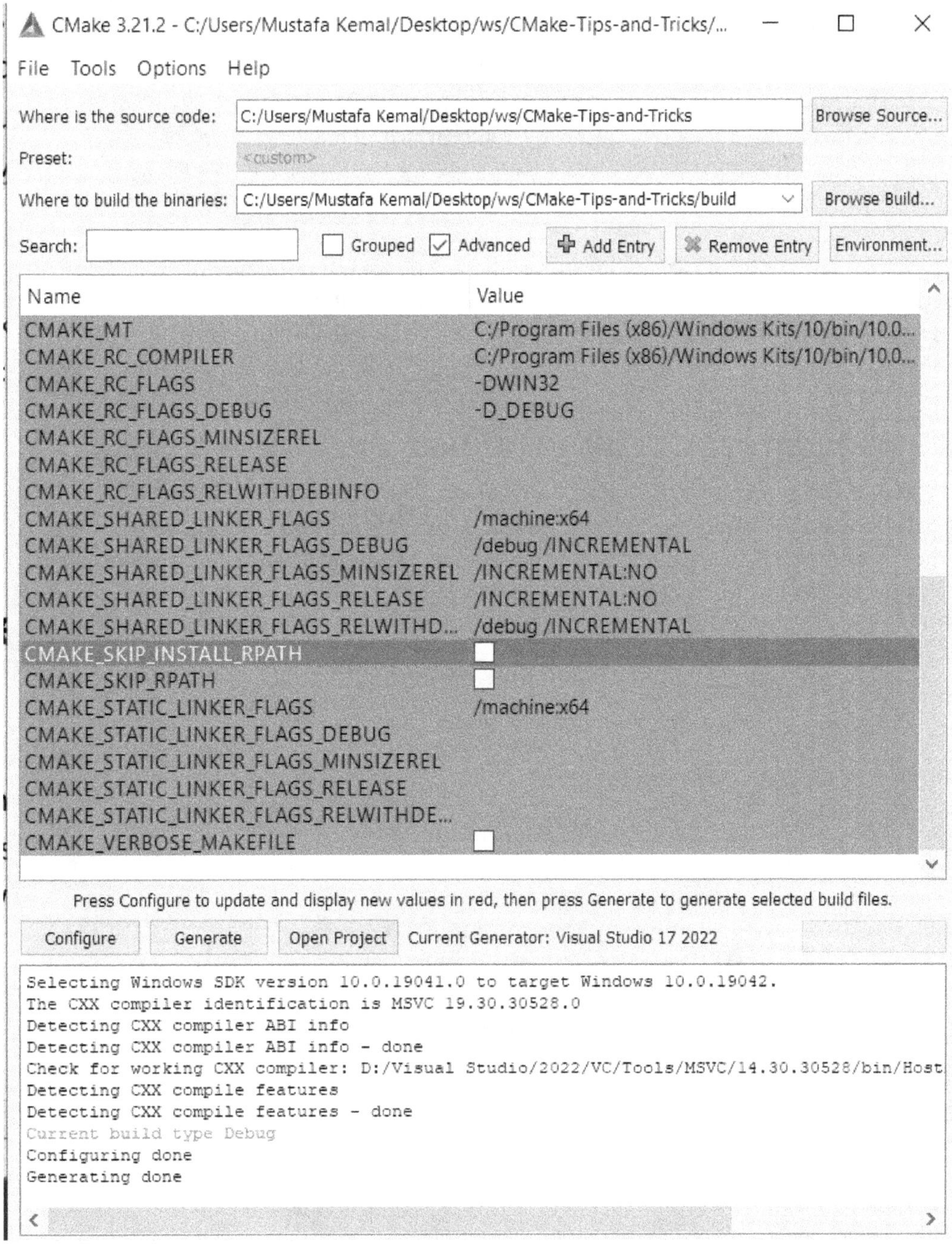

Figure 2.32 – cmake-gui in advanced mode

After tweaking any cache values, click **Configure** and then **Generate** to apply the changes.

> Tip
>
> Another useful feature is the grouping feature, which allows the grouping of cache variables into their common prefix, if there is one. Group names are determined by the first part of the variable name, up to the first underscore.

We have covered the most essential features of `cmake-gui`. Before moving to other miscellaneous stuff, if you ever need to reload cache values or delete the cache and start from scratch, you can find the **Reload Cache** and **Delete Cache** menu items in the **File** menu.

Tweaking environment variables

The CMake GUI comes with a handy environment variable editor that allows CRUD operations on environment variables. To access it, simply click the **Environment...** button on the main screen. After clicking, the **Environment Editor** window will pop up, as can be seen in the following figure:

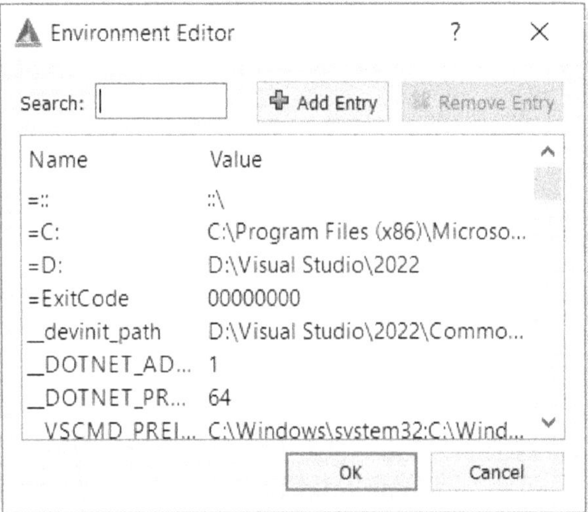

Figure 2.33 – CMake GUI environment variable editor

The **Environment Editor** window contains a list of the environment variables present in the current environment. To edit an environment variable, double-click on the value field of the desired environment variable in the table. The window also allows information to be added and deleted with the **Add Entry** and **Remove Entry** buttons.

Evaluating regular expressions with CMake

Have you ever wondered how a regular expression would get evaluated by CMake and what results it would give exactly? If so, you may have previously debugged it manually by printing out the regex match result variables with `message()`. What if I say there is a better way to do it? Let me introduce you to the `Regular Expression Explorer` tool of the CMake GUI.

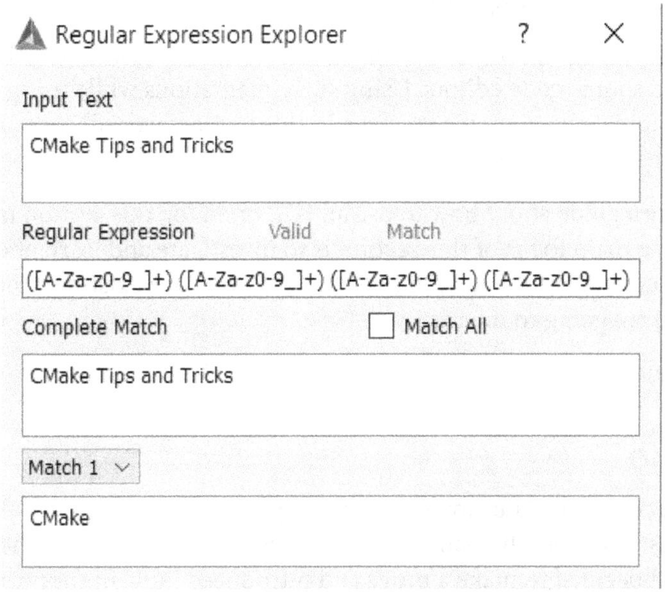

Figure 2.34 – CMake GUI Regular Expression Explorer

This hidden gem allows you to debug regular expressions using CMake's regex engine. It is located in the **Tools** menu with the name **Regular Expressions Explorer...**. Using it is pretty simple and straightforward:

1. Enter the expression in the **Regular Expression** field.

   The tool will check whether the expression is valid. If so, the **Valid** text on the screen will be green. It will turn red if CMake's regex engine did not like the expression you've given.

2. Enter the test string in the **Input Text** field. The regular expression will be matched against this text.
3. If there is any match, the **Match** word on the window will turn from red to green. The matching string will be printed in the **Complete Match** field.
4. On matching, capture groups will be assigned to **Match 1**, **Match 2**, to **Match N**, respectively, if any.

In this section, we've learned how to use CMake's native graphical interfaces. We will continue learning about using CMake by taking a look at a selection of CMake's IDE and editor integrations next.

Using CMake in Visual Studio, Visual Studio Code, and Qt Creator

Being a common tool in software development, CMake has integrations with a wide variety of IDEs and source code editors. Using such integrations while using an IDE or editor is perhaps more convenient for the user. In this section, we will cover how CMake integrates with some of the popular IDEs and editors.

If you are expecting a guide about how to use an IDE or editor, this section is not going to be about that. The main focus of this section is to investigate and learn about CMake integrations with such tools. This section assumes that you have existing experience with the IDE/editor you are going to interact with.

Let's start with Visual Studio.

Visual Studio

Visual Studio was one of the latecomers to the party when supporting CMake. Unlike other popular IDEs, Visual Studio had no native support for CMake until the year 2017. In that year, Microsoft decided to make a move and introduced built-in support for handling CMake projects, which is shipped Visual Studio 2017. Since then, it has been a solid feature of Visual Studio IDE.

To get started, obtain a copy of Visual Studio 2017 or later. For the older versions of Visual Studio, the feature is completely absent. In our examples, we'll be using Visual Studio 2022 Community Edition.

Starting a CMake project from scratch

The Visual Studio project creation feature is based on project templates. With VS2017 and upward, project templates contain a CMake project template as well. We are going to learn how to use this template to create new CMake projects.

To create a new CMake project with Visual Studio, click the **Create a new project** button on the welcome page. Alternatively, you can access it by clicking on **File | New | Project** on the main IDE window, or using the *Ctrl* + *Shift* + *N* (New Project) keyboard shortcut. The VS22 welcome screen looks like this:

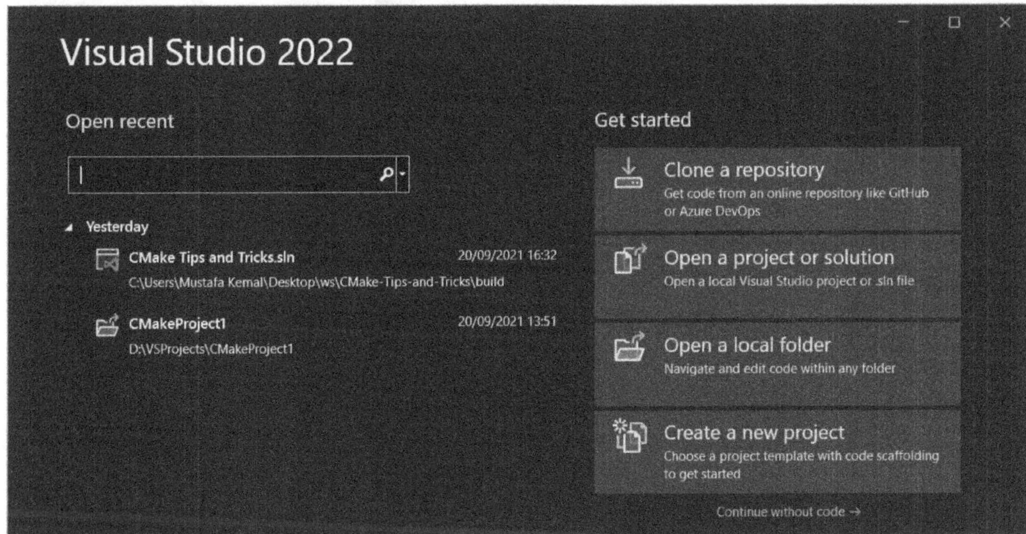

Figure 2.35 – Visual Studio 2022 welcome screen

On the **Create a new project** screen, double-click on **CMake Project** in the project template list. You can filter project templates by using the search bar located at the top of the list:

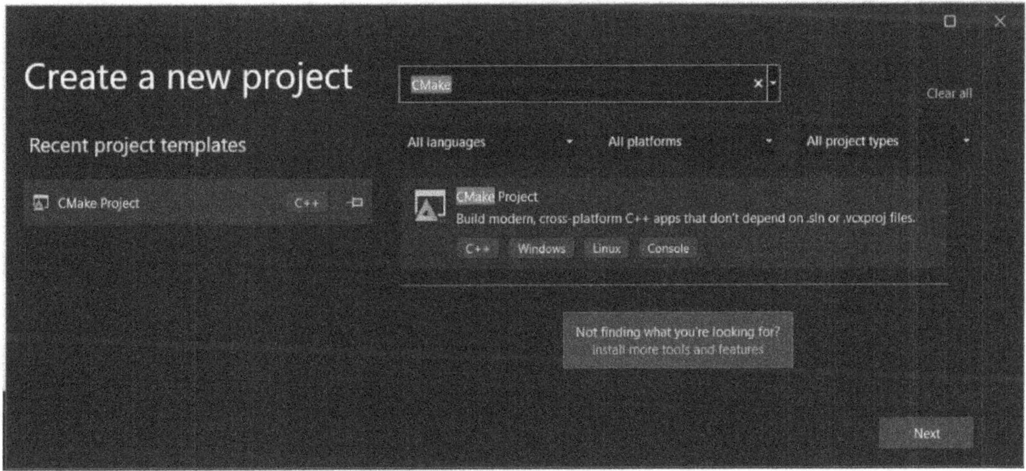

Figure 2.36 – Visual Studio 2022 Create a new project screen

After clicking **Next**, the project configuration screen will appear. On this page, you can give your new CMake project a name and choose where to put your new project. In our example, we'll go with the default project name, `CMakeProject1`.

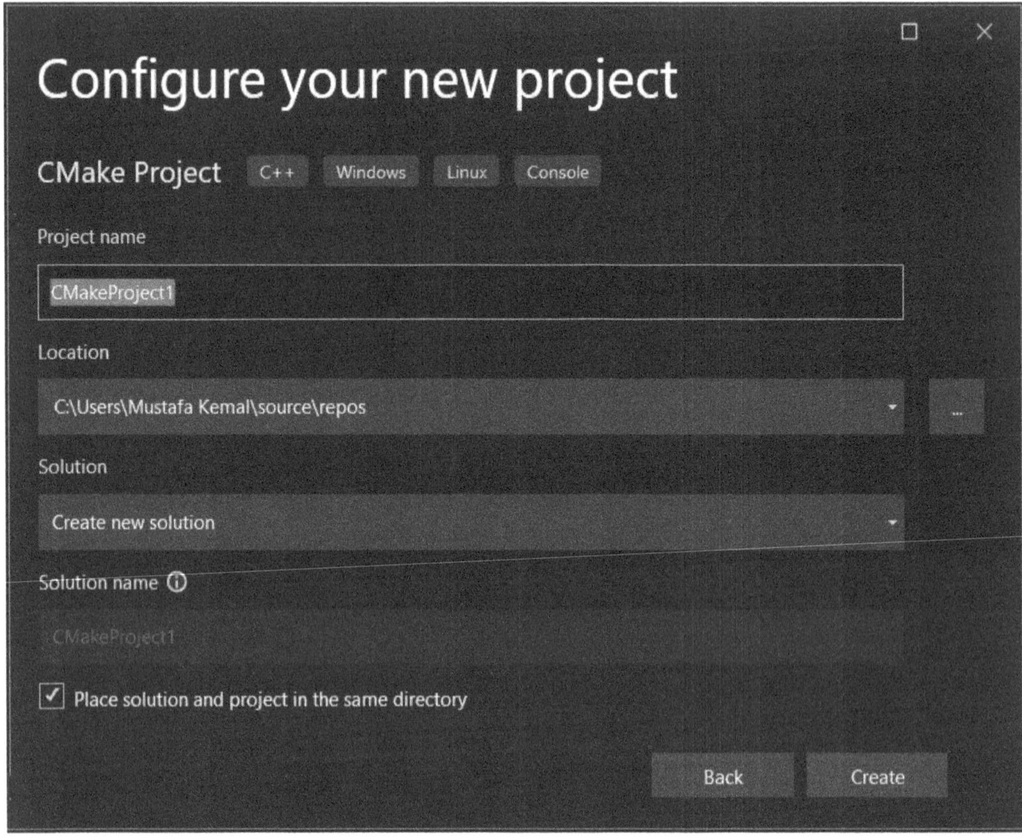

Figure 2.37 – Visual Studio 2022 new project configuration screen

After filling in the details, click **Create** to create your new CMake project. The generated project will contain a top-level `CMakeLists.txt` file, a C++ source file, and a C++ header file, named after the chosen project name. The newly created project's layout can be seen in the following figure:

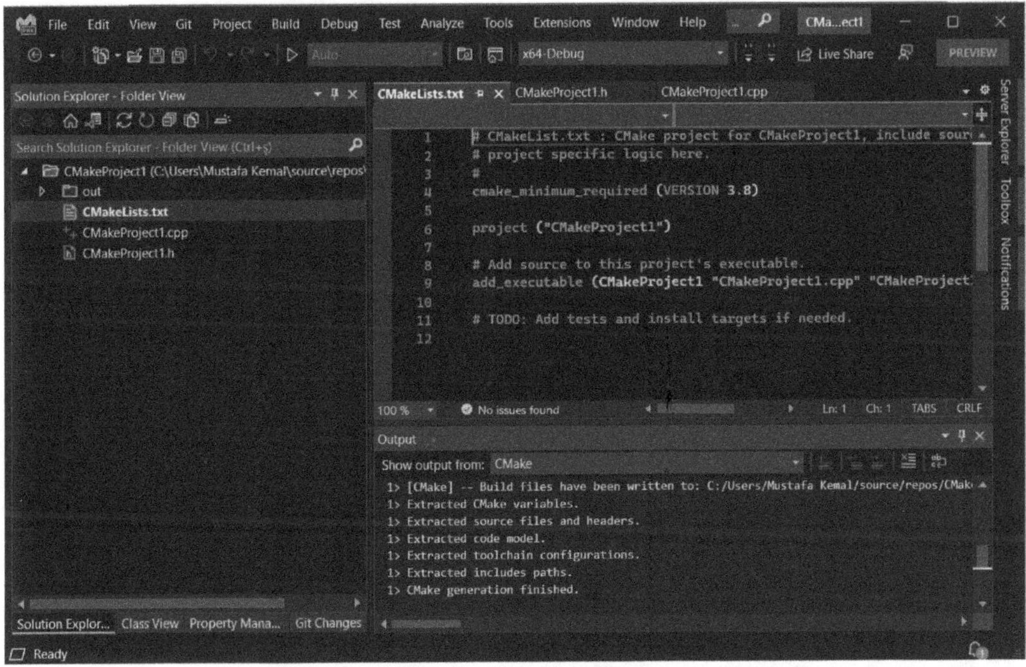

Figure 2.38 – First glance after creating a new CMake project with Visual Studio

Opening an existing CMake project

To open an existing CMake project, go to **File** | **Open** | **CMake...** and select the top-level `CMakeLists.txt` file of the project to be opened. The following figure shows what the **Open** menu looks like:

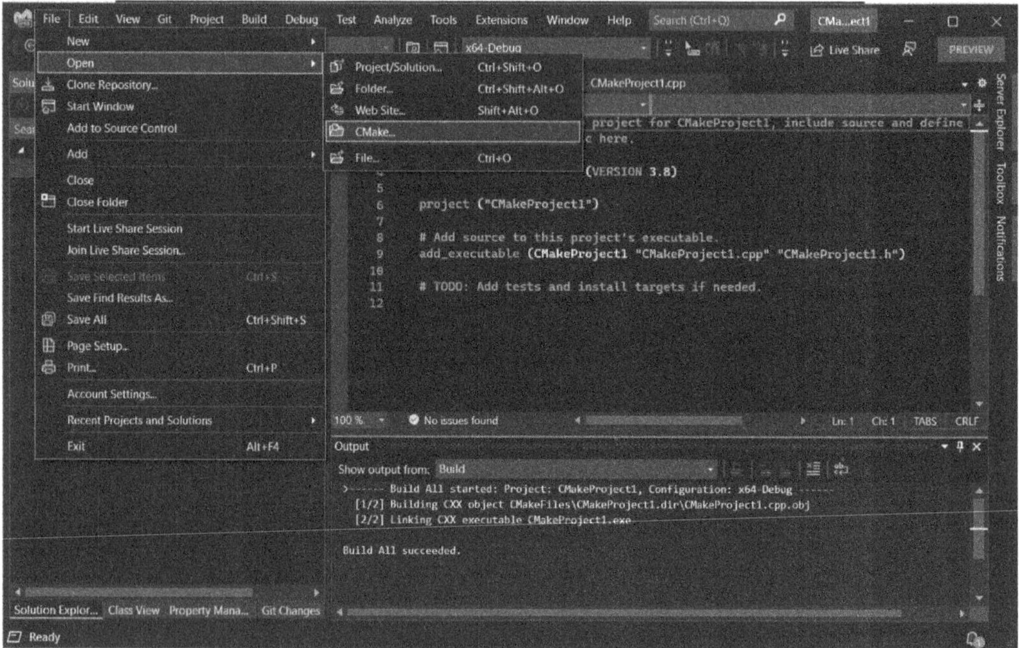

Figure 2.39 – CMake project open menu

Next, let's see how a CMake project can be configured and built.

Configuring and building a CMake project

To build a CMake project in Visual Studio, go to **Project** | **Configure** first. This will invoke the CMake `configure` step and generate the required build system files. After configuration, click **Build** | **Build All** to build the project. You can also trigger **Build All** by using the *F7* keyboard shortcut.

Note that Visual Studio will automatically invoke `configure` whenever you save a `CMakeLists.txt` file, which is a part of the project.

Executing common actions on a CMake target

Visual Studio uses a *startup target* concept for target-requiring actions such as build, debug, and launch. To set a CMake target as a startup target, use the **Select Startup Target** drop-down box located on the toolbar. Visual Studio will automatically populate this drop-down box with Cmake targets on configuration.

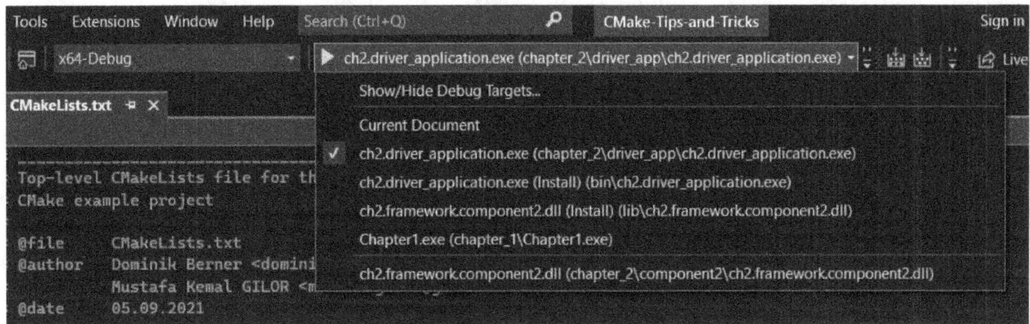

Figure 2.40 – Startup target selection drop-down menu

After setting a startup target, you can invoke actions such as **Debug**, **Build**, or **Launch**, just as you always do in Visual Studio:

1. To debug, first, click on **Debug** | **Startup Target**, and then click **Debug** | **Start Debugging** or use the *F5* keyboard shortcut.
2. To start without debugging, click on **Start without debug** or use the *Ctrl* + *F5* keyboard shortcut.
3. To build, click on **Build**, or click **Build** | **Build <target>**, or use the *Ctrl* + *B* keyboard shortcut.
4. Button locations are shown in the following figure:

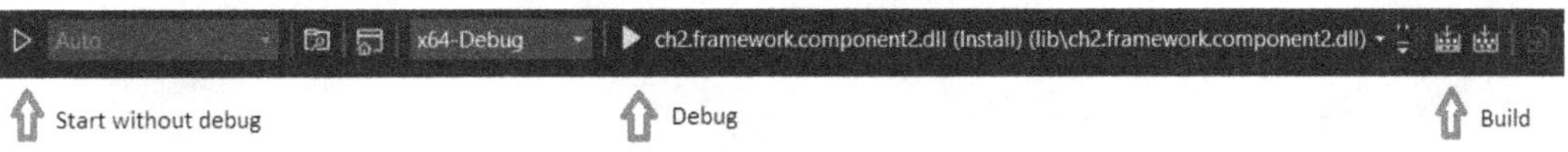

Figure 2.41 – Toolbar button locations

In this section, we've covered the basics of the Visual Studio CMake integration. In the next section, we'll continue to learn with another Microsoft product, Visual Studio Code.

Visual Studio Code

Visual Studio Code (**VSCode**) is an open source code editor developed by Microsoft. It is not an IDE but can become powerful and have IDE-like features via extensions. The extension market has a wide variety of additional content, from themes to language servers. You can find an extension for pretty much anything, which makes VSCode both powerful and liked by a wide audience. With no surprise, VSCode has an official CMake extension too. This extension was originally developed by Colby Pike (also known as *vector-of-bool*) but it is now officially maintained by Microsoft.

In this section, we are going to learn how to install the extension and perform basic CMake tasks using it.

Before going any further, VSCode must be already installed in your environment. If not, visit `https://code.visualstudio.com/learn/get-started/basics` for details on downloading and installing it.

Also, we will frequently access the Command Palette. It is strongly recommended to use it often to gain familiarity. For those asking *What the heck is the Command Palette?*, here is a screenshot:

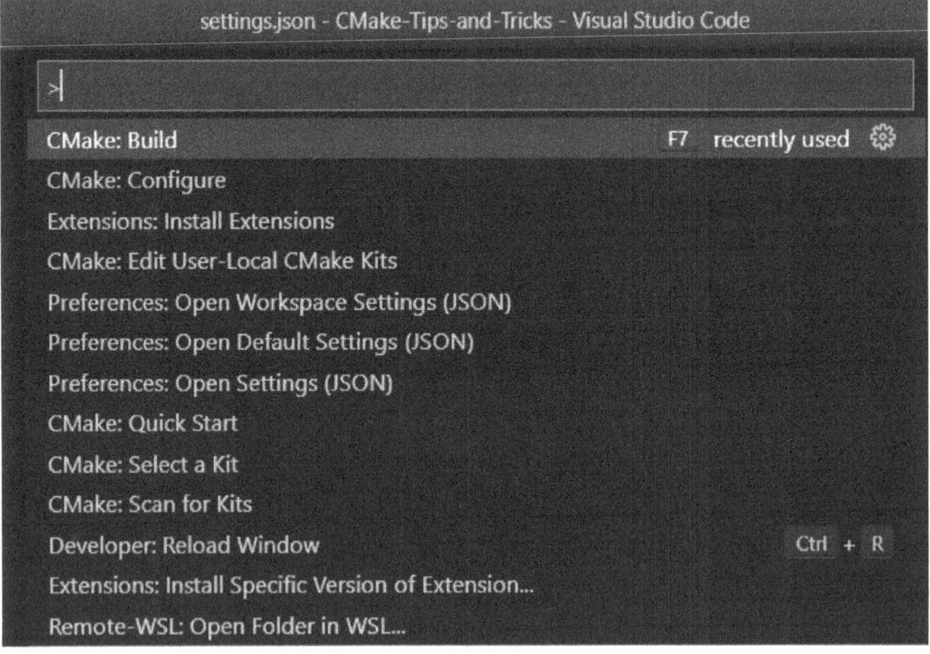

Figure 2.42 – VSCode Command Palette

Yeah, it is *that* thing. To be honest, I did not know it had a name until now. Shortcuts for accessing the Command Palette are *F1* and *Ctrl* + *Shift* + *P*. The Command Palette is the bread and butter of VSCode; it speeds up the VSCode workflow.

Installing the extension

Installing the extension is pretty straightforward and simple. To install it by using a CLI, invoke the following command (replace `code` with `code-insiders` if you're using the Insiders edition):

```
code --install-extension ms-vscode.cmake-tools
```

Alternatively, you can do the same in the VSCode GUI as well. Open VSCode and navigate to the **Extensions** page by clicking **Extensions** on the left-side navigation pane. Alternatively, you can use the *Ctrl* + *Shift* + *X* shortcut. Type `CMake Tools` into the extension search box and select **CMake Tools** from **Microsoft**. Be careful not to confuse it with the CMake extension. Press the **Install** button to install it.

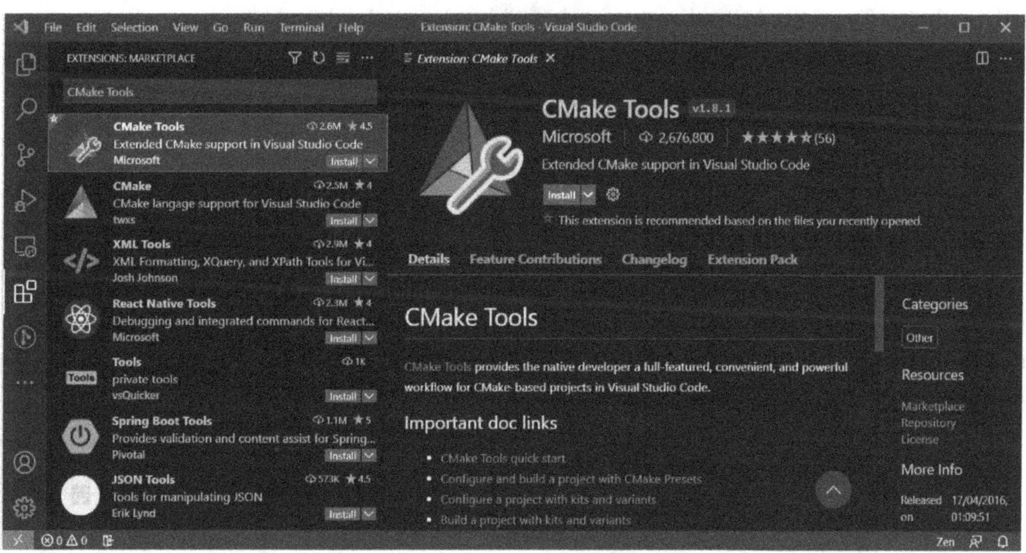

Figure 2.43 – VSCode extensions marketplace

After installation completes, the extension is ready to use.

Quick Start project

The VSCode CMake Tools extension offers a **Quick Start** option that bootstraps a CMake project with example C++ code. To use it, first open the destination folder by using the **File | Open Folder...** menu, and then press *F1* and type `cmake quick start`. Select **CMake: Quick Start** and press *Enter* on the keyboard.

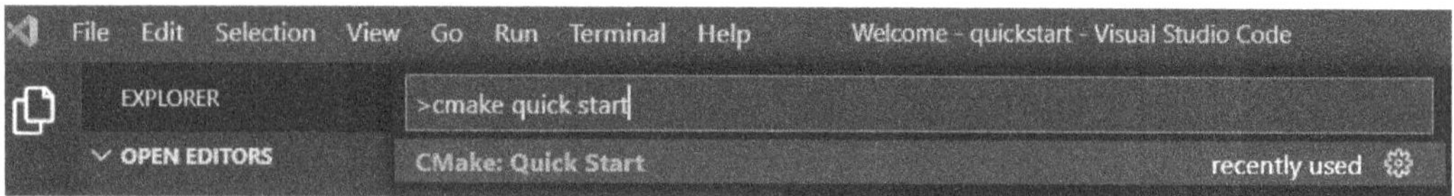

Figure 2.44 – Command Palette – Locating CMake: Quick Start

Firstly, the extension will ask which kit to use. Select the one that is appropriate for your new project. Kits will be further discussed in the *Dealing with kits* section.

After selecting a kit, you will be asked to input a project name. This will be the name of your top-level CMake project. Enter a name of your choice.

Lastly, a choice for example application code will be shown. In this choice, you will be asked to create an executable application project or a library project. Select one, and voil•! You've got yourself a working CMake project. Upon selection, `CMakeLists.txt` and `main.cpp` files will be generated. The content of these files slightly varies between executable and library choices.

Opening an existing project

There is nothing special about opening a CMake project in VSCode. Open the folder that contains the top-level `CMakeLists.txt` file of your project. The CMake Tools extension will automatically recognize this folder as a CMake project, and all CMake-related commands will become available on the VSCode Command Palette.

Configuring, building, and cleaning a project

To configure a CMake project, select the **CMake: Configure** menu item from the Command Palette. To build the project, choose a build target by selecting the **CMake: Set Build Target** menu item from the Command Palette. This will let you choose what will be built when a build is invoked. Lastly, select **CMake: Build** to build the selected build target. To build a specific target without setting it as a build target, use the **CMake: Build Target** menu item.

To clean build artifacts, use the **CMake: Clean** Command Palette item. This will run CMake's `clean` target and remove any build artifacts.

Debugging a target

To debug a target, choose a debug target by selecting the **CMake: Set Debug Target** menu item from the Command Palette. You'll see the debuggable targets listed.

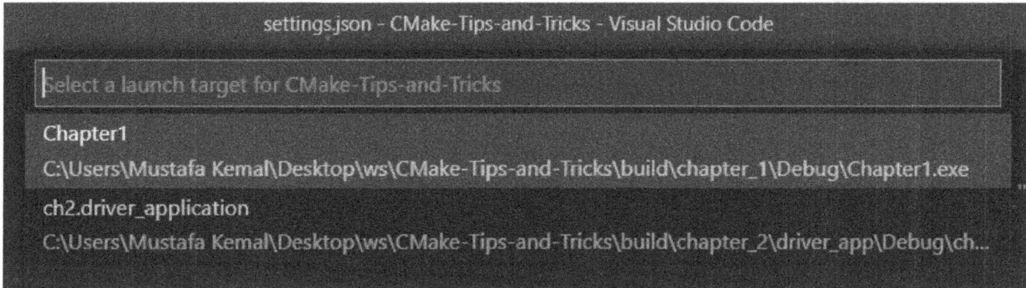

Figure 2.45 – Debug target selection

Select the target and select **CMake: Debug** (*Ctrl* + *F5*) from the Command Palette. The selected target will be started under the debugger.

If you want to run the selected target without the debugger, select **CMake: Run Without Debugging** (*Shift* + *F5*) instead.

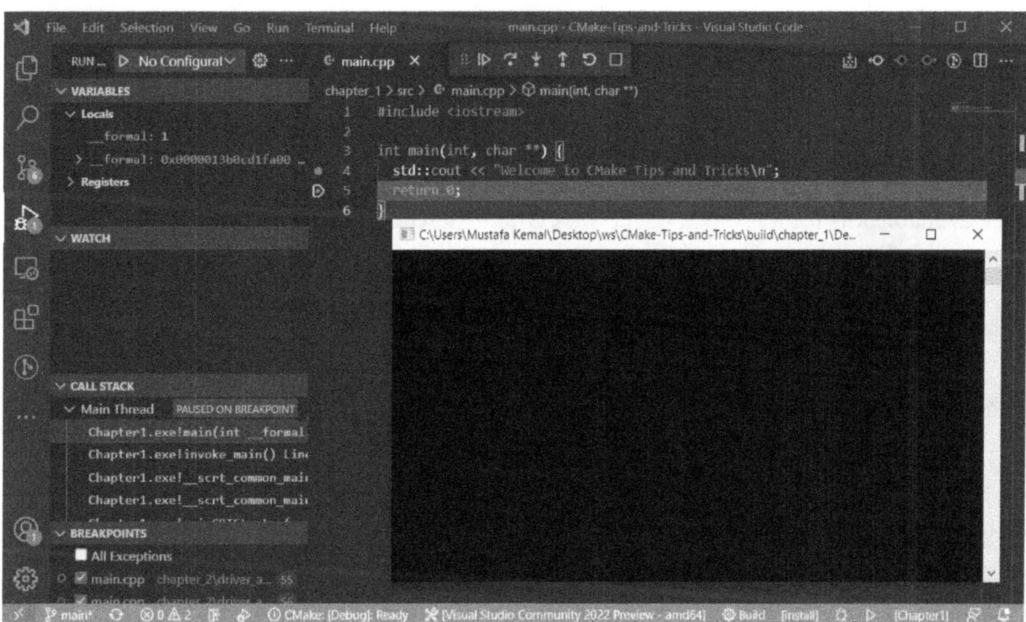

Figure 2.46 – Executable Chapter1 target being debugged

In the next section, we will look at how we can provide arguments to the debugged target.

Passing arguments to the debugged target

The target you're trying to debug might be needing command-line arguments. To pass command-line arguments to the debug target, open VSCode's `settings.json` file and append the following lines:

```
"cmake.debugConfig": {
        "args": [
            "<argument1>",
            "<argument2>"
        ]
    }
```

In the `args` JSON array, you can place any number of arguments your target requires. These arguments will be passed to all future debug targets unconditionally. If you want to have fine-grained control over the arguments, it is better to define a `launch.json` file instead.

Dealing with kits

A kit in the CMake Tools extension represents a combination of tools that can be used to build the project; hence, the term *kit* is pretty much a synonym for the toolchain. Kits make it easier to work in a multi-compiler environment, allowing a user to choose which exact compiler to work with. Kits can be discovered automatically by the extension, read from toolchain files, or defined by the user manually.

To see available kits for a project, select the **CMake: Select a Kit** menu item from the Command Palette (*F1* or *Ctrl*+*Shift*+P).

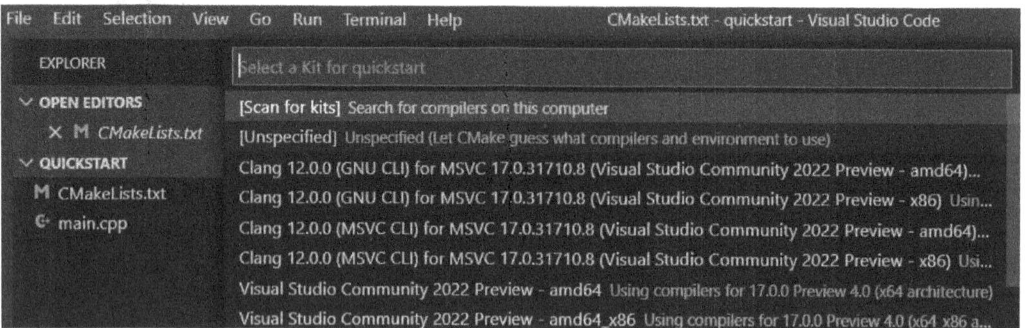

Figure 2.47 – Kit selection list

The selected kit will be used for configuring the CMake project, which means the tools that are defined in the kit will be used for compiling the project. Kit selection will automatically trigger a CMake configuration.

By default, kits are scanned by the extension automatically. As a result, discovered toolchains are listed as options in the kit selection menu. If your toolchain is not displayed here, this means CMake Tools failed to discover it. In such a scenario, try to re-scan for kits first. If it is still missing, you can always define additional kits by adding them to the `user-local cmake-tools-kits.json (1)` file manually.

Adding a new kit is not usually necessary since the extension does a good job of discovering the toolchains. In the odd case of failure, there is a kit template here, which you can customize and append to the user-local `cmake-tools-kits.json` file to define a new kit. To open the user-local kits file, select the **CMake: Edit User-Local CMake Kits** menu item from the Command Palette:

```
{
  "name":"<name of the kit>",
  "compilers" {
    "CXX":"<absolute-path-to-c++-compiler>",
    "C": "<absolute-path-to-c-compiler>"
  }
}
```

> **Note**
>
> In older versions of the CMake Tools extension, the `cmake-tools-kits.json` file may be named `cmake-kits.json` instead.

Keep in mind that if your kit name collides with an autogenerated name from CMake Tools, CMake Tools will override your entry on a scan, so, always give unique names to your kit definitions.

For further information about kits, refer to `https://github.com/microsoft/vscode-cmake-tools/blob/develop/docs/kits.md`.

Qt Creator

Qt Creator is another IDE that supports CMake projects. CMake support is decent and the support comes out of the box without the need for any extra plugins. In this section, we are going to take a quick glance into Qt Creator's CMake support.

As always, ensure that you have the IDE installed and configured properly in your environment first.

Qt Creator version 5.0.1 is used in the examples.

Adding your CMake installation

In order to use CMake with Qt Creator, the path of CMake must be defined in Qt Creator. To view and define CMake paths, navigate to **Tools | Options | Kits | CMake**.

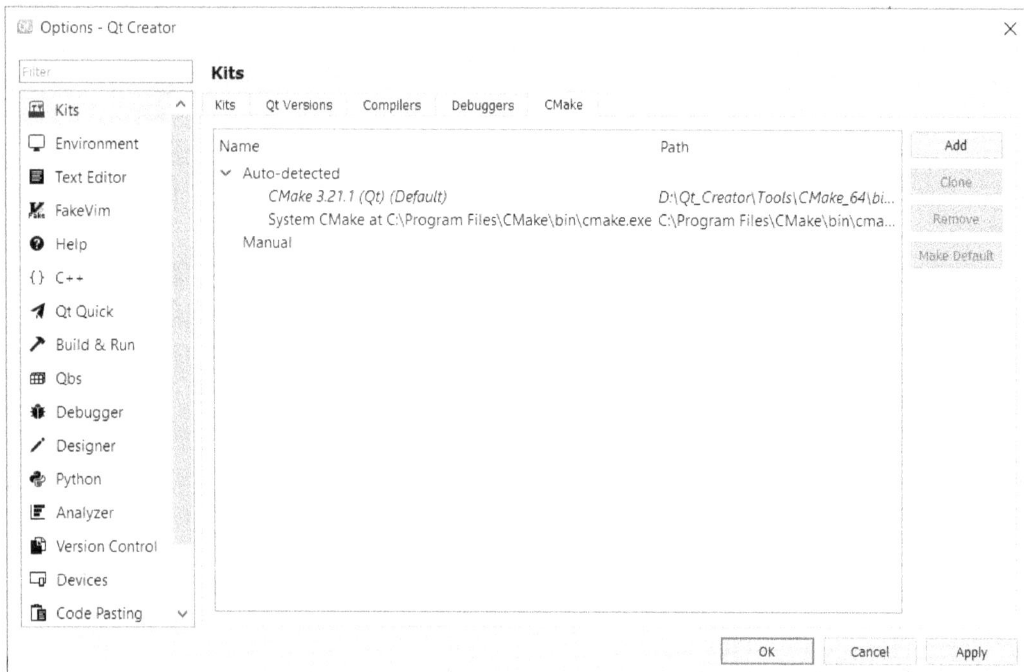

Figure 2.48 – Qt Creator CMake path settings

Albeit with a lack of any manual definition, Qt Creator was able to discover CMake installations in the system. The first entry under the **Auto-detected** section is the CMake executable shipped together with Qt Creator. The second one is the system's CMake installation. To select which CMake executable to run in Qt Creator, select the desired entry and click the **Make Default** button.

To add a new CMake executable, click **Add**. This will append a new entry in the **Manual** section and bring up a window to fill in the details for the new entry.

Figure 2.49 – Adding a new CMake executable

The window fields are described in detail here:

- **Name**: A unique name to distinguish a new CMake executable entry.
- **Path**: A CMake executable path (`cmake/cmake.exe`).
- **Version**: The version of the CMake (deduced by Qt Creator).
- **Help file**: An optional Qt Creator help file for the executable. This will allow CMake Help to appear upon pressing *F1*.
- **Autorun CMake**: Check this to run CMake automatically on any `CMakeLists.txt` file changes.

After filling in the details, click **Apply** to add the new CMake executable to Qt Creator. Don't forget to set it as default if you intend Qt Creator to use it.

Creating a CMake project

Creating a CMake project in Qt Creator follows the exact same steps as for regular project creation. Qt Creator does not treat CMake as an external build system generator. Instead, it lets its users choose between three build system generators, which are *qmake*, *cmake*, and *qbs*. Any type of Qt project can be started by any of these build system generators from scratch.

To create a CMake project in Qt Creator, click **File** | **New File or Project...** (*Ctrl* + *N*) and choose the type of project from the **New File or Project** window. We'll go with **Qt Widgets Application** for our example.

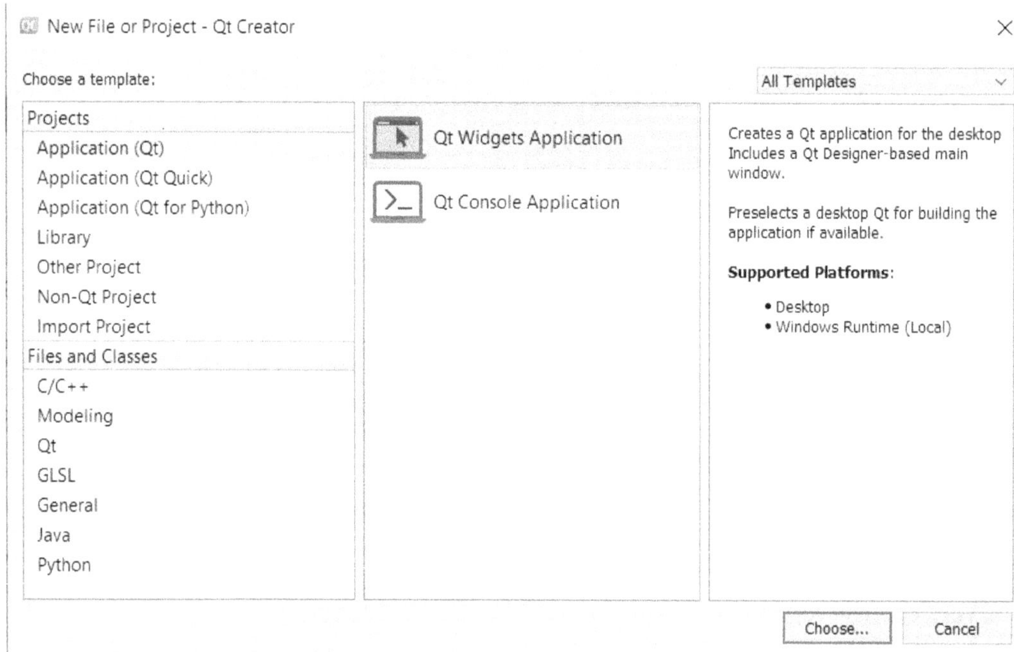

Figure 2.50 – Qt Creator New File or Project window

Upon selection, the project creation wizard will appear. Fill in the details as desired. Select **CMake** in the **Define Build System** step, as shown in the following screenshot:

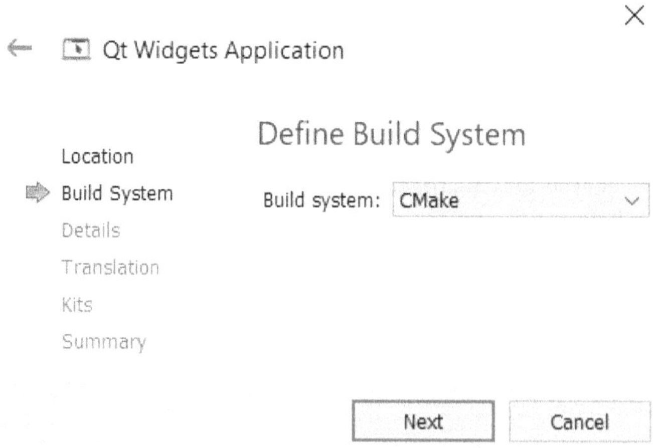

Figure 2.51 – Qt Creator new project wizard build system selection

That's it! You've got yourself a Qt application with the CMake build system.

The following figure shows a newly created CMake project:

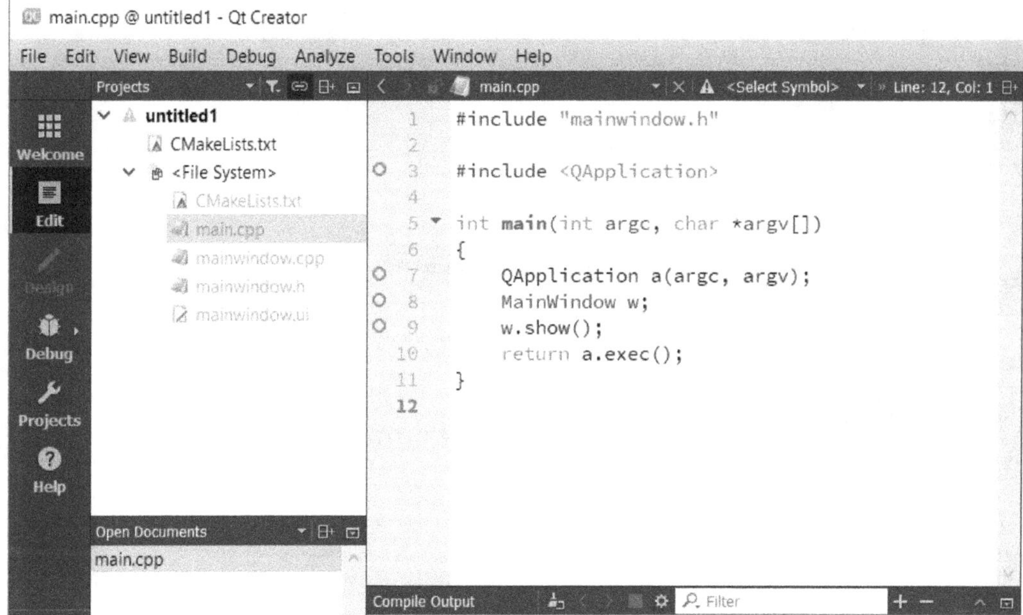

Figure 2.52 – Generated CMake-based Qt widgets application project

Opening an existing CMake project

To open an existing CMake project with Qt Creator, go to the **File | Open File or Project...** (*Ctrl* + *O*) menu item. Select the top-level `CMakeLists.txt` file of the project and then click **Open**. Qt Creator will prompt you to choose a kit for your project. Select your preferred kits and then click on the **Configure Project** button. The project will be open and the CMake configure step will be run with the selected kits.

As an example, the *CMake Best Practices* project opened with Qt Creator is shown in this figure:

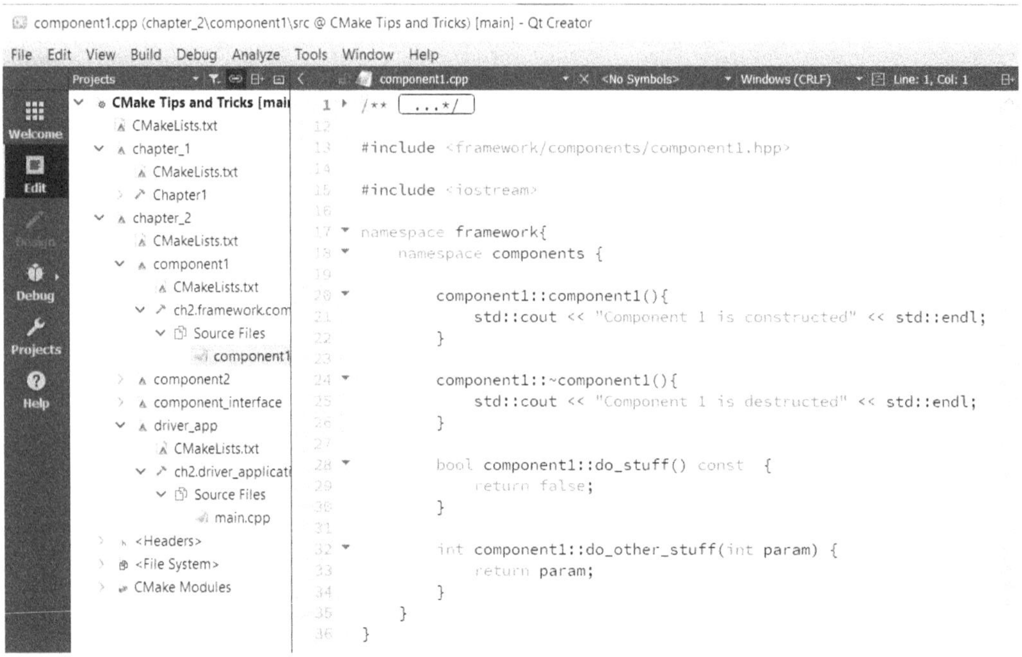

Figure 2.53 – A glance at the CMake Tips and Tricks example project in Qt Creator

After opening a CMake project for the first time, Qt Creator will create a file named `CMakeLists.txt.user` in the project's root directory. This file contains Qt-specific details that cannot be stored in the `CMakeLists.txt` file, such as kit information and editor settings.

Configuring and building

In most scenarios (for example, project opening and saving changes to `CMakeLists.txt`), Qt Creator will run CMake configuration automatically without having to run it manually. To run CMake configuration manually, click on the **Build | Run CMake** menu item.

After configuration, press the hammer icon in the left-most corner to build the project. Alternatively, the *Ctrl* + *B* keyboard shortcut can be used. This will build the whole CMake project. To build a specific CMake target only, use the **Locator** next to the **Build** button. Type `cm` and then press the space bar on your keyboard.

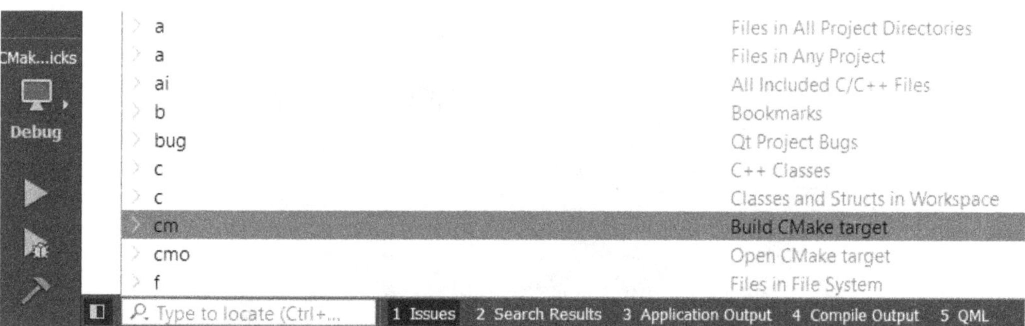

Figure 2.54 – Qt Creator locator suggestions

The locator will display CMake targets available to build. Select the desired target either by highlighting it and pressing *Enter*, or by clicking on it directly using the mouse.

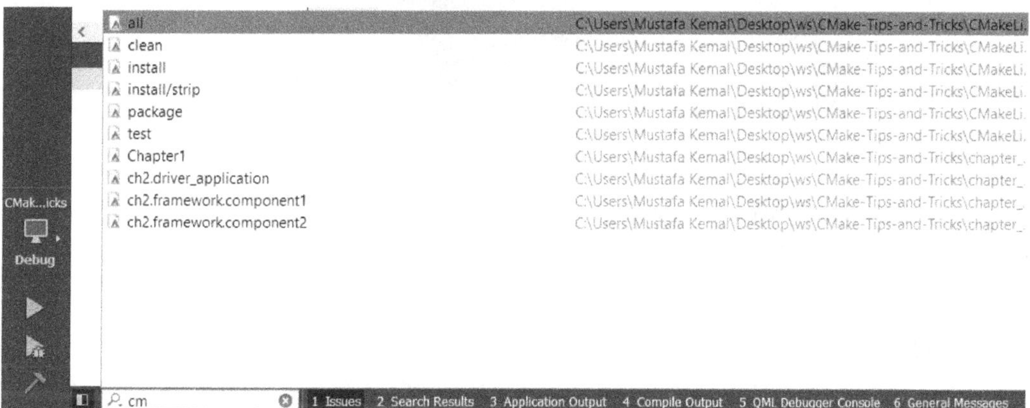

Figure 2.55 – Available CMake targets to build displayed on the locator

The selected CMake target (and naturally, its dependencies) will be built.

Run and debug

To run or debug a CMake target, press the kit selector button (the computer icon on the left navigation bar) and select the CMake target. Then, click either the run button (the *play icon* under the kit selector) to run or the debug button (the *play icon with a bug*) to debug.

The following figure shows the kit selector menu content:

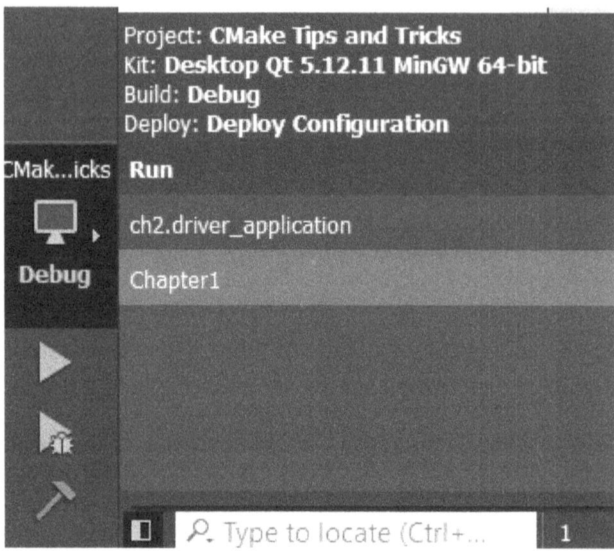

Figure 2.56 – Kit selector displaying CMake targets

Here, we conclude the basics of using CMake with Qt Creator. For more advanced topics, you can consult the resources given in the *Further reading* section.

Summary

In this chapter, we covered the fundamental methods of interacting with CMake, which are the CLI and the GUI. We also covered various IDE and editor integrations that are essential for daily workflow. Using any kind of tool requires knowledge of how to interact with it. Learning ways of interaction allows us to better utilize the tool itself and to reach our goals easier.

In the next chapter, we will be talking about the building blocks of a CMake project, which will enable you to create a well-structured, production-ready CMake project from scratch.

Questions

To reinforce what you have learned in this chapter, try to answer the following questions. If you are having a hard time answering them, go back to the relevant section and re-visit the topic:

1. Describe how a CMake project can be configured from the CLI in the build folder in the project's root directory with each of the following:

   A. A different C++ compiler, located at `/usr/bin/clang++`

   B. A Ninja generator

   C. A `-Wall` compiler flag for the `Debug` build type

2. Describe how the project previously configured in *Q1* can be built using CMake using the command line with each of the following:

   A. Eight parallel jobs

   B. The `--trace` option in the Unix Makefiles generator

3. Describe how the project previously built in *Q1* can be installed using CMake using the command line `directory/opt/project`.
4. Assuming the `CMake-Best-Practices` project is already configured and built, which command must be invoked to only install the `ch2.libraries` component?
5. What is an advanced variable in CMake?

Further reading

There are a lot of guides and documents regarding the topics we have discussed in this chapter. You can find a non-exhaustive list of recommended material to read here:

- CMake CLI documentation: `https://cmake.org/cmake/help/latest/manual/cmake.1.html`.
- CMake projects in Visual Studio: `https://docs.microsoft.com/en-us/cpp/build/cmake-projects-in-visual-studio?view=msvc-160`.
- CMake support in Visual Studio: `https://devblogs.microsoft.com/cppblog/cmake-support-in-visual-studio/`.

- CMake tools extension for VSCode: `https://devblogs.microsoft.com/cppblog/cmake-tools-extension-for-visual-studio-code/`.
- VSCode CMake Tools documentation: `https://github.com/microsoft/vscode-cmake-tools/tree/develop/docs#cmake-tools-for-visual-studio-code-documentation`.
- Debugging in VSCode: `https://code.visualstudio.com/docs/editor/debugging`.
- Qt Creator locator guide: `https://doc.qt.io/qtcreator/creator-editor-locator.html`.
- Qt Creator user interface: `https://doc.qt.io/qtcreator/creator-quick-tour.html`.

3
Creating a CMake Project

By now, you should be familiar with how to use CMake and its basic concepts, such as the two-stage build. So far, we have only looked at how CMake is used with code that already exists, but the more interesting part is when it comes to building an application using CMake. In this chapter, you will learn how to build executables and libraries and how to use them together. We will have an in-depth look at creating different kinds of libraries and we will present some good practices about how to structure CMake projects. As libraries often come with a variety of compiler settings, we will learn how we can set them and pass them onto dependent libraries if necessary. Since dependencies in projects can get quite complicated, we will also learn how to visualize the dependencies between the different targets.

In this chapter, we will cover the following topics:

- Setting up a project
- Creating a "hello world" executable
- Creating a simple library
- Bringing it together

Technical requirements

As with the previous chapters, all the examples have been tested with CMake 3.21 and run on either of the following compilers:

- GCC 9 or newer.
- Clang 12 or newer.
- MSVC 19 or newer.
- All the examples and source code for this chapter are available in this book's GitHub repository, `https://github.com/PacktPublishing/CMake-Best-Practices`, in the `chapter_3` subfolder.

Setting up a project

Although CMake can work with almost any file structure for a project, there are some good practices regarding how to organize files. The examples in this book use the following common pattern:

```
├── CMakeLists.txt
├── build
├── include/project_name
└── src
```

There are three folders and one file present in a minimal project structure. They are as follows:

- `build`: The folder where the `build` files and binaries are placed.
- `include/project_name`: This folder contains all the header files that are publicly accessible from outside the project. Adding a subfolder that contains the project's name is helpful since includes are done with `<project_name/somefile.h>`, making it easier to figure out which library a header file is coming from.
- `src`: This folder contains all the source and header files that are private.
- `CMakeLists.txt`: This is the root CMake file.

The `build` folder can be placed almost anywhere. Placing it in the project root is just very convenient. However, we strongly advise against choosing any non-empty folder as the `build` folder. Especially putting the built files into either `include` or `src` is considered as a bad practice. Additional folders such as `test` or `doc` can come in handy for organizing test projects and documentation pages.

Working with nested projects

When you're nesting projects inside each other, each project should map the file structure above and each `CMakeLists.txt` should be written so that the subproject can be built standalone. This means that each `CMakeLists.txt` file of a subproject should specify `cmake_minimum_required` and, optionally, a project definition as well. We will cover large projects and superbuilds in depth in *Chapter 9, Creating Reproducible Build Environments*.

Nested projects look something like this:

```
├── CMakeLists.txt
├── build
├── include/project_name
├── src
└── subproject
    ├── CMakeLists.txt
    ├── include
    │   └── subproject
    └── src
```

Here, the folder structure is repeated inside the `subproject` folder. Sticking to such a folder structure and making the subproject buildable on its own makes it easier to move projects around. It also allows you to only build parts of a project, which can come in handy for big projects where build times may get rather long.

Now that we have the file structure covered, let's start by creating a simple executable without any special dependencies. Later in this chapter, we will create various kinds of libraries and bring them all together.

Creating a "hello world" executable

First, we will create a simple executable from a simple hello world C++ program. The following C++ program will print out `Welcome to CMake Best Practices`:

```
#include <iostream>
int main(int, char **) {
  std::cout << "Welcome to CMake Best Practices\n";
  return 0;
}
```

To build this, we need to compile it and give the executable a name. Let's see what the `CMakeLists.txt` file to build this executable looks like:

```
cmake_minimum_required(VERSION 3.21)

project(
    hello_world_standalone
    VERSION 1.0
    DESCRIPTION"A simple C++ project"
    HOMEPAGE_URL https://github.com/PacktPublishing/CMake-Best-
      Practices
    LANGUAGES CXX
)

add_executable(hello_world)
target_sources(hello_world PRIVATE src/main.cpp)
```

With the first line, `cmake_minimum_required(VERSION 3.21)`, we tell CMake which version of CMake we are expecting to see and which features CMake will enable. For this book, we are using CMake 3.21 for all the examples, but for compatibility reasons, you could select a lower version.

For this example, version 3.1 would be the absolute minimum because, before that, the `target_sources` command is not available. It is good practice to put the `cmake_minimum_required` command at the top of each `CMakeLists.txt` file.

Next, the project is set up using the `project()` command. The first argument is the project's name – in our case, `"hello_world_standalone"`.

Next, the version of the project is set to version 1.0. What follows is a brief description and the home page's URL. Finally, the `LANGUAGES CXX` attribute specifies that we are building a C++ project. Apart from the project's name, all the arguments are optional.

Invoking the `add_executable(hello_world)` command creates a target called `hello_world`. This will also be the name of the executable file that's created by this target.

Now that the target has been created, adding the C++ source files to the target is done with `target_sources`. `Chapter3` is the target name, as specified in `add_executable`. `PRIVATE` defines that the sources are only used to build this target and not for any dependent targets. After the scope specifier, there's a list of source files that are relative to the path of the current `CMakeLists.txt` file. If needed, the location of the currently processed `CMakeLists.txt` file can be accessed with the `CMAKE_CURRENT_SOURCE_DIR` variable.

Sources can be added directly to the `add_executable` function or separately using the `target_sources` functions. Adding them with `target_sources` allows you to explicitly define where the sources can be used by using `PRIVATE`, `PUBLIC`, or `INTERFACE`. However, specifying anything other than `PRIVATE` only makes sense for library targets.

A common pattern that you often see is naming the main executable of a project after the project's name, like this:

```
project(hello_world
...
)
add_executable(${PROJECT_NAME})
```

While this seems to be convenient at first glance, it should be avoided. The project's name and the target carry a different semantic meaning, so they should be treated as separate things, so using `PROJECT_NAME` as the name for targets should be avoided.

Executables are important and quite easy to create, but unless you're building a huge monolith, you should be using libraries to modularize and distribute code. In the next section, we will learn how libraries are built and how to handle different linking methods.

Creating a simple library

Creating a library works similarly to creating an executable, although there are a few additional things to consider since library targets are usually used by other targets, either in the same project or by other projects. Since libraries usually have an internal part and a publicly visible API, we must take this into account when adding files to the project.

A simple project for a library will look like this:

```
cmake_minimum_required(VERSION 3.21)

project(
```

```
    ch3.hello_lib
    VERSION 1.0
    DESCRIPTION
      "A simple C++ project to demonstrate creating executables
        and libraries in CMake"
    LANGUAGES CXX)

add_library(hello)

target_sources(
    hello
    PRIVATE src/hello.cpp src/internal.cpp)

target_compile_features(hello PUBLIC cxx_std_17)

target_include_directories(
    hello
    PRIVATE src/hello
    PUBLIC include)
```

Again, the file starts with setting `cmake_minimum_required` and the project information, which should be familiar to you by now.

Next, the target for the library is created with `add_library` – in this case, the type of the library is not determined. We could pass `STATIC` or `SHARED` instead to determine the linking type of the library explicitly. By omitting this, we allow any downstream consumers of the library to choose how to build and link it. Generally speaking, static libraries are the easiest to handle. More information about building shared libraries can be found in the *Symbol visibility in shared libraries* subsection.

If the type of the library is omitted, the `BUILD_SHARED_LIBS` variable determines whether the libraries are built as shared or static libraries by default. This variable should not be set unconditionally in the CMake files of a project; it should always be passed by the user.

Next, the sources for the library are added with target_sources. The first argument is the target name, followed by the sources separated by the PRIVATE, PUBLIC, or INTERFACE keyword. In practice, source files are almost always added with the PRIVATE specifier. The PRIVATE and PUBLIC keywords specify where the sources should be used for compiling. Specifying PRIVATE means that the sources will only be used in the target hello itself. If PUBLIC is used, then the sources will be added to hello and any target that links to hello. As we mentioned previously, this is not usually desired. The INTERFACE keyword would mean that the sources are not added to hello but should be added to anything that links against hello. Generally, anything that's specified as PRIVATE for a target can be seen as a build requirement. Finally, the include directories for the library are set using target_include_directories. All the files inside the folders specified by this command can be accessed using #include <file.hpp> (with the angle brackets) instead of #include "", although the version with the quotes may still work.

PRIVATE includes paths that will not be included in the target property; that is, INTERFACE_INCLUDE_DIRECTORIES. CMake will read this property when targets depend on the library to determine which of the include directories are visible to the dependee.

Since the C++ code of the library uses features that are tied to a modern version of C++ such as C++ 11/14/17/20 or C++ 23 (to be released soon), we must set the cxx_std_17 property. Since it is used to compile the library itself and to interface against the library, it is set to PUBLIC. Setting it to PUBLIC or INTERFACE is only necessary if the header files contain code that requires a certain standard. If only the internal code is dependent on a certain standard, setting it to PRIVATE is preferred. Generally, try to set the public C++ standard to the lowest that is working. It is also possible to only enable certain features of one of the modern C++ standards.

A full list of the available compile features can be found at https://cmake.org/cmake/help/latest/prop_gbl/CMAKE_CXX_KNOWN_FEATURES.html.

Naming libraries

When you're creating libraries using add_library(<name>), the name of the library must be globally unique inside the project as name collisions are errors. By default, the actual filename of the library is constructed according to the conventions on the platform, such as lib<name>.so on Linux and <name>.lib or <name>.dll on Windows. The name of the file can be changed from the default behavior by setting the OUTPUT_NAME property of a target. This can be seen in the following example, where the name of the output file has been changed from ch3_hello to hello:

```
add_library(ch3_hello)
```

```
set_target_properties(
    ch3_hello
    PROPERTIES OUTPUT_NAME hello
)
```

Avoid names for libraries with the prefix or postfix of *lib* as CMake may append or prepend the appropriate string to the filename, depending on the platform.

A frequently used naming convention for shared libraries is to add the version to the filename to specify the build version and API version. By specifying the `VERSION` and `SOVERSION` properties for a library target, CMake will create the necessary filenames and symlinks when building and installing the library:

```
set_target_properties(
    hello
    PROPERTIES VERSION ${PROJECT_VERSION} # Contains 1.2.3
    SOVERSION ${PROJECT_VERSION_MAJOR} # Contains only 1
)
```

On Linux, this example will result in a filename of `libhello.so.1.0.0` with symlinks from `libhello.so` and `libhello.so.1` pointing to the actual library file. The following screenshot shows the generated file and the symbolic links pointing to it:

```
lrwxrwxrwx 1 conan    14 Sep 30 19:51 libhellod.so -> libhellod.so.1
lrwxrwxrwx 1 conan    18 Sep 30 19:51 libhellod.so.1 -> libhellod.so.1.0.0
-rwxr-xr-x 1 conan 91912 Sep 30 19:51 libhellod.so.1.0.0
```

Figure 3.1 – The library file and the generated symlinks when building with the SOVERSION property

Another convention that's often seen in projects is adding a different postfix to the filename for the various build configurations. CMake handles this by setting the `CMAKE_<CONFIG>_POSTFIX` global variable to whatever the convention is or adding the `<CONFIG>_POSTFIX` property to the targets. If this variable is set, the postfix will be automatically added to non-executable targets. As with most global variables, they should be passed to CMake over the command line or as a preset rather than hardcoded in the `CMakeLists.txt` file.

The postfix for debug libraries can also be set explicitly to a single target, as shown in the following example:

```
set_target_properties(
hello
PROPERTIES DEBUG_POSTFIX d)
```

This will result in the library file and symlinks being named `libhellod.so` when you're building in the debug configuration. Since linking libraries is done over targets rather than filenames in CMake, picking the correct filename happens automatically, so we do not have to keep track manually. However, one thing to watch out for when linking shared libraries is symbol visibility. We'll look at this in the next section.

Symbol visibility in shared libraries

To link against shared libraries, the linker has to know which symbols can be used from outside the library. These symbols can be classes, functions, types, and more, and the process of making them visible is called exporting.

Compilers have different ways and default behavior when specifying symbol visibility, which makes specifying this in a platform-independent way a bit of a hassle. It starts with the default visibility of the compilers; gcc and clang assume that all the symbols are visible, while Visual Studio compilers, by default, hide all the symbols unless they're explicitly exported. By setting `CMAKE_WINDOWS_EXPORT_ALL_SYMBOLS`, the default behavior of MSVC can be changed, but this is a brute-force approach to the problem and can only be used if all the symbols of a library should be exported.

While setting all the symbols to publicly visible is an easy way to ensure that linking is easy, it has a few downsides:

- By exporting everything, there is no way of preventing the use of internal code by dependent targets.
- Since every symbol can be used by external code, the linker cannot discard dead code, so the resulting libraries tend to be bloated. This is especially true if the library contains templates, which tend to blow up the number of symbols considerably.
- Since every symbol is exported, the only clue about what should be considered hidden or internal has to come from the documentation.
- Exposing the internal symbols of a library may expose things that should be kept hidden.

> **Setting All Symbols to Visible**
>
> Be careful when you're setting all symbols to be visible in a shared library, especially when you're concerned about security issues or when the size of the binary is important.

Changing the default visibility

To change the default visibility of the symbols, set the `<LANG>_VISIBILITY_PRESET` property to `HIDDEN`. This property can be set either globally or for a single library target. `<LANG>` is substituted for the language that the library is written in, such as CXX for C++ or C for C. If all the symbols are hidden symbols to be exported, they must be marked specially in the code. The most common way to do this is to specify a preprocessor definition that determines whether a symbol is visible or not:

```
class HELLO_EXPORT Hello {
...
};
```

The `HELLO_EXPORT` definition will contain information about whether the symbol will be exported when the library is compiled or whether it should be imported when you're linking against the library. GCC and Clang use the `__attribute__(...)` keyword to determine this behavior, while on Windows, `_declspec(...)` is used. Writing header files that handle this in a cross-platform manner is not an easy task, especially if you also have to consider that libraries might be built as static and object libraries. Luckily, CMake provides the `generate_export_header` macro, which is imported by the `GenerateExportHeader` module, to make this easier.

In the following example, the symbols for the `hello` library are set to be hidden by default. Then, they are individually enabled again with the use of the `generate_export_header` macro, which is imported by the `GenerateExportHeader` module. Additionally, this example sets the `VISIBILITY_INLINES_HIDDEN` property to `TRUE` to further reduce the export symbol table by hiding inlined class member functions. Setting the visibility for inlines is not strictly necessary, but it's often done when the default visibility is set:

```
add_library(hello SHARED)
set_property(TARGET hello PROPERTY CXX_VISIBILITY_PRESET
  "hidden")
set_property(TARGET hello PROPERTY VISIBILITY_INLINES_HIDDEN
  TRUE)
include(GenerateExportHeader)
generate_export_header(hello EXPORT_FILE_NAME export/hello/
  export_hello.hpp)

target_include_directories(hello PUBLIC "${CMAKE_CURRENT_
  BINARY_DIR}/export")
```

The call to `generate_export_header` creates a file named `export_hello.hpp` in the `CMAKE_CURRENT_BINARY_DIR/export/hello` directory that can be included in the files of the library. It is good practice to put these generated files in a subfolder of the build directory so that only part of the directory is added to the `include` path. The `include` structure of the generated files should match the `include` structure of the rest of the library. So, if, in this example, all the public header files are included by calling `#include <hello/a_public_header.h>`, the export header should also be placed in a folder called `hello`. The generated file also has to be added to the installation instructions, as explained in *Chapter 4, Packaging, Deploying, and Installing a CMake Project*. Additionally, to create the export file, the necessary compiler flags for exporting the symbols must be set to the target.

Since the generated header file must be included in the files that are declaring the classes, functions, and types to be exported, `CMAKE_CURRENT_BINARY_DIR/export/` is added to `target_include_directories`. Note that this has to be `PUBLIC` so that dependent libraries can find the file as well.

There are many more options regarding the `generate_export_header` macros, but what we have seen in this section covers the majority of use cases. Additional information about setting symbol visibility can be retrieved from the official CMake documentation at `https://cmake.org/cmake/help/latest/module/GenerateExportHeader.html`.

Interface or header-only libraries

Header-only libraries are a bit special as they are not compiled; instead, they export their headers so that they're directly included in other libraries. In most aspects, header-only libraries work like normal libraries, but their header files are exposed using the `INTERFACE` keyword, rather than the `PUBLIC` keyword.

Since header-only libraries do not need to be compiled, they do not add sources to the targets. The following code creates a minimal header-only library:

```
project(
  ch3_hello_header_only
  VERSION 1.0
  DESCRIPTION "Chapter 3 header-only example"
  LANGUAGES CXX)

add_library(hello_header_only INTERFACE)
```

```
target_include_directories(hello_header_only INTERFACE
   include/)
target_compile_features( hello_header_only INTERFACE cxx_
   std_17)
```

It is also worth noting that before CMake version 3.19, the `INTERFACE` libraries could not have any `target_sources` added. Now, header-only libraries can have no sources listed.

Object libraries - for internal use only

Sometimes, you may want to split off code so that parts of it can be reused without the need to create a full-blown library. A common practice is when you want to use some code in an executable and unit tests, without the need to recompile everything twice.

For this, CMake provides object libraries, where the sources are compiled, but not archived or linked. An object library is created by calling `add_library(MyLibrary OBJECT)`.

Since CMake 3.12, these objects can be used like normal libraries by adding them to `target_link_libraries` functions. Before version 3.12, object libraries needed to be added with a generator expression; that is, `$<TARGET_OBJECTS:MyLibrary>`. This expands to a list of objects during build system generation. This can still be done, but it is no longer recommended as it quickly becomes unmaintainable, especially if there are multiple object libraries in a project.

> **When to Use Object Libraries**
>
> Object libraries help speed up building and modularizing code without making the modules public.

With object libraries, all the different types of libraries are covered. Libraries on their own are fun to write and maintain, but unless they are integrated into a bigger project, they do not do anything. So, let's see how all the libraries we've defined so far can be used in an executable.

Bringing it together - using your libraries

So far, we have created three different libraries – a binary library to be linked either statically or dynamically, an interface or header-only library, and a precompiled but not linked object library.

Let's learn how to use them in an executable in a shared project. Installing them as system libraries or using them as external dependencies will be covered in *Chapter 5, Integrating Third-Party Libraries and Dependency Management.*

So, we can either put the `add_library` calls in the same `CMakeLists.txt` file or we can integrate them by using `add_subdirectory`. Both are valid options and depend on how the project is set up, as described in the *Setting up a project* and *Working with nested projects* sections of this chapter.

In the following example, we're assuming that three libraries have been defined with `CMakeLists.txt` files in the `hello_lib`, `hello_header_only`, and `hello_object` directories. These libraries can be included using the `add_subdirectory` command. Here, a new target called `chapter3`, which is our executable, is created. Then, the libraries are added to the executable by `target_link_libraries`:

```
add_subdirectory(hello_lib)
add_subdirectory(hello_header_only)
add_subdirectory(hello_object)

add_executable(chapter3)
target_sources(chapter3 PRIVATE src/main.cpp)
target_link_libraries(chapter3 PRIVATE hello_header_only hello
  hello_object)
```

The target of `target_link_libraries` can also be another library. Again, the libraries are linked using an access specifier, which is either of the following:

- `PRIVATE`: The library is used to link against, but it is not a part of the public interface. The linked library is only a requirement when you're building the target.
- `INTERFACE`: The library is not linked against, but it is part of the public interface. The linked library is a requirement when you're using the target somewhere else. This is usually only used when you're linking header-only libraries from other header-only libraries.
- `PUBLIC`: The library is linked against, and it is part of the public interface. So, the library is both a build dependency and a usage dependency.

> **Attention – Bad Practices**
>
> The following practices are actively discouraged by the authors of this book as they tend to create unmaintainable projects that make it hard to port between different build environments. However, we have included them for completeness.
>
> Instead of passing another target after `PUBLIC`, `PRIVATE`, or `INTERFACE`, you can also pass full paths to libraries or the filename of a library, such as `/usr/share/lib/mylib.so` or just `mylib.so`. These practices are possible but discouraged as it makes the CMake project less portable. Additionally, it is possible to pass linker flags here by passing something such as `-nolibc`, though again, this is discouraged. If special linker flags are needed for all the targets, then passing them using the command line is the preferred way. If a single library needs special flags, then using `target_link_options` is the preferred way to do this, preferably in combination with the options that are set over the command line.

In the next section, we'll look at setting compiler and linker options.

Setting compiler and linker options

C++ compilers have lots of options regarding some of the most common flags to be set and it is also a common practice to set preprocessor definitions from the outside. In CMake, these are passed using the `target_compile_options` command. Changing linker behavior is done with the `target_link_options` command. Unfortunately, compilers and linkers may have different ways of how flags are set. For instance, in GCC and Clang, options are passed with a dash (`-`), while the Microsoft compiler takes slashes (`/`) as prefixes for its options. But by using generator expressions, which we covered in *Chapter 1*, *Kickstarting CMake*, this can be conveniently handled in CMake, as shown in the following example:

```
target_compile_options(
  hello
  PRIVATE $<$<CXX_COMPILER_ID:MSVC>:/SomeOption>
          $<$<CXX_COMPILER_ID:GNU,Clang,AppleClang>:-
      someOption>
)
```

Let's look at the generator expression in detail.

$<$<CXX_COMPILER_ID:MSVC>:/SomeOption> is a nested generator expression that is evaluated inside out. Generator expressions are evaluated during build system generation. First, $<CXX_COMPILER_ID:MSVC> evaluates to true if the C++ compiler equals MSVC. If this is the case, then the outer expression will return /SomeOption, which is then passed to the compiler. If the inner expression evaluates to false, then nothing is passed on.

$<$<CXX_COMPILER_ID:GNU,Clang,AppleClang>:-fopenmp> works similarly, but instead of just checking against a single value, a list containing GNU,Clang,AppleClang is passed. If CXX_COMPILER_ID matches either of these, the inner expression evaluates to true and someOption is passed to the compiler.

Passing compiler or linker options as PRIVATE marks them as a build requirement for this target that is not needed for interfacing the library. if PRIVATE is substituted with PUBLIC, then the compile option also becomes a usage requirement and all the targets that depend on the original targets will use the same compiler options. Exposing compiler options to the dependent targets is something that needs to be done with caution. If a compiler option is only needed to use a target but not to build it, then keyword INTERFACE can be used. This is mostly the case when you're building header-only libraries.

A special case of compiler options is compile definitions, which are passed to the underlying program. These are passed with the target_compile_definitions function.

Debugging compiler options

To see all the compile options, you can look at the generated build files, such as Makefiles or Visual Studio projects. A much more convenient way is to let CMake export all the compile commands as a JSON compilation database.

By enabling the CMAKE_EXPORT_COMPILE_COMMANDS variable, a file called compile_commands.json containing the full commands for compiling is created in the build folder.

Enabling this option and running CMake will produce results similar to the following:

```
{
  "directory": "/workspaces/CMake-Best-Practices/build",
  "command": "/usr/bin/g++ -I/workspaces/CMake-Best-Practices/
  chapter_3/hello_header_only/include -I/workspaces/CMake-Best-
    Practices/chapter_3/hello_lib/include -I/workspaces/CMake-
    Best-Practices/chapter_3/hello_object_lib/include -g
      -fopenmp -o
```

```
    chapter_3/CMakeFiles/chapter3.dir/src/main.cpp.o -c /
      workspaces/CMake-Best-Practices/chapter_3/src/main.cpp",
    "file": "/workspaces/CMake-Best-Practices/chapter_3/src/
      main.cpp"
  },
```

Note the addition of the manually specified `-fopenMP` flag from the previous example. `compile_commands.json` can be used as a build system-agnostic way to load the commands. Some IDEs, such as VS Code and CLion, can interpret the JSON file and generate project information themselves. It also often comes in handy for debugging compiler options in case something did not work as expected. The full specifications of the compile commands database can be found at `https://clang.llvm.org/docs/JSONCompilationDatabase.html`.

Library aliases

Library aliases are a way to refer to a library without creating a new build target, sometimes referred to as namespaces. A common pattern is to create a library alias in the form of `MyProject::Library` for each library that is installed from a project.

They can be used to semantically group multiple targets. They also help avoid clashes in naming, especially when projects contain common targets such as libraries named `utils`, `helpers`, and similar. It is good practice to collect all the targets of the same project under the same namespace. When you're linking libraries from other projects, including the namespace avoids you accidentally including the wrong library. All the library targets in this chapter will be aliased with a namespace to group them so that they can be referenced by their namespace:

```
add_library(Chapter3::hello ALIAS hello)
...
target_link_libraries(SomeLibrary PRIVATE Chapter3::hello)
```

In addition to helping with determining the origin of a target, CMake uses namespaces to recognize imported targets and create better diagnostic messages, as we will see when we look at installing and packaging in *Chapter 4, Packaging, Deploying, and Installing a CMake Project*, as well as in *Chapter 5, Integrating Third-Party Libraries and Dependency Management*, where we will cover dependency management.

> **Always Use Namespaces**
>
> As a good practice, always `ALIAS` your targets with a namespace and reference them using the `namespace::` prefix.

Namespaces are a great way to organize build artifacts, but sometimes, they are not enough. Sometimes, we want to see the bigger picture of what is using what and which artifact depends on which library. CMake can help with creating dependency graphs that provide such insights, as we'll see in the next chapter.

Summary

Now that you've completed this chapter, you're ready to create applications and libraries with CMake and start building more complex projects than just "hello world." You have learned how to link different targets together and how compiler and linker options can be passed to targets. We also covered object libraries for internal use only and discussed symbol visibility for shared libraries. Finally, you learned how to document these dependencies automatically so that you have an overview of large projects.

In the next chapter, you will learn how to package and install your applications and libraries on different platforms.

Questions

Answer the following questions to test your knowledge of this chapter:

1. What is the CMake command for creating an executable target?
2. What is the CMake command for creating a library target?
3. How can you specify whether a library is statically or dynamically linked?
4. What is special about object libraries and where do they come in handy?
5. How can you specify the default symbol visibility for shared libraries?
6. How are compiler options specified for a target and how can you see the compile commands?

Part 2: Practical CMake – Getting Your Hands Dirty with CMake

In this part, you will be able to set up a software project using CMake in a way that suits the vast majority of use cases. Chapters 4 and 5 will cover aspects of installing and packaging your project, dependency management, and the use of package managers. Chapters 6, 7, and 8 will be about integrating external tools into your CMake project to create documentation, ensure code quality, or do almost any arbitrary task required for building software. Chapters 9 and 10 will be about creating reusable environments for building projects and chapter and handling big projects from distributed repositories. And finally, Chapter 11 will show you how to facilitate fuzzing with CMake.

This part contains the following chapters:

- *Chapter 4, Packaging, Deploying, and Installing a CMake Project*
- *Chapter 5, Integrating Third-Party Libraries and Dependency Management*
- *Chapter 6, Automatically Generating Documentation*
- *Chapter 7, Seamlessly Integrating Code-Quality Tools with CMake*
- *Chapter 8, Executing Custom Tasks with CMake*
- *Chapter 9, Creating Reproducible Build Environments*
- *Chapter 10, Handling Big Projects and Distributed Repositories in a Superbuild*
- *Chapter 11, Automated Fuzzing with CMake*

4
Packaging, Deploying, and Installing a CMake Project

Building a software project is only half the story. The other half is about delivering and presenting the software to your consumers. Consumers are the biggest stakeholders of any project, even if you are writing a hobby project for yourself. These consumers may have a variety of experiences and purposes. They might be developers, package maintainers, power users, or the *average Joes*. It is important to understand their use cases, scenarios, and requirements. Since the software is mostly abstract, let's assume that your project is baked beans instead – it may be delicious and may smell good in the factory, but improper packaging will reduce its shelf life, making it hard to transport or consume. This will make your product less likely to be desired by the consumers. Even though your product is wonderful at the start, the consumer will not notice it since their experience with your product is bad due to bad packaging. Therefore, it is important to get these things right from the start. Remember, happy consumers will bring value to a product.

CMake has good internal support and tooling for making installing and packaging easy. The good side of this is that CMake leverages the existing project code to do such things. Thus, making a project installable or packaging a project does not result in heavy maintenance costs. In this chapter, we're going to learn how to leverage CMake's existing abilities regarding installing and packaging for deployments.

In this chapter, we will cover the following topics:

- Making CMake targets installable
- Supplying configuration information for others using your project
- Creating an installable package with CPack

Prerequisites

Before you dive into this chapter, you should have a good grasp of targets in CMake (covered briefly in *Chapter 1, Kickstarting CMake*, and *Chapter 3, Creating a CMake Project*, in detail). This chapter will build on top of that knowledge.

Please obtain this chapter's examples from this book's GitHub repository at `https://github.com/PacktPublishing/CMake-Best-Practices`. This chapter's exemplary content will be available in the `chapter_4/` subfolder.

Making CMake targets installable

The most primitive way of supporting deployment in a project is by making it *installable*. The term installable here does not refer to installing a pre-made package of the software. On the contrary, the end user still has to acquire the source code for the project and build it from scratch. An installable project has extra build system code for installing the runtime or development artifacts on the system. The build system will perform the install action here, given that it has proper instructions on how to do so. Since we're using CMake to generate the build system files, CMake must generate the relevant install code. In this section, we will dive into the basics of how to instruct CMake to generate such code for the CMake targets.

The install() command

The `install(...)` command is a built-in CMake command that allows you to generate build system instructions for installing targets, files, directories, and more. CMake will not generate install instructions unless it is explicitly told to do so. Therefore, what gets installed is always under your control. Let's inspect its basic usage.

Installing CMake targets

To make a CMake target installable, the `TARGETS` parameter must be specified with at least one argument. The command signature for this usage is as follows:

```
install(TARGETS <target>... [...])
```

The `TARGETS` parameter denotes that `install` will accept a set of CMake targets to generate the installation code for. In this form, only the output artifacts of the target will be installed. The most common output artifacts for a target are defined as follows:

- `ARCHIVE` (static libraries, DLL import libraries, and linker import files):
  - Except for targets marked as `FRAMEWORK` in macOS
- `LIBRARY` (shared libraries):
  - Except for targets marked as `FRAMEWORK` in macOS
  - Except for DLLs (in Windows)
- `RUNTIME` (executables and DLLs):
  - Except for targets marked as `MACOSX_BUNDLE` in macOS

After making a target installable, CMake will generate the necessary installation code to install the output artifacts that will be produced for the target. To illustrate this, let's make a basic executable target installable together. To see the `install(...)` command in action, let's inspect the `CMakeLists.txt` file of *Chapter 4, Example 1*, which can be found in the `chapter4/ex01_executable` folder:

```
add_executable(ch4_ex01_executable)
target_sources(ch4_ex01_executable src/main.cpp)
target_compile_features(ch4_ex01_executable PRIVATE cxx_std_11)
install(TARGETS ch4_ex01_executable)
```

In the preceding code, an executable target called `ch4_ex01_executable` is being defined and its properties are being populated in the subsequent two lines. The last line, `install(...)`, is the line we're interested in. It tells CMake to create the required install code for `ch4_ex01_executable`.

To check whether `ch4_ex01_executable` can be installed, let's build the project and install it via the CLI:

```
cmake -S . -B ./build
cmake --build ./build
cmake --install ./build --prefix /tmp/install-test
```

> **Note**
>
> Instead of having to specify the `--prefix` parameter for `cmake --install`, you can also use the `CMAKE_INSTALL_PREFIX` variable to provide the non-default install prefix.
>
> Please specify the `--config` argument for the `cmake --build` and `cmake --install` commands while using CMake with multi-config generators, such as Ninja multi-config and Visual Studio:
>
> ```
> # For multi-config generators:
> cmake --build ./build --config Debug
> cmake --install ./build --prefix /tmp/install-test
> --config Debug
> ```

Let's inspect what the `cmake --install` command did:

```
-- Install configuration: ""
-- Installing: /tmp/install-test/lib/libch2.framework.
   component1.a
-- Installing: /tmp/install-test/lib/libch2.framework.
   component2.so
-- Installing: /tmp/install-test/bin/ch2.driver_application
-- Set runtime path of "/tmp/install-test/bin/
   ch2.driver_application" to ""
-- Installing: /tmp/install-test/bin/ch4_ex01_executable
```

In the last line of the preceding output, we can see the *output artifact* of the `ch4_ex01_executable` target – that is, the `ch4_ex01_executable` binary is installed. Since that was the only output artifact that `ch4_ex01_executable` target had, we can conclude that our target has indeed become installable.

Notice that ch4_ex01_executable is not directly installed in the /tmp/install-test (prefix) directory. Instead, the install command put it in the bin/ subdirectory. This is because CMake is being smart about what kind of artifact should go where. In a traditional UNIX system, binaries go into /usr/bin, while libraries go into /usr/lib. CMake knows that the add_executable() command produces an executable binary artifact and puts it into the /bin subdirectory. These directories are provided by CMake by default, depending on the target type. The CMake module that provides the default installation path information is known as the GNUInstallDirs module. The GNUInstallDirs module defines various CMAKE_INSTALL_ paths when included. The following table shows the default installation directories for the targets:

Target Type	GNUInstallDirs Variable	Built-In Default
RUNTIME	${CMAKE_INSTALL_BINDIR}	bin
LIBRARY	${CMAKE_INSTALL_LIBDIR}	lib
ARCHIVE	${CMAKE_INSTALL_LIBDIR}	lib
PRIVATE_HEADER	${CMAKE_INSTALL_INCLUDEDIR}	include
PUBLIC_HEADER	${CMAKE_INSTALL_INCLUDEDIR}	include

To override the built-in defaults, an additional <TARGET_TYPE> DESTINATION parameter is required in the install(...) command. To illustrate this, let's try to change the default RUNTIME install directory to qbin instead of bin. Doing so only requires that we make a small modification to our original install(...) command:

```
# ...
install(TARGETS ch4_ex01_executable
        RUNTIME DESTINATION qbin
)
```

After making this change, we can rerun the configure, build, and install commands. We can confirm that the RUNTIME destination has changed by inspecting the cmake --install command's output. Different from the first time, we can observe that the ch4_ex01_executable binary is put into qbin instead of the default (bin) directory:

```
# ...
-- Installing: /tmp/install-test/qbin/ch4_ex01_executable
```

Now, let's look at another example. We will be installing a `STATIC` library this time. Let's look in the `CMakeLists.txt` file of *Chapter 4, Example 2*, which can be found in the `chapter4/ex02_static` folder. Comments and the `project(...)` command have been omitted due to space reasons. Let's start inspecting the file:

```
add_library(ch4_ex02_static STATIC)
target_sources(ch4_ex02_static PRIVATE src/lib.cpp)
target_include_directories(ch4_ex02_static PUBLIC include)
target_compile_features(ch4_ex02_static PRIVATE cxx_std_11)
include(GNUInstallDirs)
install(TARGETS ch4_ex02_static)
install (
    DIRECTORY include/
    DESTINATION "${CMAKE_INSTALL_INCLUDEDIR}"
)
```

As you may have noticed, it is a little bit different from our previous example. First, there is an additional `install(...)` command with the `DIRECTORY` argument. This is required to make the header files of the static library installable. The reason for this is that CMake will not install any file that is not an *output artifact* and the `STATIC` library target only produces a binary file as an *output artifact*. Header files are not considered *output artifacts* and should be installed separately.

> **Note**
>
> The trailing slash in the `DIRECTORY` argument causes CMake to copy the folder's content instead of copying the folder by name. CMake handles trailing slashes in the same fashion with the Linux `rsync` command.

Installing files and directories

As we saw in the previous section, the things we meant to install are not always part of a target's *output artifacts*. They may be runtime dependencies of the target, such as images, assets, resource files, scripts, and configuration files. CMake provides the `install(FILES...)` and `install(DIRECTORY...)` commands for installing any specific files or directories. Let's begin with installing files.

Installing files

The `install(FILES...)` command accepts one or more files as an argument. It requires an additional `TYPE` or `DESTINATION` parameter as well. Both parameters are used for determining the destination directory of the specified files. The `TYPE` parameter is used to indicate which files will use the default path for that file type as an installation directory. Defaults can be overridden by setting the relevant `GNUInstallDirs` variable. The following table shows the valid `TYPE` values, along with their directory mappings:

Type	GNUInstallDirs Variable	Built-In Default
BIN	${CMAKE_INSTALL_BINDIR}	bin
SBIN	${CMAKE_INSTALL_SBINDIR}	sbin
LIB	${CMAKE_INSTALL_LIBDIR}	lib
INCLUDE	${CMAKE_INSTALL_INCLUDEDIR}	include
SYSCONF	${CMAKE_INSTALL_SYSCONFDIR}	etc
SHAREDSTATE	${CMAKE_INSTALL_SHARESTATEDIR}	com
LOCALSTATE	${CMAKE_INSTALL_LOCALSTATEDIR}	var
RUNSTATE	${CMAKE_INSTALL_RUNSTATEDIR}	<LOCALSTATE dir>/run
DATA	${CMAKE_INSTALL_DATADIR}	<DATAROOT dir>
INFO	${CMAKE_INSTALL_INFODIR}	<DATAROOT dir>/info
LOCALE	${CMAKE_INSTALL_LOCALEDIR}	<DATAROOT dir>/locale
MAN	${CMAKE_INSTALL_MANDIR}	<DATAROOT dir>/man
DOC	${CMAKE_INSTALL_DOCDIR}	<DATAROOT dir>/doc

If you don't wish to use the `TYPE` parameter, you can use the `DESTINATION` parameter instead. It lets you provide a custom destination for the specified files in the `install(...)` command.

An alternative form of `install(FILES...)` is `install(PROGRAMS...)`, which is the same as `install(FILES...)` except it also sets `OWNER_EXECUTE`, `GROUP_EXECUTE`, and `WORLD_EXECUTE` permissions for installed files. This makes sense for binaries or script files that must be executed by the end user.

To understand `install(FILES|PROGRAMS...)`, let's look at an example. The example we're going to look into is *Chapter 4, Example 3* (`chapter_4/ex03_file`). It essentially contains three files: `chapter4_greeter_content`, `chapter4_greeter.py`, and `CMakeLists.txt`. First, let's look at its `CMakeLists.txt` file:

```
install(FILES "${CMAKE_CURRENT_LIST_DIR}/chapter4_greeter_
  content"
  DESTINATION "${CMAKE_INSTALL_BINDIR}")
```

```
install(PROGRAMS "${CMAKE_CURRENT_LIST_DIR}/chapter4_greeter.
  py"
  DESTINATION "${CMAKE_INSTALL_BINDIR}" RENAME chapter4_
    greeter)
```

Let's digest what we have seen; in the first `install(...)` command, we're telling CMake to install the `chapter4_greeter_content` file in the current `CMakeLists.txt` directory (`chapter4/ex03_file`) in the default `BIN` directory of the system. In the second `install(...)` command, we're telling CMake to install `chapter4_greeter.py` in the default `BIN` directory with the name `chapter4_greeter`.

> **Note**
>
> The `RENAME` parameter is only valid for single-file `install(...)` calls.

With these `install(...)` instructions, CMake should install the `chapter4_greeter.py` and `chapter4_greeter_content` files in the `${CMAKE_INSTALL_PREFIX}/bin` directory. Let's build and install the project via the CLI:

```
cmake -S . -B ./build
cmake --build ./build
cmake --install ./build --prefix /tmp/install-test
```

Let's look at what the `cmake --install` command did:

```
/* … */
-- Installing: /tmp/install-test/bin/chapter4_greeter_content
-- Installing: /tmp/install-test/bin/chapter4_greeter
```

The preceding output confirms that CMake generated the required installation code for the `chapter4_greeter_content` and `chapter4_greeter.py` files. Lastly, let's check if the `chapter4_greeter` file can be executed since we used the `PROGRAMS` parameter to install it:

```
15:01 $ /tmp/install-test/bin/chapter4_greeter
['Hello from installed file!']
```

With that, we have concluded the `install(FILES|PROGRAMS...)` section. Let's continue with installing directories.

Installing directories

The `install(DIRECTORY...)` command is useful for installing directories. The directory's structure will be copied as-is to the destination. Directories can either be installed as a whole or selectively. Let's begin with the most basic directory installation example:

```
install(DIRECTORY dir1 dir2 dir3 TYPE LOCALSTATE)
```

The preceding example will install the `dir1` and `dir2` directories in the `${CMAKE_INSTALL_PREFIX}/var` directory, along with all of their subfolders and files as-is. Sometimes, installing the folder's entire content is not desirable. Luckily, CMake allows the `install` command to include or exclude directory content based on globbing patterns and regular expressions. Let's install `dir1`, `dir2`, and `dir3` selectively this time:

```
include(GNUInstallDirs)
install(DIRECTORY dir1 DESTINATION ${CMAKE_INSTALL_
  LOCALSTATEDIR} FILES_MATCHING PATTERN "*.x")
install(DIRECTORY dir2 DESTINATION ${CMAKE_INSTALL_
  LOCALSTATEDIR} FILES_MATCHING PATTERN "*.hpp"
  EXCLUDE PATTERN "*")
install(DIRECTORY dir3 DESTINATION ${CMAKE_INSTALL_
  LOCALSTATEDIR} PATTERN "bin" EXCLUDE)
```

In the preceding example, we used the `FILES_MATCHING` parameter to define criteria for file selection. `FILES_MATCHING` can be followed by either the `PATTERN` or `REGEX` argument. `PATTERN` allows you to define a globbing pattern, whereas `REGEX` allows you to define a regular expression. By default, these expressions are used for including files. If you want to exclude files that match the criteria, you can append the `EXCLUDE` argument to the pattern. Note that these filters are not applied to subdirectory names because of the `FILES_MATCHING` parameter. We also used `PATTERN` in the last `install(...)` command without `FILES_MATCHING` prepended, which allows us to filter subdirectories instead of files. This time, only the files with the `.x` extension in `dir1`, files that don't have the `.hpp` extension in `dir2`, and all the content except for the `bin` folder in `dir3` will be installed. This example is available as *Chapter 4, Example 4* in the `chapter4/ex04_directory` folder. Let's compile and install it to see whether it does the right thing:

```
cmake -S . -B ./build
cmake -build ./build
cmake -install ./build -prefix /tmp/install-test
```

The output of `cmake --install` should look like this:

```
-- Installing: /tmp/install-test/var/dir1
-- Installing: /tmp/install-test/var/dir1/subdir
-- Installing: /tmp/install-test/var/dir1/subdir/asset5.x
-- Installing: /tmp/install-test/var/dir1/asset1.x
-- Installing: /tmp/install-test/var/dir2
-- Installing: /tmp/install-test/var/dir2/chapter4_hello.dat
-- Installing: /tmp/install-test/var/dir3
-- Installing: /tmp/install-test/var/dir3/asset4
```

> **Note**
>
> `FILES_MATCHING` cannot be used after `PATTERN` or `REGEX` but it can be done vice versa.

In the output, we can see that only the files with the `.x` extension are picked from `dir1`. This is because of the `FILES_MATCHING PATTERN "*.x"` parameter in the first `install(...)` command, causing the `asset2` file to not be installed. Also, note that the `dir2/chapter4_hello.dat` file is installed and the `dir2/chapter4_hello.hpp` file is skipped. This is due to the `FILES_MATCHING PATTERN "*.hpp" EXCLUDE PATTERN "*"` parameters in the second `install(…)` command. Lastly, we can see that the `dir3/asset4` file is installed and the `dir3/bin` directory is completely skipped since the `PATTERN "bin" EXCLUDE` parameter is specified in the last `install(...)` command.

With `install(DIRECTORY...)`, we have covered the basics of the `install(...)` command. Let's continue with the `install(…)` command's other common parameters.

Other common parameters of the install() command

As we've seen, the `install()` command's first parameter indicates what to install. There are additional parameters that allow us to customize the installation. Let's inspect some of the common parameters together.

The DESTINATION parameter

This parameter allows you to specify a target directory for the files specified in the `install(...)` command. The directory path can be relative or absolute. Relative paths will be relative to the `CMAKE_INSTALL_PREFIX` variable. It is recommended to use relative paths to make the installation *relocatable*. Also, it is important to use relative paths for packaging since `cpack` requires the install paths to be relative. It is good practice to use a path that begins with the relevant `GNUInstallDirs` variable so that package maintainers can override the install destination if needed. The `DESTINATION` parameter can be used together with the `TARGETS`, `FILES`, `IMPORTED_RUNTIME_ARTIFACTS`, `EXPORT`, and `DIRECTORY` installation types.

The PERMISSIONS parameter

This parameter allows you to change the installed file permissions on supported platforms. The available permissions are `OWNER_READ`, `OWNER_WRITE`, `OWNER_EXECUTE`, `GROUP_READ`, `GROUP_WRITE`, `GROUP_EXECUTE`, `WORLD_READ`, `WORLD_WRITE`, `WORLD_EXECUTE`, `SETUID`, and `SETGID`. The `PERMISSIONS` parameter can be used with the `TARGETS`, `FILES`, `IMPORTED_RUNTIME_ARTIFACTS`, `EXPORT`, and `DIRECTORY` installation types.

The CONFIGURATIONS parameter

This allows you to limit a set of parameters to be applied when a specific build configuration has been specified.

The OPTIONAL parameter

This parameter makes the file to be installed optional so that installation does not fail when the file is not present. The `OPTIONAL` parameter can be used together with the `TARGETS`, `FILES`, `IMPORTED_RUNTIME_ARTIFACTS`, and `DIRECTORY` installation types.

In this section, we learned how to make targets, files, and directories installable. In the next section, we will learn how to generate configuration information so that we can import CMake projects directly into another CMake project.

Supplying configuration information for others using your project

In the previous section, we learned how to make our project installable so that others can consume our project by installing it on their system. But sometimes, delivering the artifacts is not enough. For example, if you are delivering a library, it must also be easy to import it into a project – especially a CMake project. In this section, we will learn how to make this importing process easier for other CMake projects.

There are convenient ways of importing a library, given that the project to be imported has the proper configuration files. One of the prominent ways of doing so is by utilizing the `find_package()` method (which we will cover in *Chapter 5, Integrating Third-Party Libraries and Dependency Management*). If you have consumers that are using CMake in their workflows, they will be happy if they can just write `find_package(your_project_name)` and start using your code. In this section, we will learn how to generate the required configuration files to make `find_package()` work for your project.

CMake's preferred way of consuming dependencies is via packages. Packages convey dependency information for CMake-based build systems. Packages can be in the form of Config-file packages, Find-module packages, or pkg-config packages. All of the package types can be found and consumed via `find_package()`. For the sake of space and simplicity, we will only be covering Config-file packages in this section. The rest of the ways are mostly workarounds for lack of config files, so they are better left in the past. Let's start discovering the Config-file packages.

Entering the CMake package world – Config-file packages

Config-file packages are based on configuration files that contain package content information. This information indicates the package's content locations, so CMake reads this file and uses the package. Thus, discovering only the package configuration files is enough to consume the package.

There are two types of configuration files – a package configuration file and an optional package version file. Both files must have a specific naming convention. Package configuration files can be named `<ProjectName>Config.cmake` or `<projectname>-config.cmake`, depending on personal preference. Both notations will be picked by CMake on `find_package(ProjectName)/find_package(projectname)` calls. The content of package configuration files looks similar to the following:

```
set(Foo_INCLUDE_DIRS ${PREFIX}/include/foo-1.2)
set(Foo_LIBRARIES ${PREFIX}/lib/foo-1.2/libfoo.a)
```

Here, `${PREFIX}` is the installation prefix of the project. It is a variable since the installation prefix can be changed based on the system's type and can also be changed by the user.

Similar to package configuration files, package version files can be named `<ProjectName>ConfigVersion.cmake` or `<projectname>-config-version.cmake` as well. CMake expects package configuration and package version files to be present in the `find_package(...)` search paths. You can create these files with the help of CMake. One of the many places that `find_package(...)` looks while searching for packages is the `<CMAKE_PREFIX_PATH>/cmake` directory. We'll be putting our Config-file package configuration files into this folder throughout our examples.

To create config-file packages, we will need to learn a few extra things, such as *exporting targets* and the *CmakePackageConfigHelpers* module. To learn about these things, let's start diving into an actual example. We'll be following *Chapter 4, Example 5* to learn how to structure a CMake project to make it a config-file package. It is located in the `chapter4/ex05_config_file_package` folder. Let's start by inspecting the `CMakeLists.txt` file in the `chapter4/ex05_config_file_package` directory (comments and project commands have been omitted in favor of space; also, note that the lines that are not relevant to the topic will not be mentioned):

```
include(GNUInstallDirs)
set(ch4_ex05_lib_INSTALL_CMAKEDIR cmake CACHE PATH
"Installation directory for config-file package cmake files")
/*...*/
```

The `CMakeLists.txt` file is quite similar to `chapter4/ex02_static`. This is because it is the same example, except it supports the config-file packaging. The first line, `include(GNUInstallDirs)`, is used to include the `GNUInstallDirs` module. This provides the `CMAKE_INSTALL_INCLUDEDIR` variable, which will be used later. `set(ch4_ex05_lib_INSTALL_CMAKEDIR...)` is a user-defined variable for setting the target installation directory of the config-file packaging configuration files. It is a relative path that should be used in the `install(...)` directives, so it is implicitly relative to `CMAKE_INSTALL_PREFIX`:

```
/*...*/
target_include_directories(ch4_ex05_lib PUBLIC
        $<BUILD_INTERFACE:${CMAKE_CURRENT_SOURCE_DIR}/include>
)
target_compile_features(ch4_ex05_lib PUBLIC cxx_std_11)
/*...*/
```

The `target_include_directories(...)` call is quite different than the usual calls. It uses `generator expressions` to distinguish between build-time `include` directories and install-time `include` directories since the build-time `include` path will not be present when the target is imported into another project. The following set of commands will make the target installable:

```
/*…*/
install(TARGETS ch4_ex05_lib
        EXPORT ch4_ex05_lib_export
        INCLUDES DESTINATION ${CMAKE_INSTALL_INCLUDEDIR}
)
install (
     DIRECTORY ${PROJECT_SOURCE_DIR}/include/
     DESTINATION ${CMAKE_INSTALL_INCLUDEDIR}
)

/*…*/
```

`install(TARGETS...)` is a bit different than usual as well. It contains an extra `EXPORT` parameter. This `EXPORT` parameter is used to create an export name from the given `install(…)` targets. These targets can then be exported using this export name. The path that's specified with the `INCLUDES DESTINATION` parameter will be used to populate the `INTERFACE_INCLUDE_DIRECTORIES` property of the exported target and will be automatically prefixed with the install prefix path. Here, the `install(DIRECTORY...)` command is used to install the target's header files and is located in `${PROJECT_SOURCE_DIR}/include/`, in the `${CMAKE_INSTALL_PREFIX}/${CMAKE_INSTALL_INCLUDEDIR}` directory. The `${CMAKE_INSTALL_INCLUDEDIR}` variable is used to give consumers the ability to override the `include` directory for this installation. Now, let's create an export file from the export name we created in the preceding example:

```
/*…*/
install(EXPORT ch4_ex05_lib_export
        FILE ch4_ex05_lib-config.cmake
        NAMESPACE ch4_ex05_lib::
        DESTINATION ${ch4_ex05_lib_INSTALL_CMAKEDIR}
)
/*…*/
```

`install(EXPORT...)` is perhaps the most important piece of code in this file. It is the code that is doing the actual target exporting. It generates a CMake file that contains all the exported targets in the given export name. The `EXPORT` parameter accepts an existing export name to perform the export. It is referring to the `ch4_ex05_lib_export` export name that we created by the previous `install(TARGETS...)` call. The `FILE` parameter is used to determine the export's filename and is set to `ch4_ex05_lib-config.cmake`. The `NAMESPACE` parameter is used to prefix all the exported targets with a namespace. This allows you to connect all the exported targets under a common namespace and avoid collisions with packages that have similar target names. Lastly, the `DESTINATION` parameter determines the installation path of the generated export file. This is set to `${ch4_ex05_lib_INSTALL_CMAKEDIR}` to allow `find_package()` to discover it.

> **Note**
>
> Since we are not providing any extras other than exported targets, the name of the export file is `ch4_ex05_lib-config.cmake`. It is the package configuration file name that's required for this package. We've done this because the example project does not require any extra dependencies to be satisfied first and can be directly imported as-is. If any extra action is required, it is recommended to have an intermediate package configuration file that satisfies those dependencies and includes the exported file after.

With the `install(EXPORT...)` command, we obtained the `ch4_ex05_lib-config.cmake` file. This means that our target can be consumed via `find_package(..)`. One additional step is required to achieve full support for `find_package(...)`, which is obtaining the `ch4_ex05_lib-config-version.cmake` file:

```
/*...*/
include(CMakePackageConfigHelpers)

write_basic_package_version_file(
  "ch4_ex05_lib-config-version.cmake"
  # Package compatibility strategy. SameMajorVersion is
    essentially 'semantic versioning'.
  COMPATIBILITY SameMajorVersion
)
install(FILES
  "${CMAKE_CURRENT_BINARY_DIR}/ch4_ex05_lib-config-version.
    cmake"
```

```
  DESTINATION "${ch4_ex05_lib_INSTALL_CMAKEDIR}"
)
/* end of the file */
```

In the last few lines, you can find the code that's required to generate and install the `ch4_ex05_lib-config-version.cmake` file. With the `include(CMakePackageConfigHelpers)` line, the `CMakePackageConfigHelpers` module is imported. This module provides the `write_basic_package_version_file(…)` function. `write_basic_package_version_file(…)` is used to automatically generate package version files, depending on the given parameters. The first positional argument is the output's filename. The `VERSION` parameter is used to specify the version of the package we're generating in `major.minor.patch` form. It is opted out to allow `write_basic_package_version_file` to get it from the project version automatically. The `COMPATIBILITY` parameter allows you to specify compatibility policies, depending on the version's value. `SameMajorVersion` denotes that this package is compatible with any version that has the same major version value of this package. The other possible values are `AnyNewerVersion`, `SameMinorVersion`, and `ExactVersion`.

Now, let's test whether this works. To test the package configuration, we must install the project regularly:

```
cmake -S . -B ./build
cmake --build ./build
cmake --install ./build --prefix /tmp/install-test
```

The `cmake --install` command's output should look as follows:

```
/* … */
-- Installing: /tmp/install-test/cmake/ch4_ex05_lib-config.
  cmake
-- Installing: /tmp/install-test/cmake/ch4_ex05_lib-config-
  noconfig.cmake
-- Installing: /tmp/install-test/cmake/ch4_ex05_lib-config-
  version.cmake
/*…*/
```

Here, we can see that our package configuration files have been successfully installed in the `/tmp/install-test/cmake` directory. I'll leave inspecting the content of those files to you as an exercise. So, we have a consumable package in our hands. Let's switch sides and try to consume our freshly baked package. To do that, we'll look at the `chapter4/ex05_consumer` example. Let's inspect the `CMakeLists.txt` file together:

```
if(NOT PROJECT_IS_TOP_LEVEL)
  message(FATAL_ERROR "The chapter-4, ex05_consumer project is
    intended to be a standalone, top-level project. Do not
      include this directory.")
endif()
find_package(ch4_ex05_lib 1 CONFIG REQUIRED)
add_executable(ch4_ex05_consumer src/main.cpp)
target_compile_features(ch4_ex05_consumer PRIVATE cxx_std_11)
target_link_libraries(ch4_ex05_consumer ch4_ex05_lib::ch4_ex05_
  lib)
```

In the first few lines, we can see verification regarding whether the project is a top-level project or not. Since this example is intended to be an external application, it should not be part of the root example project. Thus, we can guarantee that we will use the targets that are exported by the package, not the root project's targets. The root project also does not include the `ex05_consumer` folder. Next, there's a `find_package(...)` call, where `ch4_ex05_lib` is given as a package name. It is also explicitly requested that the package should have a major version of 1; `find_package(...)` must only consider `CONFIG` packages and packages specified in this `find_package(...)` call are required. In the consequent lines, a regular executable is defined called `ch4_ex05_consumer` that's linked against `ch4_ex05_lib` in the `ch4_ex05_lib` namespace (`ch4_ex05_lib::ch4_ex05_lib`). `ch4_ex05_lib::ch4_ex05_lib` is the actual target we have defined in our package. Let's look at the source file, `src/main.cpp`:

```
#include <chapter4/ex05/lib.hpp>
int main(void){
    chapter4::ex05::greeter g;
    g.greet();
}
```

This is a simple application that includes chapter4/ex05/lib.hpp, creates an instance of the greeter class, and calls the greet() function. Let's try to compile and run the application:

```
cd chapter_4/ex05_consumer
cmake -S . -B build/ -DCMAKE_PREFIX_PATH:STRING=/tmp/install-
test
cmake --build build/
./build/ch4_ex05_consumer
```

Since we have installed the package using a custom prefix (/tmp/install-test), we can indicate this by setting the CMAKE_PREFIX_PATH variable. This causes find_package(...) to search /tmp/install-test for packages as well. For default prefix installations, this parameter setting is not required. We should see the infamous Hello, world! message if everything goes well:

```
 ./build/ch4_ex05_consumer
Hello, world!
```

Here, our consumers can use our little **greeter** and everybody is happy. Now, let's conclude this section by learning how to package with **CPack**.

Creating an installable package with CPack

So far, we have seen how CMake can structure software projects. Although CMake is the star of the show, CMake has some powerful friends too. It is time to introduce you to CPack, the packaging tool of CMake. It is shipped with CMake installations by default. It allows you to leverage existing CMake code to generate platform-specific installations and packages. CPack is similar to CMake in concept. It is based on generators that generate packages instead of build system files. The following table shows the available CPack generator types, as of CPack version 3.21.3:

Generator Name	Description
7Z	7-zip archive
DEB	Debian package
External	CPack external package
IFW	Qt Installer Framework
NSIS	Null Soft Installer
NSIS64	Null Soft Installer (64-bit)

Generator Name	Description
NuGet	NuGet packages
RPM	RPM packages
STGZ	Self-extracting TAR gzip archive
TBZ2	Tar BZip2 archive
TGZ	Tar GZip archive
TXZ	Tar XZ archive
TZ	Tar Compress archive
TZST	Tar Zstandard archive
ZIP	Zip archive

CPack uses CMake's installation mechanism to populate the content of the packages. CPack uses the configuration details that are present in the `CPackConfig.cmake` and `CPackSourceConfig.cmake` files to generate packages. These files can either be populated manually or generated automatically by CMake with the help of the CPack module. Using CPack on an existing CMake project is as easy as including the CPack module, given that the project already has proper `install(...)` commands. Including the CPack module will cause CMake to generate the `CPackConfig.cmake` and `CPackSourceConfig.cmake` files, which are the CPack configurations that are needed to pack the project. Also, an additional `package` target will become available for the build step. This step will build the project and run CPack so that it starts packaging. CPack can be used when the CPack configuration files have been populated properly, either by CMake or the user. The CPack module allows you to customize the packaging process. A large amount of CPack variables can be set. These variables are separated into two groups – common variables and generator-specific variables. Common variables affect all of the package generators, whereas generator-specific variables only affect a specific type of generator. We'll be inspecting the most basic and prominent ones and we will mostly deal with the common variables. The following table shows the most common CPack variables we will use in our examples:

Variable Name	Description	Default Value
CPACK_PACKAGE_NAME	Package name	Project name
CPACK_PACKAGE_VENDOR	Package vendor name	"Humanity"
CPACK_PACKAGE_VERSION_MAJOR	Package major version	Project major version
CPACK_PACKAGE_VERSION_MINOR	Package minor version	Project minor version

Variable Name	Description	Default Value
CPACK_PACKAGE_VERSION_PATCH	Package patch version	Project patch version
CPACK_GENERATOR	List of CPack generators to use	N/A
CPACK_THREADS	Number of threads to use when parallelism is supported	1

Any changes that must be made to the variables must be made before you include the CPack module. Otherwise, the defaults will be used. Let's dive into an example to see CPack in action. We will be following the *Chapter 4, Example 6* (`chapter4/ex06_pack`) example. This example is structured as a standalone project and is not part of the root example project. It is a regular project with two subdirectories named `executable` and `library`. The `CMakeLists.txt` file of the `executable` directory looks as follows:

```
add_executable(ch4_ex06_executable src/main.cpp)
target_compile_features(ch4_ex06_executable PRIVATE cxx_std_11)
target_link_libraries(ch4_ex06_executable PRIVATE ch4_ex06_
  library)
install(TARGETS ch4_ex06_executable)
```

The `CMakeLists.txt` file of the `library` directory looks as follows:

```
add_library(ch4_ex06_library STATIC src/lib.cpp)
target_compile_features(ch4_ex06_library PRIVATE cxx_std_11)
target_include_directories(ch4_ex06_library PUBLIC include)
set_target_properties(ch4_ex06_library PROPERTIES PUBLIC_HEADER
  include/chapter4/ex06/lib.hpp)
include(GNUInstallDirs) # Defines the ${CMAKE_INSTALL_
  INCLUDEDIR} variable.
install(TARGETS ch4_ex06_library)
install (
    DIRECTORY ${PROJECT_SOURCE_DIR}/include/
    DESTINATION ${CMAKE_INSTALL_INCLUDEDIR}
)
```

The `CMakeLists.txt` files for these folders do not contain anything out of the ordinary. They contain regular, installable CMake targets and declare nothing about CPack. Let's take a look at the top-level `CMakeLists.txt` file as well:

```
cmake_minimum_required(VERSION 3.21)
project(
  ch4_ex06_pack
  VERSION 1.0
  DESCRIPTION "Chapter 4 Example 06, Packaging with CPack"
  LANGUAGES CXX)
if(NOT PROJECT_IS_TOP_LEVEL)
  message(FATAL_ERROR "The chapter-4, ex06_pack project is
    intended to be a standalone, top-level project.
    Do not include this directory.")
endif()

add_subdirectory(executable)
add_subdirectory(library)

set(CPACK_PACKAGE_VENDOR "CTT Authors")
set(CPACK_GENERATOR "DEB;RPM;TBZ2")
set(CPACK_THREADS 0)
set(CPACK_DEBIAN_PACKAGE_MAINTAINER "CTT Authors")
include(CPack)
```

The top-level `CMakeLists.txt` file is pretty much a regular, top-level `CMakeLists.txt` file, except for the last four lines. It sets three CPack-related variables and then includes the CPack module. These four lines are enough to provide basic CPack support. The `CPACK_PACKAGE_NAME` and `CPACK_PACKAGE_VERSION_*` variables are not set to let CPack deduce them from the top-level project's name and version parameters. Let's configure the project to see whether it works:

```
cd chapter_4/ex06_pack
cmake -S . -B build/
```

After configuring the project, the `CpackConfig.cmake` and `CpackConfigSource.cmake` files should be generated by the CPack module to the `build/CPack*` directory. Let's check if they're present:

```
$ ls build/CPack*
build/CPackConfig.cmake  build/CPackSourceConfig.cmake
```

Here, we can see that the CPack configuration files are automatically generated. Let's build this and try to package the project with CPack:

```
cmake --build build/
cpack --config build/CPackConfig.cmake -B build/
```

The `--config` argument is the main input of the CPack command. The `-B` argument overrides the default package directory that CPack will write its artifacts to. Let's look at CPack's output:

```
CPack: Create package using DEB
/*…*/
CPack: - package: /home/user/workspace/personal/CMake-Best-
Practices/chapter_4/ex06_pack/build/ch4_ex06_pack-1.0-Linux.deb
generated.
CPack: Create package using RPM
/*…*/
CPack: - package: /home/user/workspace/personal/CMake-Best-
Practices/chapter_4/ex06_pack/build/ch4_ex06_pack-1.0-Linux.rpm
generated.
CPack: Create package using TBZ2
/*…*/
CPack: - package: /home/user/workspace/personal/CMake-Best-
Practices/chapter_4/ex06_pack/build/ch4_ex06_pack-1.0-Linux.tar.
bz2
generated.
```

Here, we can see that CPack has used the `DEB`, `RPM`, and `TBZ2` generators to generate the `ch4_ex06_pack-1.0-Linux.deb`, `ch4_ex06_pack-1.0-Linux.rpm`, and `ch4_ex06_pack-1.0-Linux.tar.bz2` packages, respectively. Let's try to install the generated Debian package on a Debian environment:

```
sudo dpkg -i build/ch4_ex06_pack-1.0-Linux.deb
```

If the packaging was correct, we should be able to invoke `ch4_ex06_executable` directly on the command line:

```
13:38 $ ch4_ex06_executable
Hello, world!
```

It works! As an exercise, try to install the RPM and `tar.bz2` packages as well.

In this section, we learned how to use CPack to pack our project. This is by no means an exhaustive guide. CPack itself deserves several chapters so that it can be covered in detail. With that, we've successfully reached the end of this chapter.

Summary

In this chapter, we learned the basics of making a target installable, as well as how to package a project for development and consumer environments. Deployment is a very important aspect of professional software projects and with the help of the things we have covered in this chapter, you will be able to tackle such deployment requirements easily.

In the next chapter, we will learn how to integrate third-party libraries into CMake projects.

Questions

Answer the following questions to test your knowledge of this chapter:

1. How can we instruct CMake to make a CMake target installable?
2. Which files are installed when a target is installed via the `install(TARGETS)` command?
3. For library targets, are header files installed by the `install(TARGETS)` command? Why? If not, what can be done to install them as well?
4. What does the `GNUInstallDirs` CMake module provide?
5. How can you selectively install a directory's content in a destination directory?
6. Why should we use relative paths when specifying install destination directories?
7. What are the essential files that are required for a `config-file` package?
8. What does exporting a target mean?
9. How can you make a CMake project packageable with CPack?

5
Integrating Third-Party Libraries and Dependency Management

So far, in this book, we have covered how to build and install our own code with CMake. In this chapter, we will take a look at how to use files, libraries, and programs that are not part of a CMake project. The first part of the chapter will be about how to find those things in general, while the latter part will focus on how to manage dependencies to build your CMake project.

One of the biggest advantages of using CMake is that it has built-in dependency management for the discovery of many third-party libraries. In this chapter, we will look at how to integrate libraries that are installed on your system and locally downloaded dependencies. Additionally, you will learn how third-party libraries can be downloaded and used as binaries and, alternatively, how they can be built from source directly out of a CMake project.

We will look at how to write instructions for CMake to reliably find almost any library on your system. Finally, we will take a look at how to use package managers such as Conan and vcpkg with CMake. The practices for dependency management, as covered in this chapter, will help you to create stable and portable builds with CMake. It doesn't matter if you are using precompiled binaries or compiling them in place from scratch, setting up CMake to handle dependencies in a structured and consistent way will reduce the time spent fixing broken builds in the future. Here's the list of main topics that we'll cover in this chapter:

- Finding files, programs, and paths with CMake
- Using third-party libraries in your CMake project
- Using package managers with CMake
- Getting the dependencies as source code

Technical requirements

As with the previous chapters, all the examples are tested with CMake 3.21 and run on either of the following compilers:

- GCC 9 or newer
- Clang 12 or newer
- MSVC 19 or newer

Additionally, some examples will need OpenSSL 1.1 installed to be able to compile. Some examples pull dependencies from various online locations, so an internet connection is also required. All of the examples and source code are available from the GitHub repository to this book, which can be found at `https://github.com/PacktPublishing/CMake-Best-Practices`.

The examples for the external package managers require Conan (version 1.40 or newer) and vcpkg installed on your system to run. You can get the software here:

- **Conan**: `https://conan.io/`
- **Vcpkg**: `https://github.com/microsoft/vcpkg`

Finding files, programs, and paths with CMake

Most projects quickly grow to a size and complexity where they depend on files, libraries, and perhaps even programs that are managed outside the project. CMake provides built-in commands to find these things. At first glance, the process of searching and finding things appears to be quite simple. However, on closer analysis, there are quite a few things to consider. First, we have to handle the search order of where to look for files. Then, we might want to add additional locations where the file might be, and finally, we have to account for the differences between different operating systems.

On an abstraction level higher than the individual files, CMake has the ability to find whole packages that define targets, include paths, and package specific variables. Refer to the *libraries in your CMake project* section for more detail.

There are five `find_...` commands that share very similar options and behaviors:

- `find_file`: This locates a single file.
- `find_path`: This finds a directory containing a specific file.
- `find_library`: This finds library files.
- `find_program`: This finds executable programs.
- `find_package`: This finds complete sets of packages.

All of these commands work in a similar way, but there are some small but important differences when it comes to where they look for things. In particular, `find_package` does more than just locate files; it not only looks for packages but makes the file content available for easy use in the CMake project. In this chapter, first, we will look at the simpler `find` functions before we cover how to find complex packages.

Finding files and paths

The most low-level and basic things to find are files and paths. The `find_file` and `find_path` functions share the same signature. The only difference between them is that `find_path` stores the directory in which a file is found in the result, while `find_file` will store the full path, including the filename. The signature of the `find_file` command is shown as follows:

```
find_file (
          <VAR>
          name | NAMES name1 [name2 ...]
          [HINTS [path | ENV var]... ]
          [PATHS [path | ENV var]... ]
```

```
        [PATH_SUFFIXES suffix1 [suffix2 ...]]
        [DOC "cache documentation string"]
        [NO_CACHE]
        [REQUIRED]
        [NO_DEFAULT_PATH]
        [NO_PACKAGE_ROOT_PATH]
        [NO_CMAKE_PATH]
        [NO_CMAKE_ENVIRONMENT_PATH]
        [NO_SYSTEM_ENVIRONMENT_PATH]
        [NO_CMAKE_SYSTEM_PATH]
        [CMAKE_FIND_ROOT_PATH_BOTH |
         ONLY_CMAKE_FIND_ROOT_PATH |
         NO_CMAKE_FIND_ROOT_PATH]
      )
```

The preceding command either searches for a single file, if the name has been passed directly, or for a list of likely names if the `NAMES` option has been used. The resulting path is stored in the variable passed as `<VAR>`. If the file cannot be found, the variable will contain `<VARIABLENAME>-NOTFOUND`.

Passing a list of names is useful if the files being searched for have variations in their names such as different capitalizations or naming conventions that may or may not include version numbers and so on. When passing a list of names, the names should be ordered in a preferred way, as the search stops once the first file has been found.

> **Searching for Files Containing Version Numbers**
> It is recommended that you search for filenames without version numbers before searching for those that contain some form of version numbering. This is so that locally built files are preferred to the ones installed by the operating system.

The `HINTS` and `PATHS` options contain additional locations to the default locations where the file is searched for. `PATH_SUFFIXES` could contain a number of subdirectories that are searched below each of the other locations.

The find_... commands search for things in defined places and within a defined order. The NO_..._PATH arguments of the commands can be used to skip the respective location. The following table shows the order of the search locations and the options for skipping a location:

Location	The skip option in the command
Package root variables	NO_PACKAGE_ROOT_PATH
CMake-specific cache variables	NO_CMAKE_PATH
CMake-specific environment variables	NO_CMAKE_ENVIRONMENT_PATH
Paths from the HINTS option	
System-specific environment variables	NO_SYSTEM_ENVIRONMENT_PATH
System-specific cache variables	NO_CMAKE_SYSTEM_PATH
Paths from the PATHS option	

Let's look at the search order more closely along with what the different locations mean:

- **Package root variables**: This is only used when find_file is used as part of the find_package command. Please refer to the *Using third-party libraries in your CMake project* section for an in-depth discussion.
- **CMake-specific cache variables**: The locations derived from the CMAKE_PREFIX_PATH, CMAKE_INCLUDE_PATH, and CMAKE_FRAMEWORK_PATH cache variables for macOS. Generally, setting the CMAKE_PREFIX_PATH cache variable is preferred over the other two types, as this is used for all of the find_ commands. The prefix path is the base point for any searches under which the common file structures such as bin, lib, include, and more are located. CMAKE_PREFIX_PATH is a list of paths, and for each entry, find_file will search <prefix>/include or <prefix>/include/${CMAKE_LIBRARY_ARCHITECTURE} if the respective variable has been set. Generally, CMake sets the variable automatically, and they should not be changed by developers. Architecture-specific paths take precedence over generic paths.
- The CMAKE_INCLUDE_PATH and CMAKE_FRAMEWORK_PATH cache variables should only be used if the standard directory structure is not applicable. They do not add additional include suffixes to the paths.
- Searching these paths can be skipped by passing the NO_CMAKE_PATH option to the command or, globally, by setting the CMAKE_FIND_USE_PATH variable to false.

- **System-specific environment variables**: These are derived from the `CMAKE_PREFIX_PATH`, `CMAKE_INCLUDE_PATH`, and `CMAKE_FRAMEWORK_PATH` system environment variables. The variables work in the same way as the cache variables, but they are usually set from outside the call to CMake.
- Note that, on Unix platforms, the lists are separated by colons (:) instead of semicolons (;) in order to conform to the platform-specific environment variables.
- **Paths from the `HINTS` option**: These are the additional search locations that are manually specified. They could be constructed from other values such as property values, or they could depend on a previously found file or path.
- **System-specific environment variables**: The `INCLUDE` and `PATH` environment variables could each contain a list of directories to be searched. Again, on Unix platforms, the list is separated by colons (:) instead of semicolons (;).
- On Windows, the `PATHS` entries are handled in a more complex manner. For each entry, a base path is extracted by dropping any trailing `bin` or `sbin` directories. If `CMAKE_LIBRARY_ARCHITECTURE` is set, the `include/${CMAKE_LIBRARY_ARCHITECTURE}` subdirectory is added as the first priority for each path. After that, `include` (without a postfix) is searched. Only then, the original path, which might or might not end in `bin` or `sbin`, is searched. Passing either the `NO_SYSTEM_ENVIRONMENT_PATH` variable or setting the `CMAKE_FIND_USE_CMAKE_SYSTEM_PATH` variable to `false` will skip the locations in the environment variables.
- Assuming that the `PATH` option contains `C:\myfolder\bin;C:\yourfolder`, and `CMAKE_LIBRARY_ARCHITECTURE` is set to `x86_64`, the search order will be as follows:

  I. `C:\myfolder\include\x86_64`

  II. `C:\myfolder\include\`

  III. `C:\myfolder\bin`

  IV. `C:\yourfolder\include\x86_64`

  V. `C:\yourfolder\include\`

  VI. `C:\yourfolder\`

- **System-specific cache variables**: Here, the `CMAKE_SYSTEM_PREFIX_PATH` and `CMAKE_SYSTEM_FRAMEWORK_PATH` variables work similarly to the CMake-specific cache variables. These variables are not supposed to be changed by the developer but are configured when CMake sets up the platform toolchain. One exception here is if an own toolchain file is provided, such as when using sysroots or cross-compiling, as explained in *Chapter 11, Automated Fuzzing with CMake*.

- In addition to the NO_CMAKE_SYSTEM_PATH option, the CMAKE_FIND_USE_CMAKE_SYSTEM_PATH variable can be set to `false` to skip searching in locations provided by the system-specific cache variables.
- **Paths specified in the PATHS option**: In the same way as the HINTS option, these are additional search locations that are manually provided. Although not technically prevented, it is the convention that the PATHS variables should be fixed paths and not dependable on other values.

If only the locations provided by HINTS or PATHS are to be searched, adding the NO_DEFAULT_PATH option skips all the other locations.

Occasionally, you might want to ignore particular paths for searching. In such cases, a list of paths might be specified in CMAKE_IGNORE_PATH or CMAKE_SYSTEM_IGNORE_PATH. Both of these variables were designed with cross-compiling scenarios in mind and are rarely used in other circumstances.

Searching for files when cross-compiling

When cross-compiling, the process of searching for files is often different because cross-compilation toolchains are collected under their own self-contained directory structure, which does not mix with the system toolchain. Generally, first, you will want to look inside the toolchain's directory for files. By setting the CMAKE_FIND_ROOT variable, the origin for all searches can be changed to a new location.

Additionally, the CMAKE_SYSROOT, CMAKE_SYSROOT_COMPILE, and CMAKE_SYSROOT_LINK variables affect the search locations, but they should only be set in a toolchain file, not by a project itself. If any of the regular search locations are already under the sysroot or the location specified by CMAKE_FIND_ROOT, they will not be changed. Any path that starts with a tilde (~), and is passed to the `find_` commands, will not be changed to avoid skipping directories that are under the user's home directory.

By default, first, CMake searches in the locations provided by any of the variables from the preceding paragraph and then continues to search the host system. This behavior can be changed globally by setting the CMAKE_FIND_ROOT_PATH_MODE_INCLUDE variable to either BOTH, NEVER, or ONLY. Alternatively, you can set the CMAKE_FIND_ROOT_PATH_BOTH option, the ONLY_CMAKE_FIND_ROOT_PATH option, or the NO_CMAKE_FIND_ROOT_PATH option to `find_file`.

The following table shows the search order when setting either of the options or the variables in the different search modes:

Mode	Option	Search order
BOTH	CMAKE_FIND_ROOT_PATH_BOTH	• CMAKE_FIND_ROOT_PATH • CMAKE_SYSROOT_COMPILE • CMAKE_SYSROOT_LINK • CMAKE_SYSROOT • All regular search locations
NEVER	NO_CMAKE_FIND_ROOT_PATH	• All regular search locations
ONLY	ONLY_CMAKE_FIND_ROOT_PATH	CMAKE_FIND_ROOT_PATH • CMAKE_SYSROOT_COMPILE • CMAKE_SYSROOT_LINK • CMAKE_SYSROOT • Any regular paths, one of the other locations, or under CMAKE_STAGING_PREFIX

The CMAKE_STAGING_PREFIX variable is used to provide installation paths for cross-compiling. The CMAKE_SYSROOT variable should not be changed by installing things on it. Setting up cross-compilation toolchains will be covered, in detail, in *Chapter 11, Automated Fuzzing with CMake*, where we talk about cross-compiling.

Finding programs

Finding executables is very similar to finding files and paths, and the find_program command has almost the same signature as find_file. Additionally, find_program has the NAMES_PER_DIR option, which tells the command to search one directory at a time and search for all provided filenames in each directory instead of searching through each directory for each file.

On Windows, the .exe and .com file extensions are automatically added to the filenames provided, but not .bat or .cmd.

The cache variables used by `find_program` are slightly different from the ones used by `find_file`:

- `find_program` automatically adds `bin` and `sbin` to the search locations provided by `CMAKE_PREFIX_PATH`.
- Values in `CMAKE_LIBRARY_ARCHITECTURE` are ignored and have no effect.
- `CMAKE_PROGRAM_PATH` is used instead of `CMAKE_INCLUDE_PATH`.
- `CMAKE_APPBUNDLE_PATH` is used instead of `CMAKE_FRAMEWORK_PATH`.
- `CMAKE_FIND_ROOT_PATH_MODE_PROGRAM` is used to change the mode for searching programs.

As with the other `find` commands, `find_program` will set the `<varname>-NOTFOUND` variable if CMake is unable to find the program. This is often handy to determine whether a custom build step that depends on a certain external program should be enabled.

Finding libraries

Finding libraries is a special case of finding files, so the `find_library` command supports the same set of options as `find_file`. And similar to the `find_program` command, it has the additional `NAMES_PER_DIR` option that checks for all filenames first, before moving to the next directory. The difference between finding regular files and finding libraries is that `find_library` automatically applies the platform-specific naming conventions to the filenames. On Unix platforms, the names will be prefixed with `lib`, while on Windows, the `.dll` or `.lib` extensions will be added.

Again, the cache variables are slightly different from the ones used in `find_file` and `find_program`:

- `find_library` adds `lib` to the search locations by `CMAKE_PREFIX_PATH`, and it uses `CMAKE_LIBRARY_PATH` instead of `CMAKE_INCLUDE_PATH` to find libraries. The `CMAKE_FRAMEWORK_PATH` variable is used similarly to `find_file`. The `CMAKE_LIBRARY_ARCHITECTURE` variable works the same as in `find_file`.
- This is done by appending the respective folders to the search paths. `find_library` searches the locations in the `PATH` environment variable in the same way as `find_file`, but it appends `lib` to each prefix. Also, it uses the `LIB` environment variable if this has been set instead of the `INCLUDE` variable.
- `CMAKE_FIND_ROOT_PATH_MODE_LIBRARY` is used to change the mode for searching libraries.

CMake is generally aware of conventions regarding 32-bit and 64-bit search locations such as platforms using the `lib32` and `lib64` folders for different libraries of the same name. The behavior is controlled by the `FIND_LIBRARY_USE_LIB[32|64|X32]_PATHS` variables, which control what should be searched first. Additionally, projects can define their own suffix using the `CMAKE_FIND_LIBRARY_CUSTOM_LIB_SUFFIX` variable, which overrides the behavior of the other variables. However, the need to do this is very rare, and tampering with the search order inside a `CMakeLists.txt` file quickly makes projects hard to maintain and heavily impacts the portability between systems.

Finding static or shared libraries

In most cases, simply passing the base name of a library to CMake works well enough, but sometimes, the behavior has to be overridden. One reason for this is if, on some platform, the static version of a library should be preferred over the shared one or vice versa. The best way to do this is to split up the `find_library` call into two calls instead of trying to achieve this in a single call. It is more robust if the static library is in a different directory from the dynamic one:

```
find_library(MYSTUFF_LIBRARY libmystuff.a)
find_library(MYSTUFF_LIBRARY mystuff)
```

On Windows, this approach cannot be used, as static libraries and import libraries for DLLs do have the same `.lib` suffix, so they are not distinguishable by name. The `find_file`, `find_path`, `find_program`, and `find_library` commands are often handy when looking for specific things. On the other hand, finding dependencies happens on a higher level. This is where CMake really excels by providing the `find_package` methods. With `find_package`, we do not need to, first, search for all the include files, followed by all the library files, and then add them manually to each target and, in the end, account for all platform-specific behaviors. Let's dive into the process of how to find dependencies next.

Using third-party libraries in your CMake project

If you're writing software in earnest, sooner or later, you will hit the point where your project will rely on libraries from outside your project. Instead of looking for individual library files or header files, the recommended way by the CMake community and us authors is to do this with the `find_package` command. The preferred approach for finding dependencies in CMake is using packages.

Packages provide a set of information about dependencies for CMake and the generated build systems. They can be integrated into a project in two forms. This is either by their configuration details (also called config-file packages), which are provided by the upstream project, or as so-called find module packages, which are usually defined somewhere that is unrelated to the package, either by CMake itself or by the project using the package. Both types can be found by using `find_package`, and the result is a set of imported targets and/or a set of variables containing information that is relevant to the build system.

The `findPkgConfig` module, which uses find-pkg to find the relevant meta-information for a dependency, also provides indirect support for packages.

Typically, *Find modules* are used for locating dependencies, for instance, when the upstream does not provide the necessary information for package configuration. They are not to be confused with CMake utility modules, which are used with `include()`. Whenever possible, using a package provided by the upstream should be used instead of the `find` modules. If possible, fixing the upstream project to provide the necessary information is preferred over writing a find module.

Note that the `find_package` command has two signatures: a basic or short signature and a full or long signature. In almost all scenarios, using the short signature is sufficient to find the packages we're looking for, and it should be preferred because it is easier to maintain. The short form supports both the module and config packages, but the long form only supports configuration mode.

The signature of the short mode is as follows:

```
find_package(<PackageName> [version] [EXACT] [QUIET] [MODULE]
             [REQUIRED] [[COMPONENTS] [components...]]
             [OPTIONAL_COMPONENTS components...]
             [NO_POLICY_SCOPE])
```

Let's suppose we want to write a program that converts a string into a sha256 hash by using the appropriate functionality of the OpenSSL library. To compile and link this example, we have to inform CMake that this project needs the OpenSSL library and then attach it to the target. For the moment, let's assume that the necessary libraries have been installed at a default location on your system; for example, by using a regular package manager such as apt, RPM, or similar for Linux, chocolatey for Windows, or brew for macOS.

A sample CMakeLists.txt file might look like this:

```
find_package(OpenSSL REQUIRED COMPONENTS SSL)
add_executable(find_package_example)
target_link_libraries(find_package_example PRIVATE
  OpenSSL::SSL)
```

The preceding example does the following things:

1. On the first line in the example, there is a `find_package(OpenSSL REQUIRED COMPONENTS SSL)` call. This tells CMake that we're looking for a set of libraries and header files for OpenSSL. Specifically, we're looking for the SSL component and ignoring the crypto component. The `REQUIRED` keyword tells CMake that it is required to build this project. This means that CMake will stop the configuration process with an error if the library is not found.
2. Once the package has been found, we tell CMake to link the library to the target using `target_link_libary`. Specifically, we tell CMake to link the `OpenSSL::SSL` target provided by the package OpenSSL.

If a dependency has to be of a certain version, it can be specified either as a single version of the `major[.minor[.patch[.tweak]]]` format or as a version range with the `versionMin..[<]versionMax` format. For version ranges, both `versionMin` and `versionMax` should have the same format, and by specifying <, the upper version will be excluded.

Unfortunately, as of version 3.21, CMake cannot query the modules for the available components. So, we have to rely on the documentation of the modules or library providers to find out which components are available. The available modules can be queried with the following commands:

```
cmake --help-module-list #< lists all available modules
cmake --help-module <mod> #< prints the documentation for
  module
  <mod>
cmake --help-modules #< lists all modules and their
  documentation
```

A list of modules shipped with CMake can be found at `https://cmake.org/cmake/help/latest/manual/cmake-modules.7.html`.

> **Finding Individual Libraries and Files**
>
> It is possible to look for individual libraries and files, but the preferred way is to use packages. Finding individual files and making them available to CMake will be covered in the *Writing your own find module* section.

When run in module mode, the `find_package` command searches for files called `Find<PackageName>.cmake`; this occurs, first, in the paths specified by `CMAKE_MODULE_PATH` and then among the find modules provided by the CMake installation. If you wish to learn how to create CMake packages, head over to *Chapter 4, Packaging, Deploying, and Installing a CMake Project.*

When run in config mode, `find_package` searches for files called after either of the following patterns:

- `<lowercasePackageName>-config.cmake`
- `<PackageName>Config.cmake`
- `<lowercasePackageName>-config-version.cmake` (if the version details were specified)
- `<PackageName>ConfigVersion.cmake` (if the version details were specified)

All searches will be conducted over a set of locations in a well-defined order; if needed, some of the locations can be skipped by passing the respective option to CMake. `find_package` contains a few more options than the other `find_` commands. The following table shows the search order from a high level:

Location	Skip option in the command
Package root variables	`NO_PACKAGE_ROOT_PATH`
CMake-specific cache variables	`NO_CMAKE_PATH`
CMake-specific environment variables	`NO_CMAKE_ENVIRONMENT_PATH`
Paths specified in the `HINTS` option	
System-specific environment variables	`NO_SYSTEM_ENVIRONMENT_PATH`
User package registry	`NO_CMAKE_PACKAGE_REGISTRY`
System-specific cache variables	`NO_CMAKE_SYSTEM_PATH`
System package registry	`NO_CMAKE_SYSTEM_PACKAGE_REGISTRY`
Paths specified in the `PATHS` option	

Let's look at the search order and search locations more closely:

- **Package root variables**: The package root for each `find_package` call is stored in a variable called `<PackageName>_ROOT`. They are the first priority for searching files belonging to a package. The package root variables work in the same way as `CMAKE_PREFIX_PATH`, not just for the call to `find_package` but for all other `find_` calls that might happen inside the find module belonging to the package.
- **CMake-specific cache variables**: These are the locations derived from `CMAKE_PREFIX_PATH`. For macOS, the `CMAKE_FRAMEWORK_PATH` variable is also considered a search location.
- By setting the `CMAKE_FIND_USE_CMAKE_PATH` variable to `false`, the locations from the CMake-specific cache variables will be skipped.
- **CMake-specific environment variables**: In addition to specifying `CMAKE_PREFIX_PATH` and `CMAKE_FRAMEWORK_PATH` as cache variables, CMake will also consider them if they are set as environment variables.
- Setting the `CMAKE_FIND_USE_ENVIRONMENT_PATH` variable to `false` will disable this behavior.
- `HINTS` from `find_package`: These are optional paths passed to `find_package`.
- **System-specific environment variables**: The `PATH` environment variable is used to look for packages and files and the trailing `bin` and `sbin` directories are removed. The default locations for each system, such as `/usr`, `/lib`, and similar locations, are usually searched at this point.
- **User package registry**: Often, packages are found either in the standard locations or in the locations passed to CMake by using the `CMAKE_PREFIX_PATH` option. Package registries are another way to tell CMake where to look for dependencies. Package registries are special locations where collections of packages reside. The user registry is valid for the current user account, while the system package registry is valid system-wide. On Windows, the location for the user package registry is stored in the Windows registry under the following:
- `HKEY_CURRENT_USER\Software\Kitware\CMake\Packages\<packageName>\`
- On the Unix platform, it is stored in the user's home directory as follows:

  `~/.cmake/packages/<PackageName>`

- **Platform-specific cache variables**: For `find_package`, the platform-specific cache variables, `CMAKE_SYSTEM_PREFIX_PATH`, `CMAKE_SYSTEM_FRAMEWORK_PATH`, and `CMAKE_SYSTEM_APPBUNDLE_PATH`, work in a similar way to the other find calls. They are set by CMake itself and should not be changed by the project.
- **System package registry**: Similar to the user package registry, this is a location where CMake is looking for packages. On Windows, it is stored under `HKEY_LOCAL_MACHINE\Software\Kitware\CMake\Packages\<packageName>\`.
- Unix systems do not provide a system package registry.
- `PATHS` from `find_package`: These are optional paths that are passed to `find_package`. Usually, the `HINTS` options are computed from other values or depend on variables, whereas `PATHS` options are fixed paths.

Specifically, when looking for packages in config mode, CMake will look for the following file structure under the various prefixes:

```
<prefix>/
<prefix>/(cmake|CMake)/
<prefix>/<packageName>*/
<prefix>/<packageName>*/(cmake|CMake)/
<prefix>/(lib/<arch>|lib*|share)/cmake/<packageName>*/
<prefix>/(lib/<arch>|lib*|share)/<packageName>*/
<prefix>/(lib/<arch>|lib*|share)/<packageName>*/(cmake|CMake)/
<prefix>/<packageName>*/(lib/<arch>|lib*|share)/cmake/
  <packageName>*/
<prefix>/<packageName>*/(lib/<arch>|lib*|share)/<packageName>*/
<prefix>/<packageName>*/(lib/<arch>|lib*|share)/<packageName>*/
  (cmake|CMake)/
```

On macOS platforms, the following folders are also searched:

```
<prefix>/<packageName>.framework/Resources/
<prefix>/<packageName>.framework/Resources/CMake/
<prefix>/<packageName>.framework/Versions/*/Resources/
<prefix>/<packageName>.framework/Versions/*/Resources/CMake/
<prefix>/<packageName>.app/Contents/Resources/
<prefix>/<packageName>.app/Contents/Resources/CMake/
```

You can find out more about packages in the official CMake documentation at `https://cmake.org/cmake/help/latest/manual/cmake-packages.7.html`.

In terms of modules, so far, we've only covered how to find existing modules. But what happens if we want to look for dependencies that are neither integrated into CMake, nor in the standard places, or they do not provide configuration instructions for CMake? Well, let's find out about that in the next section.

Writing your own find module

While CMake is almost an industry standard, there are still lots of libraries out there that are not managed with CMake or that are managed with CMake but do not export a CMake package. If they can be installed in the default location of a system, finding these libraries is usually not a problem, but this is not always possible or wanted. A common case is when using a proprietary third-party library that is only needed for a certain project or that uses a different version of a library to build from the one that is installed by the systems package manager.

If you're developing multiple projects side by side, you might want to handle the dependencies locally for each project. Either way, it is good practice to set up your project in a way so that dependencies are managed locally and do not depend too much on what is installed on the system. Creating fully reproducible builds is described in *Chapter 11, Automated Fuzzing with CMake*; however, for now, let's focus on finding dependencies.

If no module and no configuration file exists for a dependency, often, writing your so-called find module is the solution. The goal is to provide enough information so that, later, we can use any package by using `find_package`.

`Find` modules are instructions for CMake on how to find the necessary header and binary files for a library and which create imported targets for CMake to use. As described earlier in this chapter, when invoking `find_package` in module mode, CMake searches for files called `Find<PackageName>.cmake` in `CMAKE_MODULE_PATH`.

Let's assume that we're building a project where the dependencies have already been downloaded or built and have been placed in a folder called `dep` before we use them. So, the project structure might look like this:

```
├── dep <-- The folder where we locally keep dependencies
├── cmake
│   └── FindLibImagePipeline.cmake <-- This is what we need to
      write
```

```
├── CMakeLists.txt <-- Main CmakeLists.txt
├── src
|   ├── *.cpp files
```

The first thing we have to do is to add the `cmake` folder to the `CMAKE_MODULE_PATH`, which is a list. So, first, we add the following line to the `CMakeLists.txt` file:

```
list(APPEND CMAKE_MODULE_PATH "${CMAKE_CURRENT_SOURCE_DIR}/
  cmake")
```

This tells CMake that it should look for `find` modules in the `cmake` folder. Typically, a `find` module does things in the following order:

1. It looks for files belonging to the package.
2. It sets up variables containing the `include` and `library` directories for the packages.
3. It sets up targets for the imported package.
4. It sets properties for the targets.

A simple `FindModules.cmake` for a library called `obscure` might look like this:

```
cmake_minimum_required(VERSION 3.21)
find_library(
    OBSCURE_LIBRARY
    NAMES obscure
    HINTS ${PROJECT_SOURCE_DIR}/dep/
    PATH_SUFFIXES  lib  bin  build/Release  build/Debug
)

find_path(
    OBSCURE_INCLUDE_DIR
    NAMES obscure/obscure.hpp
    HINTS ${PROJECT_SOURCE_DIR}/dep/include/
)

include(FindPackageHandleStandardArgs)

```

```
find_package_handle_standard_args(
    Obscure
    DEFAULT_MSG
    OBSCURE_LIBRARY
    OBSCURE_INCLUDE_DIR
)

mark_as_advanced(OBSCURE_LIBRARY OBSCURE_INCLUDE_DIR)

if(NOT TARGET Obscure::Obscure)
    add_library(Obscure::Obscure UNKNOWN IMPORTED )
    set_target_properties(Obscure::Obscure  PROPERTIES
               IMPORTED_LOCATION "${OBSCURE_LIBRARY}"
               INTERFACE_INCLUDE_DIRECTORIES
                 "${OBSCURE_INCLUDE_DIR}"
               IMPORTED_LINK_INTERFACE_LANGUAGES "CXX"
)
endif()
```

When looking at the example, we can observe that the following things happen:

1. First, the actual library file belonging to the dependency is searched for using the `find_library` command. If found, the path to it, including the actual filename, is stored in the `OBSCURE_LIBRARY` variable. It is a common practice to name the `<PACKAGENAME>_LIBRARY` variable. The `NAMES` argument is a list of possible names for the library. The names are automatically extended with common prefixes and extensions. So, although, in the preceding example, we look for the name "obscure," a file named `libobscure.so` or `obscure.dll` will be found. More details about the search order, hints, and paths will be covered later in this section.
2. Next, the `Find` module attempts to locate the `include` path. This is done by finding a known path pattern of the library, usually one of the public header files. The result is stored in the `OBSCURE_INCLUDE_DIR` variable. Again, the common practice is to name this variable as `<PACKAGENAME>_INCLUDE_DIR`.

3. Since handling all the requirements for a find module can be tedious and is often very repetitive, CMake provides the `FindPackageHandleStandardArgs` module, which provides a handy function to handle all the common cases. It provides the `find_package_handle_standard_args` function, which handles `REQUIRED`, `QUIET`, and the version-related arguments of `find_package`. `find_package_handle_standard_args` has a short signature and a long signature. In the example, the short signature is used:

```
find_package_handle_standard_args(<PackageName>
  (DEFAULT_MSG|<custom-failure-message>)
  <required-var>...
  )
```

4. For most cases, the short form of `find_package_handle_standard_args` is sufficient. In the short form, the `find_package_handle_standard_args` function takes the package name as the first argument and a list of variables that are required for the package. The `DEFAULT_MSG` argument tells it to print default messages in the event of success or failure, depending on whether `find_package` was invoked with `REQUIRED` or `QUIET`. The message can be customized, but we recommend that you stick to the default messages whenever possible. That way, the messages are consistent for all the `find_package` commands. In the preceding example, `find_package_handle_standard_args` checks whether the `OBSCURE_LIBRARY` and `OBSCURE_INCLUDE_DIR` variables that have been passed are valid. If that is the case, the `<PACKAGENAME>_FOUND` variable is set.
5. If all goes well, the find module defines the target. Before we do this, it is helpful to check whether the target we are trying to create does not already exist (to avoid overwriting it in the case that we have multiple calls to `find_package` for the same dependency). Creating the target is done with `add_library`. Since we cannot be sure whether it is a static or dynamic library, the type is `UNKNOWN` and the `IMPORTED` flag is set.
6. Finally, the properties for the library are set. The minimum setting that we recommend is the `MPORTED_LOCATION` property and the location of the `include` files in `INTERFACE_INCLUDE_DIR`.

If everything works as expected, the library can then be used like this:

```
find_package(Obscure PRIVATE REQUIRED)
...
target_link_libraries(find_module_example Obscure::Obscure)
```

So, now we understand how other libraries are added to your projects if they are already available for use. But how do we get the libraries into our system in the first place? Let's find that out in the next section.

Using package managers with CMake

The easiest way to get dependencies into your project is to regularly install them using apt-get, brew, or Chocolatey. The downside of installing everything is that you can pollute your system with many different versions of libraries and the version you are looking for might not be available at all. This is especially true if you are working on multiple projects with different requirements regarding the dependencies side by side. Often, a developer downloads the dependencies locally for each project so that each project can work independently. A very good way to handle dependencies is by using package managers such as Conan or vcpkg.

Using a dedicated package manager has many advantages when it comes to dependency management. Two of the more popular ones for handling C++ dependencies are Conan and vcpkg. Both of them can handle complex build systems, and mastering them will require whole books on their own, so we will only cover the bare necessities to start working with them here. In this book, we will focus on using packages that are already available in your CMake project, rather than on creating your own packages.

Getting dependencies from Conan

Over the last few years, the Conan package manager gained much popularity, mostly because it integrates very well with CMake. Conan is a decentralized package manager that has been built on a client/server architecture. This means that the local client fetches or uploads packages to one or more remote servers.

One of the most powerful features of Conan is that it can create and manage binary packages for multiple platforms, configurations, and versions. When creating packages, they are described with a `conanfile.py` file that lists all dependencies, sources, and build instructions.

The packages are built and uploaded to the remote server with the Conan client. This has an additional benefit where, if no binary package that fits your local configuration can be found, the package can be built locally from its sources.

A very convenient way to use Conan with CMake is to use Conan from CMake itself. However, if you do not desire to do this, calling Conan externally also works. While not strictly necessary, we recommend that you check for the Conan program using `find_program` before using Conan.

To call Conan out of CMake directly, Conan offers a CMake wrapper for downloading. The following example downloads the `conan-cmake` wrapper and then pulls the `fmt` formatting library from ConanCenter to be used as a regular library in the project:

```
if(NOT EXISTS "${CMAKE_CURRENT_BINARY_DIR}/conan.cmake")
  message(STATUS "Downloading conan.cmake from
    https://github.com/conan-io/cmake-conan")
  file( DOWNLOAD
  "https://raw.githubusercontent.com/conan-io/cmake-
    conan/0.17.0/conan.cmake"
    "${CMAKE_CURRENT_BINARY_DIR}/conan.cmake"
    EXPECTED_HASH
    SHA256=3bef79da16c2e031dc429e1dac87a08b9226418b300ce00
      4cc125a82687baeef
    STATUS download_status
    )

  if(NOT download_status MATCHES "^0;")
    message(FATAL_ERROR "Downloading conan.cmake failed with
      ${download_status}")
  endif()
endif()
include(${CMAKE_CURRENT_BINARY_DIR}/conan.cmake)

conan_cmake_autodetect(CONAN_SETTINGS)

conan_cmake_configure(REQUIRES fmt/6.1.2 GENERATORS cmake_find_
  package_multi)

conan_cmake_install(PATH_OR_REFERENCE .
                    BUILD missing
                    SETTINGS ${CONAN_SETTINGS}
                   )
```

```
list(APPEND CMAKE_PREFIX_PATH ${CMAKE_CURRENT_BINARY_DIR})
find_package(fmt 6.1 REQUIRED)
add_executable(conan_example)
target_link_libraries(conan_example PRIVATE fmt::fmt)
```

The preceding CMake code does the following things:

1. If not already downloaded, the `conan.cmake` file is downloaded to the current binary directory.
2. Next, it is included to make the Conan functions available.
3. Once included, Conan is told to detect the settings, such as the compiler, the platform, and more, from the current CMake configuration and to store them in the `CONAN_SETTINGS` variable.
4. The `conan_cmake_configure` functions define the requirements for `fmt` and set the generator for Conan so that we can use `find_package` to include the dependencies. This will generate a `conanfile.txt` file that contains the necessary instructions for Conan in the current build directory.
5. And, finally, `conan_cmake_install` installs the dependencies.
6. `PATH_OR_REFERENCE` tells us where the definition of the dependencies is located. This command runs in the same build directory as `conan_cmake_configure`, so passing a single dot will search the same directory. `BUILD missing` tells Conan to build the packages locally if they are not available as binaries from the remote server.
7. `SETTINGS` passes the retrieved settings along to Conan.
8. Since the generated find modules will be located in the current binary directory, they have to be added to `CMAKE_MODULE_PATH`.
9. Once downloaded, the dependencies can be included by using `find_package` and then added to the existing targets as usual.

Instead of directly declaring the dependencies inside CMake, it is also possible to provide the information as a `conanfile.txt` file. It might look something like this:

```
[requires]
fmt/6.1.2
[generators]
cmake_find_package
```

Here, instead of running conan_cmake_configure and conan_cmake_install, Conan is invoked through conan_cmake_run. Adapting the preceding example to use a conanfile.txt file will look similar to the following (the part that downloads the conan.cmake file stays the same):

```
include(${CMAKE_CURRENT_BINARY_DIR}/conan.cmake)
conan_cmake_autodetect(CONAN_SETTINGS)
conan_cmake_run(CONANFILE ${CMAKE_CURRENT_LIST_DIR}/conanfile.
  txt
                BASIC_SETUP
                BUILD missing
                SETTINGS ${CONAN_SETTINGS})
list(APPEND CMAKE_PREFIX_PATH ${CMAKE_CURRENT_BINARY_DIR})
find_package(fmt 6.1 REQUIRED)
add_executable(conan_conanfile_example)
target_link_libraries(conan_conanfile_example  PRIVATE
  fmt::fmt)
```

This setup is similar to the previous example, except that Conan is pointed to the conanfile.txt file instead of generating it from the CMake instructions. BASIC_SETUP will tell Conan to create the necessary CMake variables automatically. The conan_cmake_run command can also be used to run almost any Conan command.

Of course, Conan can also be called from outside CMake manually. While some find this to be a cleaner approach, because of the separation of concerns between Conan and CMake, it can be tedious to maintain. Indeed, the information about the dependency has to be tracked in two places, and the build configurations, such as the compilers, the libc version, the platform, and more, have to be configured not just for CMake but for Conan too.

The full Conan documentation can be found at https://docs.conan.io/en/latest/.

Using vcpkg for dependency management

Another popular open source package manager is *vcpkg* from Microsoft. It works similarly to Conan in the way that it is set up as a client/server architecture. It was originally built to work with the Visual Studio compiler environment, and CMake was added later. Packages can either be installed manually, by calling vcpkg in the so-called *classic mode*, or directly out of CMake in the so-called *manifest mode*. The command to install packages with vcpkg in classic mode is as follows:

```
vcpkg install [packages]
```

When run in manifest mode, the dependencies of a project are defined in a `vcpkg.json` file in the root of the project. Manifest mode has a big advantage in that it integrates better with CMake, so, whenever possible, use manifest mode. A vcpkg manifest might look like this:

```
{
        "name" : "vcpkg-example",
        "version-semver" : "0.0.1",
        "dependencies" :
        [
        "someLibrary",
        "anotherLibrary",
]
}
```

In order for CMake to find the packages, a vcpkg toolchain file has to be passed to CMake, so the call to CMake will be as follows:

```
cmake -S <source_dir> -D <binary_dir> -DCMAKE_TOOLCHAIN_
  FILE=[vcpkg root]/scripts/buildsystems/vcpkg.cmake
```

If they are run in manifest mode, the packages specified in the vcpkg.json file will be automatically downloaded and installed locally. If they are run in classic mode, the packages have to be manually installed before running CMake. When passing the vcpkg toolchain file, the installed packages can be used, as usual, by using `find_package` and `target_link_libraries`.

Microsoft recommends that you install vcpkg as a submodule in the repository at the same level as the CMake root project, but it can be installed almost everywhere.

Setting the toolchain file might cause problems when cross-compiling, as `CMAKE_TOOLCHAIN_FILE` might point to a different file already. In this case, a second toolchain file can be passed with the `VCPKG_CHAINLOAD_TOOLCHAIN_FILE` variable. In this case, the call to CMake would look something like this:

```
cmake -S <source_dir> -D <binary_dir> -DCMAKE_TOOLCHAIN_
  FILE=[vcpkg root]/scripts/buildsystems/vcpkg.cmake -DVCPKG
    _CHAINLOAD_TOOLCHAIN_FILE=/path/to/other/toolchain.cmake
```

Conan and vcpkg are just two package managers for C++ and CMake that are popular. Of course, there are many more, but it would require a separate book to describe them all. Especially when projects get more complex, we highly advise that you use package managers.

Which package manager you choose will depend on the context a project is being developed in and your personal preference. Conan has a slight advantage over vcpkg in that it is supported on more platforms as it runs everywhere Python runs. In terms of features and ability for cross-compiling, both are more or less equal. Overall, Conan offers more advanced configuration options and control over the packages, which comes at the price of more complex handling. Another way to work with local dependencies is to create fully isolated build environments by using containers, sysroots, and more. This will be covered in *Chapter 12, Cross-Platform Compiling and Custom Toolchains*. For the moment, let's assume we're running CMake with your standard system installation.

Using packages managers for dependency management is the recommended way when working with project-specific dependencies. However, sometimes, package managers are not an option. This might be because of mysterious company policies or some other reason. In these cases, CMake also supports downloading dependencies as the source and integrating them into the project as external targets.

Getting the dependencies as source code

There are several ways to get dependencies as sources into your project. A relatively straightforward but dangerous way is to manually download or clone them into a subfolder inside your project and then add this folder with `add_subdirectory`. While this works and is pretty fast, it quickly becomes tedious and hard to maintain. So, this should be automated as soon as possible.

> **Note**
>
> The practice of downloading and integrating a copy of a third-party software directly into a product is called vendoring. While it has the advantage that it often makes building software easy, it creates issues with packaging libraries. Vendoring is avoided by either using either a package manager or by installing the third-party software in a location on your system.

Downloading dependencies as the source using pure CMake

At the base of getting external content is the CMake `ExternalProject` module and the more sophisticated `FetchContent` module, which is built on `ExternalProject`. While `ExternalProject` offers more flexibility, `FetchContent` is often more convenient to use, especially if the downloaded project is also built using CMake. Both of them download projects as source files and can be used to build them.

Using FetchContent

For external projects that use CMake to build, using the `FetchContent` module is the way of choice to add source dependencies. For binary dependencies, using `find_package` and find modules is still the preferred way. One of the main differences between `ExternalProject` and `FetchContent` is that `FetchContent` downloads and configures external projects during configuration time, while `ExternalProject` does everything during the build step. The drawback to this is that the source and its configuration are not available during configuration time.

Before `FetchContent`, you would have used Git submodules to manually download the dependencies and then add them using `add_subdirectory`. This works in some cases, but it can be rather inconvenient and cumbersome to maintain.

`FetchContent` provides a list of functions to pull in source dependencies, mainly `FetchContent_Declare`, which defines the parameters for downloading and building `FetchContent_MakeAvailable`, which populates the targets of the dependency and makes them available for the build. In the following example, the `bertrand` library for design by contract is pulled from Git using GitHub and made available for use:

```
include(FetchContent)
FetchContent_Declare(
  bertrand
  GIT_REPOSITORY https://github.com/bernedom/bertrand.git
  GIT_TAG 0.0.17)

FetchContent_MakeAvailable(bertrand)

add_executable(fetch_content_example)
target_link_libraries(
```

```
    fetch_content_example
    PRIVATE bertrand::bertrand
)
```

`FetchContent_MakeAvailable` has been available since version 3.14, and it is recommended that you overpopulate the project manually by using `FetchContent_Populate`. This should be used whenever possible because it makes code bases very maintainable due to its simplicity. As `ExternalProject`, `FetchContent` can download from HTTP(S), Git, SVN, Mercurial, and CVS, and the same good practices, such as specifying MD5 hashes for the downloaded content or using Git hashes, apply.

`FetchContent_MakeAvailable` is the recommended way to make external CMake-based projects available, but if you want to have more control over your external projects, it is also possible to populate the projects manually. The following example does the same as the previous example but in a more verbose way:

```
FetchContent_Declare(
  bertrand
  GIT_REPOSITORY https://github.com/bernedom/bertrand.git
  GIT_TAG 0.0.17)
if(NOT bertrand_POPULATED)
FetchContent_Populate(bertrand)
add_subdirectory(${bertrand_SOURCE_DIR} ${bertrand_BINARY_DIR})
endif()
```

`FetchContent_Populate` has additional options to be specified to control the build more closely. The signature is as follows:

```
FetchContent_Populate( <name>
  [QUIET]
  [SUBBUILD_DIR <subBuildDir>]
  [SOURCE_DIR <srcDir>]
  [BINARY_DIR <binDir>]
  ...
)
```

Let's look at the options of `FetchContent_Populate`:

- `QUIET`: This can be specified to suppress the output of the population if it succeeds. If the command fails, the output will be shown even if the option is specified to allow for debugging.
- `SUBBUILD_DIR`: This specifies where the external project will be located. The default is `${CMAKE_CURRENT_BINARY_DIR}/<name>-subbuild`. Generally, this option should be left as it is.
- `SOURCE_DIR` and `BINARY_DIR`: These change where the source and build directories of the external project are located. The default settings are `${CMAKE_CURRENT_BINARY_DIR}/<lcName>-src` for `SOURCE_DIR` and `${CMAKE_CURRENT_BINARY_DIR}/<lcName>-build` for `BINARY_DIR`
- Any additional parameters added will be passed on to the underlying `ExternalProject_Add`. However, `FetchContent` prohibits you from editing the commands for the different steps, so attempts to tamper with `CONFIGURE_COMMAND`, `BUILD_COMMAND`, `INSTALL_COMMAND`, and `TEST_COMMAND` will cause `FetchContent_Populate` to fail with an error.

> **Note**
> If you find yourself in a situation where you need to pass options to the underlying `ExternalProject_Add`, consider using `ExternalProject` directly instead of first going through `FetchContent`.

Information about the source and build directories, along with whether a project has been populated, can be retrieved either by reading the `<name>_SOURCE_DIR`, `<name>_BINARY_DIR`, and `<name>_POPULATED` variables or by calling `FetchContent_GetProperties`. Note that `<name>` will always be available in all caps and all lowercase. This is so that CMake can identify the packages despite different capitalizations.

Another big advantage of `FetchContent` is that it can handle cases where the external projects share common dependencies and prevent them from being downloaded and built multiple times. The first time a dependency is defined over `FetchContent` the details are cached, and any further definitions will be silently ignored. The benefit of this is that a parent project can overrule the dependencies of the child projects.

Let's assume we have a top project called `MyProject` that fetches two external projects, `Project_A` and `Project_B`, with each depending on a third external project called `AwesomeLib`, but on a different minor version. In most cases, we do not want to download and use two versions of `AwesomeLib` but only one to avoid conflicts. The following diagram, shows what the dependency graph might look like:

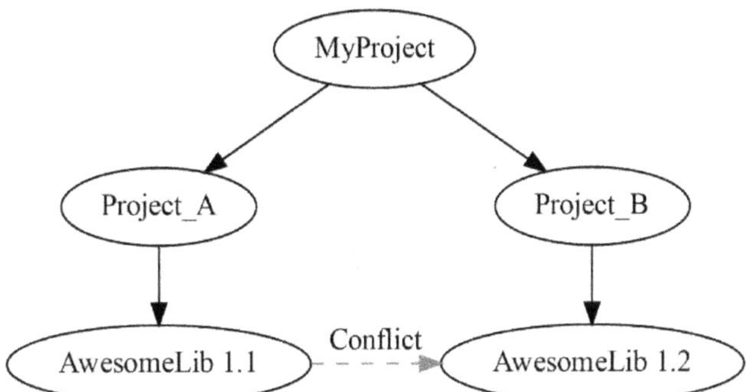

Figure 5.1 – Both Project_A and Project_B depend on different versions of AwesomeLib

To resolve this, we can specify which version of `AwesomeLib` to pull by placing a `FetchContent_Declare` call for `AwesomeLib` in the top-level `CMakeLists.txt` file. The order of the declaration inside the `CMakeLists.txt` file is not relevant here, only the level on which it is declared. Since both `Project_A` and `Project_B` contain the code to populate `AwesomeLib`, the top-level project does not need to use `FetchContent_MakeAvailable` or `FetchContent_Populate`. The resulting top-level `CMakeLists.txt` file might appear as follows:

```
include(FetchContent)
FetchContent_Declare(Project_A GIT_REPOSITORY ... GIT_TAG ...)
FetchContent_Declare(Project_B GIT_REPOSITORY ... GIT_TAG ...)
# Force AwesomeLib dependency to a certain version
FetchContent_Declare(AwesomeLib
GIT_REPOSITORY … GIT_TAG 1.2 )
FetchContent_MakeAvailable(Project_A)
FetchContent_MakeAvailable(Project_B)
```

This will force `AwesomeLib` to be pinned to version 1.2 for all projects. Of course, this only works if the interface between the versions required by `Project_A` and `Project_B` are compatible, resulting in a dependency graph, as illustrated in the following diagram:

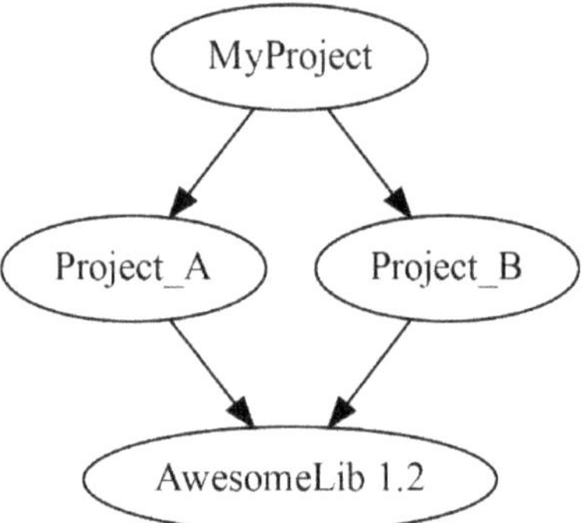

Figure 5.2 – The corrected dependency graph after MyProject declares the version of AwesomeLib

Adding dependencies as sources has some advantages, but it comes with major drawbacks in that it increases configuration and build time considerably. In *Chapter 9, Creating Reproducible Build Environments*, we will tackle superbuilds with distributed repositories and provide more information about how to handle source dependencies.

At the beginning of the chapter, we looked at `find_package`, which can be used to include binary dependencies, but we did not talk about how to conveniently download local binary dependencies using CMake. While `FetchContent` and `ExternalProject` can be used for that, it is not their purpose. Instead, dedicated package managers such as Conan and vcpkg will be better suited. Let's learn more about them next.

Using ExternalProject

The `ExternalProject` module is used to download and build external projects that are not fully integrated into the main project. When building an external project, the build is fully isolated, meaning that it will not automatically take over any settings regarding architecture or platforms. This isolation can come in handy to avoid clashes in the naming of targets or components. The external project creates a primary target and several child targets that contain the following isolated build steps:

1. **Download**: `ExternalProject` can download content in several ways, such as through pure HTTPS downloads or by accessing versioning systems such as Git, Subversion, Mercurial, and CVS. If the contents are archived, the download step will also unpack them.

2. **Updating and patching**: The downloaded source code can either be patched or updated to the newest version if the content is pulled from a **Server Configuration Monitor (SCM)**.
3. **Configure**: If the downloaded source uses CMake, the configure step is executed on it. For non-CMake projects, a custom command that does the configuration can be provided.
4. **Build**: By default, the same build tool that is used in the main project is used to build the dependency, but a custom command can be provided if this is not desired. If a custom build command is supplied, it is up to the user to ensure that the necessary compiler flags are passed on so that the results are ABI-compatible.
5. **Install**: The isolated build can be installed locally, usually somewhere in the build tree of the main project.
6. **Test**: If the external content comes with a set of tests, the main project might choose to run them. By default, the tests are not run.

All the steps, including downloading, run at build time. So, depending on the external project, this can increase the build time quite significantly. CMake caches the downloads and builds, so unless the external project has been changed, the overhead is primarily for the first run. The possibility of adding more steps to the external build does exist, but for most projects, the default steps are sufficient. The steps can be customized or omitted, as we will discover later.

In the following example, the `bertrand` library for using the design by contract is downloaded over HTTPS and locally installed in the current `build` directory:

```
include(ExternalProject)
ExternalProject_Add(
  bertrand
  URL https://github.com/bernedom/bertrand/archive
    /refs/tags/0.0.17.tar.gz
  URL_HASH MD5=354141c50b8707f2574b69f30cef0238

  INSTALL_DIR ${CMAKE_CURRENT_BINARY_DIR}/bertrand_install
   CMAKE_CACHE_ARGS -DBERTRAND_BUILD_TESTING:BOOL=OFF
-DCMAKE_INSTALL_PREFIX:PATH=<INSTALL_DIR>

)
```

Note that the `ExternalProject` module is not available by default and has to be included in the first line with `include(ExternalProject)`. As the external library is installed in the local build directory, the `INSTALL_DIR` option is specified. Since `bertrand` itself is a CMake project, the installation directory is passed as `<INSTALL_DIR>` by using the `CMAKE_INSTALL_PREFIX` variable to build the project. `<INSTALL_DIR>` is a placeholder that points back to the `INSTALL_DIR` option. `ExternalProject` knows placeholders for the various directories, such as `<SOURCE_DIR>`, `<BINARY_DIR>`, and `<DOWNLOAD_DIR>`. For a complete list, please consult the module documentation at https://cmake.org/cmake/help/latest/module/ExternalProject.html.

> **Verify Your Downloads**
>
> It is highly recommended that you add the download hash to any URL, as this sends you a notification if the contents of an artifact change.

For this to work, any target that depends on `bertrand` has to be built after the external dependency. As `bertrand` is a header-only library, we want to add the `include` path to a target. Using the external project for another target in CMake could look similar to the following:

```
ExternalProject_Get_Property(bertrand INSTALL_DIR)
set(BERTRAND_DOWNLOADED_INSTALL_DIR "${INSTALL_DIR}")

# Create a target to build an executable
add_executable(external_project_example)
# make the executable to be built depend on the external
  project
# to force downloading first
add_dependencies(external_project_example bertrand)

# make the header file for bertrand available
target_include_directories(external_project_example PRIVATE
  ${BERTRAND_DOWNLOADED_INSTALL_DIR}/include)
```

In the first line, the installation directory is retrieved with `ExternalProject_Get_Property` and stored in the `INSTALL_DIR` variable. Unfortunately, the variable name is always the same as the property, so it is recommended that you store it immediately after retrieval in a variable with a unique name that expresses its use better.

Next, the target we want to build is created and made dependent on the target created by `ExternalProject_Add`. This is necessary to enforce the correct build order.

And, finally, the path to the local installation is added to the target with `target_include_directories`. Additionally, we could import the CMake targets provided by the external library, but the purpose of this is to illustrate how this could work if the external project is not built by CMake.

Downloading from source code management systems happens with the respective options. For Git, this usually looks like the following:

```
ExternalProject_Add(MyProject GIT_REPOSITORY
  https://github.com/PacktPublishing/SomeRandomProject.git
    GIT_TAG 56cc1aaf50918f208e2ff2ef5e8ec0111097fb8d )
```

Note that `GIT_TAG` can be any valid revision number for Git, including tag names and long and short hashes. If `GIT_TAG` is omitted, the latest version of the default branch—usually called main or master—is downloaded. We highly recommend that you always specify the version to download. The most robust way is to define a commit hash, as tags can be moved around, although they rarely are in practice. Downloading from SVN is similar to downloading from Git. For additional details, please consult the official documentation for `ExternalProject`.

Using non-CMake projects and cross-compiling

A common use case for `ExternalProject` is to build dependencies that are not handled by CMake but by autotools or automake instead. In that case, you would need to specify the configuration and build commands as follows:

```
find_program(MAKE_EXECUTABLE NAMES nmake gmake make)
ExternalProject_Add(MyAutotoolsProject
   URL     someUrl
   INSTALL_DIR ${CMAKE_CURRENT_BINARY_DIR}/myProject_install
   CONFIGURE_COMMAND <SOURCE_DIR>/configure --prefix=<INSTALL_
    DIR>
    BUILD_COMMAND ${MAKE_EXECUTABLE}
)
```

Note that the first `find_program` command is used to find a version of make and store it in the `MAKE_EXECUTABLE` variable. A common issue with external projects is that you have to closely control where the dependencies are installed. Most projects want to install to a default system location, which often requires root privileges and could accidentally pollute a system. So, passing the necessary options to the configuration or a build step is often necessary. Another way to handle this is to avoid the installation process entirely by replacing `INSTALL_COMMAND` with an empty string as follows:

```
ExternalProject_Add(MyAutotoolsProject
   URL    someUrl
   CONFIGURE_COMMAND <SOURCE_DIR>/configure
    BUILD_COMMAND ${MAKE_EXECUTABLE}
    INSTALL_COMMAND ""
)
```

One problem with using non-CMake projects such as this is that they do not define the necessary targets for using the dependency directly. So, in order to use an externally built library in another target, you often have to add the full library name to the `target_link_libraries` calls. The major drawback of this is that you have to manually maintain the different names and locations of the files for the various platforms. The `find_library` or `find_file` calls are of little use because they happen at configuration time, while `ExternalProject` only creates the necessary files at build time.

Another common use case is to use `ExternalProject` to build the contents of an existing source directory for a different target platform. In this case, the parameter that handles the downloading is simply omitted. If the external project is using CMake to build, the toolchain file can be passed as a CMake option to the external project. More information about toolchain files is available in *Chapter 11, Automated Fuzzing with CMake*. A very common pitfall here is that `ExternalProject` will not recognize any changes to the sources of the external projects, so CMake might not rebuild them. For this reason, the `BUILD_ALWAYS` option should be passed, which has the downside of often making the build time considerably longer:

```
ExternalProject_Add(ProjectForADifferentPlatform
SOURCE_DIR $
    {CMAKE_CURRENT_LIST_DIR}/ProjectForADifferentPlatform
INSTALL_DIR ${CMAKE_CURRENT_BINARY_DIR}/
  ProjectForADifferentPlatform-install
CMAKE_ARGS
```

```
-D CMAKE_TOOLCHAIN_FILE=${CMAKE_CURRENT_LIST_DIR}/fwtoolchain.
  cmake
-D CMAKE_BUILD_TYPE=Release
-D CMAKE_INSTALL_PREFIX=<INSTALL_DIR>
BUILD_ALWAYS YES
)
```

Managing the steps in ExternalProject

As mentioned in the preceding section, the steps of `ExternalProject` can be configured further and used in a more granular way. `ExternalProject` can be told to create regular targets for each step either by passing the `STEP_TARGETS` option or by calling `ExternalProject_Add_StepsTargets`. The following calls expose both the configure step and the build step of an external project as targets:

```
ExternalProject_Add(MyProject
   # various options
   STEP_TARGETS configure build
)
ExternalProject_Add_StepTargets(MyProject configure build)
```

The targets are named after `<mainName>-step`. In the preceding example, two additional targets, `MyProject-configure` and `MyProject-build`, will be created. Creating step targets has two main uses: you are able to create custom steps that are sorted in the order of the download, configure, build, install, or test sequence, or you can make the steps dependent on other targets. These can either be regular targets, created by `add_executable`, `add_library`, or `add_custom_target`, or targets from other add executables. A common case is when external projects depend on each other, so the configuration step of one has to depend on another. In the next example, the configure step of project B will depend on the completion of project A:

```
ExternalProject_Add(ProjectA
... # various options
        STEP_TARGETS install
)
ExternalProject_Add(ProjectB
... # various options
)
ExternalProject_Add_StepDependencies(ProjectB configure
  ProjectA)
```

Finally, we can also create custom steps to be interjected into an external project. The process of adding steps is done with the `ExternalProject_Add_Step` command. Custom steps cannot be named the same as any of the predefined steps (such as `mkdir`, `download`, `update`, `patch`, `configure`, `build`, `install`, or `test`). The following example will create a step that adds the license information of an external project to a specific `tar` file after building:

```
ExternalProject_Add_Step(bertrand_downloaded copy_license
     COMMAND ${CMAKE_COMMAND} -E tar "cvzf" ${CMAKE_CURRENT_
       BINARY_DIR}/licenses.tar.gz <SOURCE_DIR>/LICENSE
         DEPENDEES build
)
```

All in all, `ExternalProject` is a very powerful tool; however, it can become very complex to manage. Often, it is that flexibility that also makes `ExternalProject` hard to use. While it can help isolate builds, it often forces the project maintainer to manually expose any information from the inner workings of the external project to CMake, which, ironically, is what CMake is supposed to help with in the first place.

Summary

In this chapter, we covered a general approach for finding files, libraries, and programs, along with the more complex search for CMake packages. You learned how to create an imported package definition if it cannot be found automatically by providing your own find module. We looked over source code-based dependencies with `ExternalProject` and `FetchContent` and how even non-CMake projects can be built using CMake.

Additionally, if you want to become even more sophisticated with your dependency management, we briefly introduced Conan and vcpkg as two package handlers that integrate very well with CMake.

Dependency management is a tough topic to cover and can be tedious at times. Nevertheless, it pays off to take the time to set it up correctly with the techniques described in this chapter. The versatility of CMake and its various ways of finding dependencies are its greatest strengths but also its greatest weaknesses. By using the various `find_` commands, `FetchContent`, `ExternalProject`, or integrating any of the available package managers with CMake, almost any dependency can be integrated into a project. However, with so many methods to choose from, finding the best one can be tough. Nevertheless, we recommend using `find_package` whenever possible. The more popular CMake becomes, the better the chances are that other projects can be seamlessly integrated.

In the next chapter, you will learn how to automatically generate and package documentation for your code.

Questions

Answer the following questions to test your knowledge of this chapter:

1. Which `find_` programs exist in CMake?
2. Which properties should be set for targets imported by a `find` module?
3. When finding things, which option takes precedence, `HINTS` or `PATHS`?
4. At what stage does `ExternalProject` download the external content?
5. At what stage does `FetchContent` download the external content?

6
Automatically Generating Documentation with CMake

Documentation is—without a doubt—an essential part of all projects. Documentation conveys information that is not implicitly available to the user. It is a way of sharing the intent, functionality, capabilities, and restrictions regarding a project and it enables both technical and non-technical people to work on a project. But it is indeed a time-consuming process to write documentation, thus it is crucial to make use of the tools available for generating documentation.

This chapter will look into ways of integrating Doxygen, `DOT`, and PlantUML into CMake to speed up the documentation process. These tools will allow us to lessen the context switch between code and documentation and also ease the maintenance burden of documentation.

To understand the skills shared in this chapter, we'll cover the following main topics:

- Generating documentation from your code
- Packaging and distributing documentation with CPack
- Creating dependency graphs of CMake targets

Let's begin with the technical requirements.

Technical requirements

Before you dive further into this chapter, you should have a good grasp of the content covered in *Chapter 4, Packaging, Deploying, and Installing a CMake Project*, and *Chapter 5, Integrating Third-Party Libraries and Dependency Management*. The techniques that will be used in this chapter are all covered in these two chapters. Additionally, it is recommended to obtain this chapter's example content from `https://github.com/PacktPublishing/CMake-Best-Practices/tree/main/chapter_6`. All of the examples assume that you will be using the development environment container provided by the project found at the following link: `https://github.com/PacktPublishing/CMake-Best-Practices`. This is a Debian-like environment that contains all the prerequisites installed beforehand. Commands and outputs may differ slightly if a different environment is used. If you are not using the provided Docker container, ensure that you have installed Doxygen, PlantUML, and Graphviz in your environment. Consult your package manager's index for installation details.

Let's dive into the realm of documentation by learning ways of generating documentation from existing code.

Generating documentation from your code

Most people, either knowingly or not, structure their software projects in an organized way. The organization is a positive side effect of methodologies and procedures such as **object-oriented** (**OO**) design, programming language rules, personal preference, habits, or dictated by project rules. While rules and conventions tend to be boring, adhering to them results in a more understandable project structure. As we can all agree, organized material is better for both human perception and computers. When procedure, rules, order, and organization exist, the computer can make sense of it. Documentation generation software leverages that fact to our benefit.

We will now introduce you to one of the most prominent documentation generation software for C and C++ programming languages: Doxygen. We will learn how we can integrate Doxygen with CMake to automatically generate documentation for CMake projects.

Up next, the basics of Doxygen await.

Understanding what Doxygen is

Doxygen is a very popular documentation software for C++ projects that allows the generation of documentation from code. Doxygen understands C and C++ grammar and can see the code structure in a way that a compiler would see it. This allows Doxygen to dive into the structure of a software project and look into all class definitions, namespaces, anonymous functions, encapsulation, variables, inheritance relations, and so on. Doxygen combines this information with inline code documentation written by the programmer. The final result is human-readable documentation in various format that is compatible for both online and offline reading.

But there is no such thing as a free lunch, as is the saying. In order to be able to make sense of code comments, Doxygen requires comments to be in a predefined set of formats. For the sake of not distracting from the focus of this chapter, please take a look at `https://www.doxygen.nl/manual/docblocks.html` to find out the comment formats compatible with Doxygen. We will be using **Javadoc**-style comments in our examples. An example Javadoc comment for a C++ function is provided here:

```
/**
 * Does foo with @p bar and @p baz
 *
 * @param [in] bar Level of awesomeness
 * @param [in] baz Reason of awesomeness
 */
void foo(int bar, const char* baz){}
```

Doxygen also requires a **Doxyfile**, which essentially contains all the parameters of documentation generation, such as output format, excluded file patterns, project name, and so on. Because of the sheer number of configuration parameters, configuring Doxygen may be intimidating at the start, but fear not—CMake will generate a Doxyfile for you too.

As we dive further into this chapter, you will start to see the benefits of using documentation generation software for your project. Through this, it is way easier to keep the documentation consistent with code, and the ability to read the code structure also makes graphing easier.

Enough theory for now. Let's begin using Doxygen together with CMake.

Using Doxygen with CMake

CMake, being a C++-oriented build system generator, has good support for integrating external tools that are commonly used in C++ projects. As you would expect, integrating Doxygen with CMake is pretty straightforward. We'll utilize the `FindDoxygen.cmake` module of CMake to integrate Doxygen into our projects. This module is, by default, provided by the CMake installation and requires no extra setup.

`FindDoxygen.cmake` is, as the name suggests, a module package file designated to be consumed by the `find_package()` CMake function. Its primary use is locating Doxygen in the environment, and also providing some extra utility functions to enable documentation generation in a CMake project. To illustrate Doxygen's abilities, we will be following the *Chapter 6 - Example 01* example for this section. Our goal is to generate documentation for a simple calculator library along with its `README` file. The interface definition of this library looks like this:

```
class calculator : private calculator_interface {
public:
  /**
    * Calculate the sum of two numbers, @p augend lhs and
      @p addend
    *
    * @param [in] augend The number to which @p addend is
      added
    * @param [in] addend The number which is added to
      @p augend
    *
    * @return double Sum of two numbers, @p lhs and @p rhs
    */
  virtual double sum(double augend, double addend)
    override;
   /**
```

```
   * Calculate the difference of @p rhs from @p lhs
   *
   * @param [in] minuend    The number to which @p
     subtrahend is subtracted
   * @param [in] subtrahend The number which is to be
     subtracted from @p minuend
  *
   * @return double Difference of two numbers, @p minuend
     and @p subtrahend
  */
 virtual double sub(double minuend, double subtrahend)
   override;
 /*...*/}; // class calculator
```

The `calculator` class implements the class interface defined in the `calculator_interface` class. It is properly documented in Javadoc format. We will expect Doxygen to generate **application programming interface** (**API**) documentation for the `calculator` and `calculator_interface` classes, together with an inheritance diagram. The class definition is in the `calculator.hpp` file and can be found under the `include/chapter6/ex01` subdirectory of the `chapter6/ex01_doxdocgen` directory. Additionally, we have a Markdown file named `README.md` in the `chapter6/ex01_doxdocgen` directory—this contains essential information about the example project's layout. We expect this file to be the main page of the documentation. As our input material is ready, let's start diving into the example further by inspecting the example's `CMakeLists.txt` file, `chapter6/ex01_doxdocgen/CMakeLists.txt`, as usual. The `CMakeLists.txt` file begins with finding the Doxygen package, as can be seen here:

```
find_package(Doxygen)
set(DOXYGEN_OUTPUT_DIRECTORY"${CMAKE_CURRENT_BINARY_DIR}
  /docs")
set(DOXYGEN_GENERATE_HTML YES)
set(DOXYGEN_GENERATE_MAN YES)
set(DOXYGEN_MARKDOWN_SUPPORT YES)
set(DOXYGEN_AUTOLINK_SUPPORT YES)
set(DOXYGEN_HAVE_DOT YES)
set(DOXYGEN_COLLABORATION_GRAPH YES)
set(DOXYGEN_CLASS_GRAPH YES)
```

```
set(DOXYGEN_UML_LOOK YES)
set(DOXYGEN_DOT_UML_DETAILS YES)
set(DOXYGEN_DOT_WRAP_THRESHOLD 100)
set(DOXYGEN_CALL_GRAPH YES)
set(DOXYGEN_QUIET YES)
```

The `find_package(...)` call will utilize the `FindDoxygen.cmake` module provided by the CMake installation to find Doxygen in the environment if present. The `REQUIRED` parameter is omitted in order to allow package maintainers to pack the project without having to install. This ensures that Doxygen in their environment is found before proceeding any further. The subsequent lines are setting several Doxygen configurations. These configurations will be placed into the Doxyfile that will be generated by CMake. Detailed descriptions for each option are listed here:

- `DOXYGEN_OUTPUT_DIRECTORY`: Sets the output directory for Doxygen.
- `DOXYGEN_GENERATE_HTML`: Instructs Doxygen to emit **HyperText Markup Language** (**HTML**) output.
- `DOXYGEN_GENERATE_MAN`: Instructs Doxygen to emit `MAN` page output.
- `DOXYGEN_AUTOLINK_SUPPORT`: Allows Doxygen to automatically link language symbols and filenames to relevant documentation pages if available.
- `DOXYGEN_HAVE_DOT`: Tells Doxygen the environment has the `dot` command available, which can be utilized for generating graphs. This will enable Doxygen to enrich generated documentation with diagrams such as dependency, inheritance, and collaboration diagrams.
- `DOXYGEN_COLLABORATION_GRAPH`: Tells Doxygen to generate collaboration diagrams for classes.
- `DOXYGEN_CLASS_GRAPH`: Tells Doxygen to generate class diagrams for classes.
- `DOXYGEN_UML_LOOK`: Instructs Doxygen to generate **Unified Modeling Language** (**UML**)-like diagrams.
- `DOXYGEN_DOT_UML_DETAILS`: Adds type and parameter information to UML diagrams.
- `DOXYGEN_DOT_WRAP_THRESHOLD`: Sets the line wrapping threshold for UML diagrams.
- `DOXYGEN_CALL_GRAPH`: Instructs Doxygen to generate call graphs for functions in function documentation.
- `DOXYGEN_QUIET`: Suppresses Doxygen output generated to **standard output** (**stdout**).

Doxygen's set of options is pretty extensive and offers a lot more than just the options we've covered. If you want to customize documentation generation further, take a look at the full list of parameters that can be used in Doxyfiles at `https://www.doxygen.nl/manual/config.html`. To set any Doxygen option in CMake, prefix the variable name with `DOXYGEN_` and set the desired value using `set()`. With that side note written down, let's go back to the example code. The CMake code shown before is followed by the target declarations. The following lines of code define a regular static library that contains our example code for the documentation:

```
add_library(ch6_ex01_doxdocgen_lib STATIC)
target_sources(ch6_ex01_doxdocgen_lib PRIVATE
  src/calculator.cpp)
target_include_directories(ch6_ex01_doxdocgen_lib PUBLIC
  include)
target_compile_features(ch6_ex01_doxdocgen_lib PRIVATE
  cxx_std_11)
```

Subsequently, the following lines of code define an executable that consumes the static library target defined before:

```
add_executable(ch6_ex01_doxdocgen_exe src/main.cpp)
target_compile_features(ch6_ex01_doxdocgen_exe PRIVATE
  cxx_std_11)
target_link_libraries(ch6_ex01_doxdocgen_exe PRIVATE
  ch6_ex01_doxdocgen_lib)
```

Lastly, the `doxygen_add_docs(...)` function is called to specify code that we wish to generate documentation from, as can be seen next:

```
doxygen_add_docs(
  ch6_ex01_doxdocgen_generate_docs
  "${CMAKE_CURRENT_LIST_DIR}"
  ALL
  COMMENT "Generating documentation for Chapter 6 - Example
    01 with Doxygen"
)
```

The doxygen_add_docs(...) function is a function provided by the FindDoxygen.cmake module. Its sole purpose is to provide a convenient way to create a CMake target for documentation generation without explicitly dealing with Doxygen. The signature for the doxygen_add_docs(...) function is shown here (non-relevant parameters are omitted):

```
doxygen_add_docs(targetName
    [filesOrDirs...]
    [ALL]
    [COMMENT comment])
```

The first parameter of the targetName function is the name for the documentation target. The function will generate a custom target named targetName. This target will trigger Doxygen and create documentation from the code when built. The next list of parameters, filesOrDirs, is a list of files or directories that contain the code we want to generate from the documentation. The ALL parameter is used to make CMake's ALL meta-target depend on the documentation target created by the doxygen_add_docs(...) function, so documentation generation is automatically triggered when the ALL meta-target is built. Lastly, the COMMENT parameter is for making CMake print a message to the output when the target is being built. COMMENT is primarily useful for diagnostic purposes so that we can quickly know whether documentation is being generated or not.

After this brief introduction to doxygen_add_docs(...), let's go back to the example code and explain what the doxygen_add_docs(...) function call does in our scenario. It creates a target named ch6_ex01_doxdocgen_generate_docs, adds ${CMAKE_CURRENT_LIST_DIR} to the documentation generation directory, requests the ALL meta-target to depend on it, and specifies a COMMENT parameter that will be printed when the target is built.

All right—it's time to test whether this works or not. Go into the chapter_6/ directory and configure the project in the build/ directory with the following command:

```
cd chapter_6/
cmake -S . -B build/
```

Check the CMake output to see whether the configuration succeeded or not. If the configuration was successful, that means CMake succeeded to find Doxygen in the environment. You should be able to see that in the CMake output, as can be seen next:

```
Found Doxygen: /usr/bin/doxygen (found version "1.9.1")
   found components: doxygen dot
```

After a successful configuration, let's try to build it with the following command:

```
cmake --build build/
```

In the build output, you should be able to see that the text we've given into the `COMMENT` parameter is being printed to the CMake output. This means the documentation target is being built and Doxygen is running. Notice that we did not specify a `--target` argument to the CMake build command, which effectively causes CMake to build the `ALL` meta-target. Since we have given the `ALL` argument to the `doxygen_add_docs(...)` function, a `ch6_ex01_doxdocgen_generate_docs` target is being built too. The output of the build command should look similar to the output given here:

```
[ 20%] Generating documentation for Chapter 6 - Example 01
  with Doxygen
Doxygen version used: 1.9.1
Searching for include files...
Searching for example files...
/*…*/
Running dot...
Running dot for graph 1/1
lookup cache used 9/65536 hits=13 misses=9
finished...
[ 20%] Built target ch6_ex01_doxdocgen_generate_docs
[ 40%] Building CXX object ex01_doxdocgen/CmakeFiles
  /ch6_ex01_doxdocgen_lib.dir/src/calculator.cpp.o
[ 60%] Linking CXX static library
  libch6_ex01_doxdocgen_lib.a
[ 60%] Built target ch6_ex01_doxdocgen_lib
[ 80%] Building CXX object ex01_doxdocgen/CmakeFiles
  /ch6_ex01_doxdocgen_exe.dir/src/main.cpp.o
[100%] Linking CXX executable ch6_ex01_doxdocgen_exe
[100%] Built target ch6_ex01_doxdocgen_exe
```

It seems we have succeeded in building the project and the documentation. Let's inspect the generated documentation in the `${CMAKE_CURRENT_BINARY_DIR}/docs` output folder, as follows:

```
14:27 $ ls build/ex01_doxdocgen/docs/
html  man
```

Here, we can see Doxygen has emitted the HTML and `MAN` page output into the `html/` and `man/` directories. Let's inspect the final result for each type. To inspect the generated `MAN` page, simply type the following:

```
man build/ex01_doxdocgen/docs/man/man3
  /chapter6_ex01_calculator.3
NAME
       chapter6::ex01::calculator - The 'calculator' class
         interface.
SYNOPSIS
       #include <calculator.hpp>
   Static Public Member Functions
       static double sum (double augend, double addend)
       static double sub (double minuend, double
         subtrahend)
       static double mul (double multiplicand, double
         multiplier)
       static double div (double dividend, double divisor)
Detailed Description
       The 'calculator' class interface.
Member Function Documentation
    double chapter6::ex01::calculator::div (double
      dividend, double divisor) [static]
       Divide dividend with divisor
       Parameters
           dividend The number to be divided by divisor
           divisor The number by which divisor is to be
             divided
       Returns
           double Quotient of two numbers, dividend and
             divisor

 Manual page chapter6_ex01_calculator.3 line 1 (press h for
  help or q to quit)
```

Great! Our code comments turned into a `MAN` page. Similarly, let's inspect the HTML output as well. Use a browser of your choice to open the `build/ex01_doxdocgen/docs/html/index.html` file, as shown next:

```
google-chrome build/ex01_doxdocgen/docs/html/index.html
```

This will display the main page of your documentation, as shown in the following screenshot:

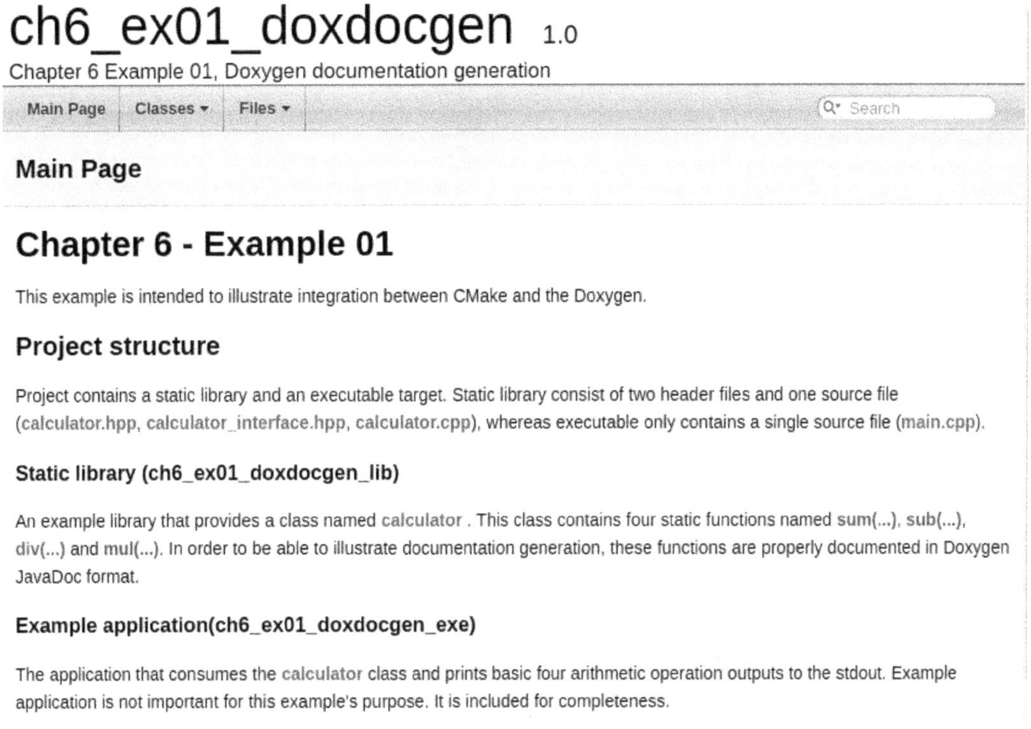

ch6_ex01_doxdocgen 1.0

Chapter 6 Example 01, Doxygen documentation generation

Main Page | Classes ▾ | Files ▾ | Search

Main Page

Chapter 6 - Example 01

This example is intended to illustrate integration between CMake and the Doxygen.

Project structure

Project contains a static library and an executable target. Static library consist of two header files and one source file (calculator.hpp, calculator_interface.hpp, calculator.cpp), whereas executable only contains a single source file (main.cpp).

Static library (ch6_ex01_doxdocgen_lib)

An example library that provides a class named calculator . This class contains four static functions named sum(...), sub(...), div(...) and mul(...). In order to be able to illustrate documentation generation, these functions are properly documented in Doxygen JavaDoc format.

Example application(ch6_ex01_doxdocgen_exe)

The application that consumes the calculator class and prints basic four arithmetic operation outputs to the stdout. Example application is not important for this example's purpose. It is included for completeness.

Figure 6.1 – The main page of the documentation

In the preceding screenshot, we can see that Doxygen has rendered the `README.md` Markdown file content into the main page. Note that the main page is just provided as an example. Doxygen can embed an arbitrary number of Markdown files into generated documentation. It even replaced filenames, class names, and function names with links to the relevant documentation. This is achieved by Doxygen's `AUTOLINK` feature and the `@ref` Doxygen command. Click on the `calculator` link under the **Static library** section of the main page to access API documentation for the `calculator` class. The `calculator` class documentation page should look similar to this:

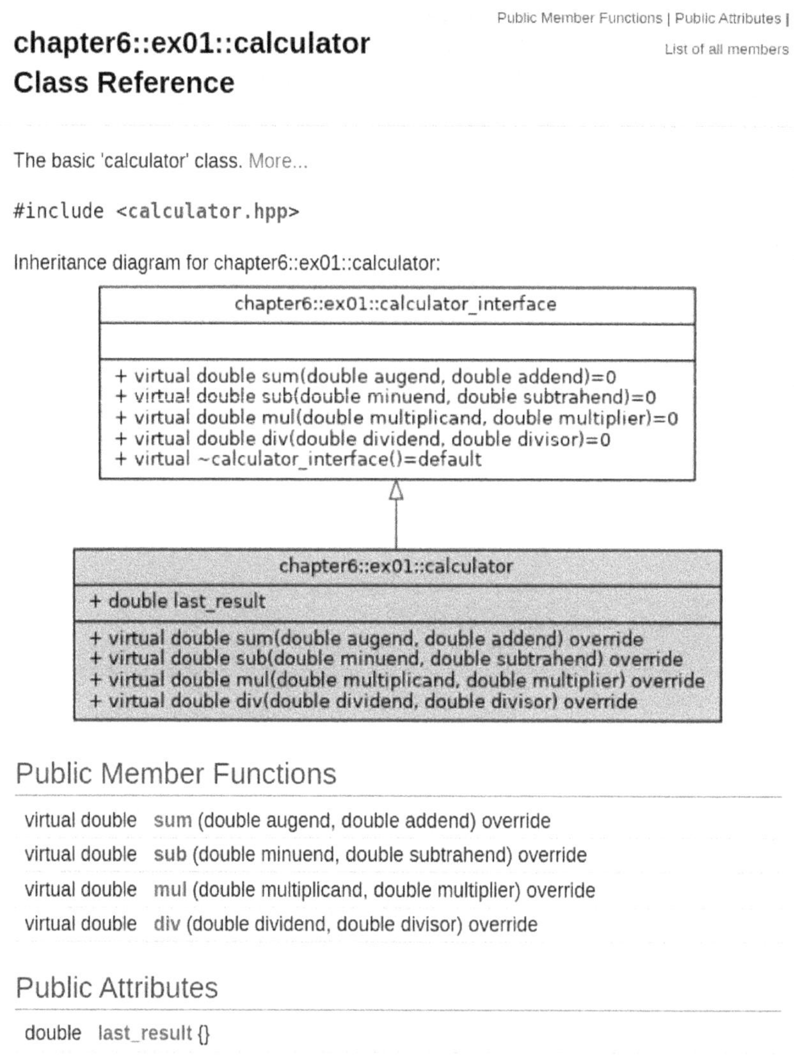

Figure 6.2 – Generated HTML documentation for the calculator class (basic layout)

In the preceding screenshot, we can see that Doxygen is aware that the `calculator` class inherits from `calculator_interface` and draws an inheritance diagram for the `calculator` class.

> **Note**
>
> Doxygen requires the `dot` tool to render diagrams. `dot` is available in the Graphviz software package.

Also, the generated diagrams contain function names and encapsulation symbols in UML style. Let's take a look at the detailed member function documentation shown in the following screenshot:

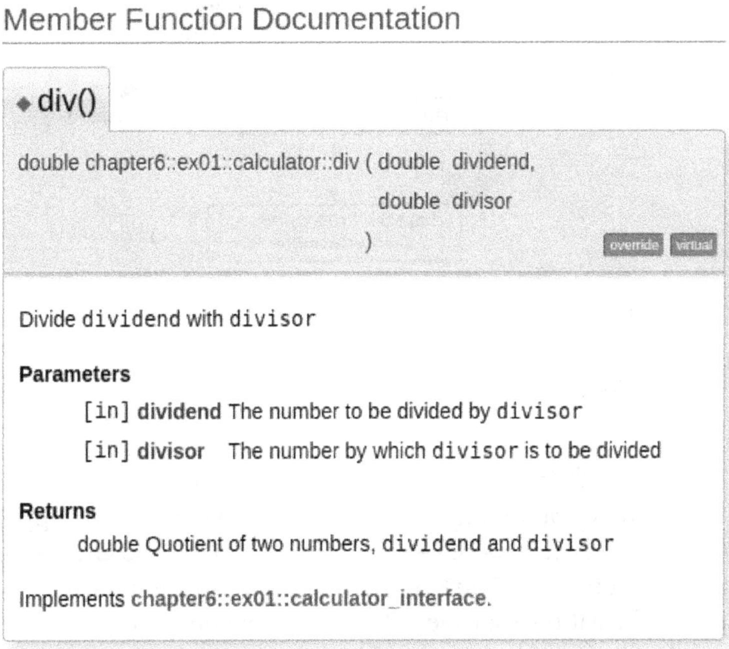

Member Function Documentation

◆ div()

double chapter6::ex01::calculator::div (double dividend,
double divisor
) override virtual

Divide `dividend` with `divisor`

Parameters

`[in]` **dividend** The number to be divided by `divisor`

`[in]` **divisor** The number by which `divisor` is to be divided

Returns

double Quotient of two numbers, `dividend` and `divisor`

Implements **chapter6::ex01::calculator_interface**.

Figure 6.3 – Generated documentation for the div() function of the calculator class

As we can see in *Figure 6.3*, Doxygen did a pretty good job of putting the contents into a clear, readable layout. If you were reading this API documentation, you would be happy. Lastly, let's navigate to **Files** | **File List** | `main.cpp` to look into the documentation of `main.cpp` to illustrate what a dependency graph looks like. You can see a representation of the documentation page in the following screenshot:

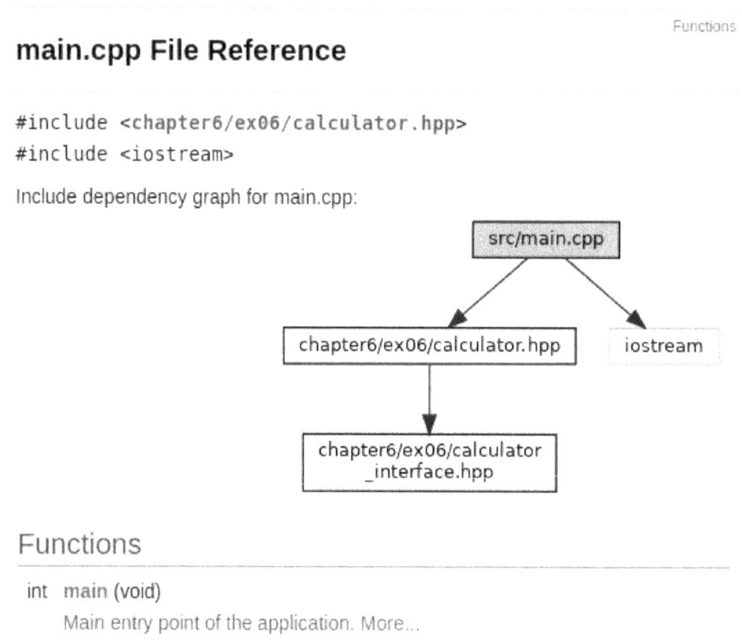

Figure 6.4 – main.cpp documentation page

The dependency graph in the preceding screenshot shows that the `main.cpp` file directly depends on `iostream` and `chapter6/ex06/calculator.hpp` and indirectly depends on `chapter6/ex06/calculator_interface.hpp` files. It is pretty useful to have dependency information available in the documentation. The consumers will know exactly what the file dependencies are without having to dive into the code. If you scroll down a bit more, you will see a call graph for the `main()` function as well, as can be seen in the following screenshot:

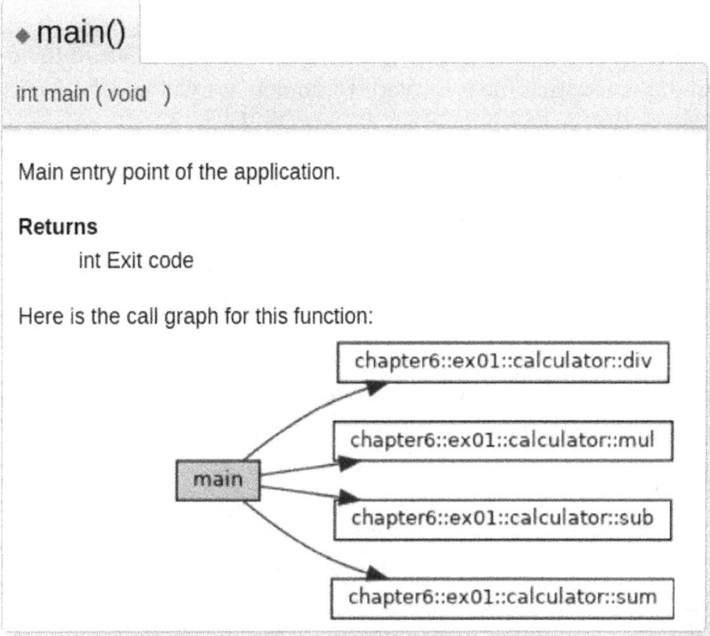

Figure 6.5 – main() function call graph

Splendid! We have generated documentation with graphs for our code in two different formats with fewer than 20 lines of extra CMake code. How cool is that? Now, with that power in your hands, it is hard to find an excuse to avoid documentation. But our journey in this chapter is not over yet. The upcoming section will enrich our knowledge by teaching us how to embed custom UML diagrams into documentation. Let's go!

Embedding custom UML diagrams into documentation

In the previous section, we learned how to utilize Doxygen to generate diagrams and documentation for our CMake project, but not every diagram is inferable from the code. We might want to draw custom diagrams to illustrate the elaborate relationships of an entity with an external system that is not available in the code context. To tackle this, the obvious choice would be somehow making this context available in the code or comments to utilize documentation generation again. Well, as could be expected, this too is possible with Doxygen. Doxygen allows the embedding of PlantUML diagrams into comments, which will enable us to draw any diagram that PlantUML supports. But before starting to put PlantUML code in Doxygen comments, there's a small thing that must be taken care of: enabling PlantUML support in Doxygen. We have got ourselves a starting point. Let's roll!

Enabling PlantUML support in Doxygen is pretty easy. Doxygen requires a `PLANTUML_JAR_PATH` variable to be set to the location of the `plantuml.jar` file in the environment. So, we have to find out where that file is located. To do that, we will use the `find_path(...)` CMake function. `find_path(...)` is similar to `find_program(...)`, except it is designated for locating the path of files instead of program locations. So, that means we should be able to locate the path of `plantuml.jar` with `find_path(...)`, give this path to Doxygen, and ... profit! Let's put that theory to the test. We will be following *Chapter 6 - Example 02* for this section. As ever, let's dive into the `CMakeLists.txt` file of the example code, located at the `chapter_6/ex02_doxplantuml/CMakeLists.txt` path. Let's start inspecting from the `find_path(...)` call, as follows:

```
find_path(PLANTUML_JAR_PATH NAMES plantuml.jar HINTS
  "/usr/share/plantuml" REQUIRED)
find_package(Doxygen REQUIRED)
set(DOXYGEN_OUTPUT_DIRECTORY "${CMAKE_CURRENT_BINARY_DIR}
  /docs")
set(DOXYGEN_GENERATE_HTML YES)
set(DOXYGEN_AUTOLINK_SUPPORT YES)
set(DOXYGEN_PLANTUML_JAR_PATH "${PLANTUML_JAR_PATH}")
set(DOXYGEN_QUIET YES)
```

In the `find_path(...)` call here, `PLANTUML_JAR_PATH` is the name of the output variable. `NAMES` is the filenames that will be searched in the search locations. `HINTS` is extra paths in addition to the default search locations. These are useful for discovering stuff in non-standard locations. Lastly, the `REQUIRED` parameter is used to make finding `plantuml.jar` a requirement, so CMake will fail and exit when `plantuml.jar` cannot be located. The following Doxygen configuration section is exactly the same as our previous example, *Chapter 6 - Example 01*, except we are setting `DOXYGEN_PLANTUML_JAR_PATH` to the PlantUML directory path we found with the `find_path(...)` call. Also, variables that are not required for this example are omitted too. Doxygen should be able to use PlantUML right now. Let's test that with an example PlantUML graph, embedded into the `src/main.cpp` source file, as follows:

```
/**
* @brief Main entry point of the application
  @startuml{system_interaction.png} "System Interaction
    Diagram"
  user -> executable : main()
```

```
    user -> stdin
    executable -> executable: read_stdin()
    executable -> stdout
    @enduml
 * @return int Exit code
 */
int main(void) {
    std::cout << "Greetings from the echo application!" <<
      std::endl;
    std::string input;
    while (std::getline(std::cin, input)) {
      std::cout << input;
    }
}
```

The `@startuml` and `@enduml` Doxygen comment keywords are for indicating the beginning and the end of a PlantUML diagram, respectively. Regular PlantUML code can be placed inside of a `@startuml - @enduml` block. In our example, we have a simple system interaction diagram of the application. If everything goes as expected, we should see the embedded PlantUML diagram in the `main()` function's documentation. Let's generate documentation by building the example with the code shown here:

```
cd chapter_6/
cmake -S ./ -B build/
cmake --build build/
```

The documentation for the second example should be built now. Open the generated `build/ex02_doxplantuml/docs/html/index.html` HTML documentation with the browser of your choice, by running the following command:

```
google-chrome build/ex02_doxplantuml/docs/html/index.html
```

This gives you the following output:

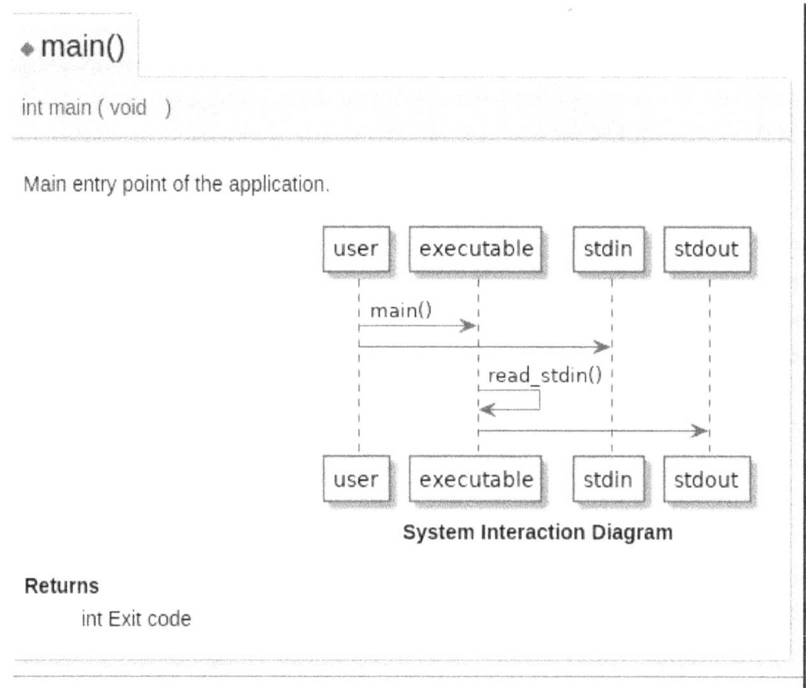

Figure 6.6 – Embedded PlantUML diagram in the main() function documentation

In *Figure 6.6*, we can see that Doxygen has generated a PlantUML diagram and embedded it into the documentation. With that, we're now able to embed custom diagrams into our generated documentation as well. This will allow us to explain complicated systems and relationships without having to interact with external graphing tools.

Now, as we have the right tools for generating documentation, it's time to learn how to package and deliver them. In the next section, we will learn about ways of delivering documentation, together with the software involved.

Packaging and distributing documentation with CPack

Packaging documentation is no different than packaging software and its artifacts—documentation is an artifact of a project, after all. Thus, we will use the techniques we learned in *Chapter 4, Packaging, Deploying, and Installing a CMake Project,* to package our documentation.

> **Note**
>
> If you have not read *Chapter 4, Packaging, Deploying, and Installing a CMake Project,* yet, it is strongly recommended to do so before reading this section.

For illustrating this section, we will return to *Chapter 6 - Example 01*. We will make the documentation we have generated in this example installable and packageable. Let's dive back into the `CMakeLists.txt` file located in the `chapter_6/ex01_doxdocgen/` folder. With the following code, we will make the HTML and `MAN` documentation installable:

```
include(GNUInstallDirs)
install(DIRECTORY "${CMAKE_CURRENT_BINARY_DIR}/docs/html/"
  DESTINATION "${CMAKE_INSTALL_DOCDIR}" COMPONENT
    ch6_ex01_html)
install(DIRECTORY "${CMAKE_CURRENT_BINARY_DIR}/docs/man/"
  DESTINATION "${CMAKE_INSTALL_MANDIR}" COMPONENT
    ch6_ex01_man)
```

Remember how we used `install(DIRECTORY...)` to install any kind of folder while preserving its structure in *Chapter 4, Packaging, Deploying, and Installing a CMake Project*? This is exactly what is happening here. We are making the generated documentation installable by installing `docs/html` and `docs/man` to the default documentation and man page directories provided by the `GNUInstallDirs` module. Also, recall that if a thing is installable, it means it is also packageable since CMake is able to generate the required packaging code from `install(...)` calls. So, let's include the `CPack` module to enable packaging for this example too. The code is illustrated in the following snippet:

```
set(CPACK_PACKAGE_NAME cbp_chapter6_example01)
set(CPACK_PACKAGE_VENDOR "CBP Authors")
set(CPACK_GENERATOR "DEB;RPM;TBZ2")
set(CPACK_DEBIAN_PACKAGE_MAINTAINER "CBP Authors")
include(CPack)
```

And there we have it! Simple as that. Let's try building and packaging the example project by invoking the following commands:

```
cd chapter_6/
cmake -S . -B build/
cmake --build build/
cpack --config  build/CPackConfig.cmake -B build/pak
```

So, let's summarize what is happening here. We have configured and built the `chapter_6/` code and invoked CPack to package the project into the `build/pak` folder using the generated `CPackConfig.cmake` file. To check whether everything is in order, let's extract the contents of the generated package into the `/tmp/ch6-ex01` path by invoking the following command:

```
dpkg -x build/pak/cbp_chapter6_example01-1.0-Linux.deb
  /tmp/ch6-ex01
export MANPATH=/tmp/ch6-ex01/usr/share/man/
```

After the extraction completes, the documentation must become available under the `/tmp/ch6-ex01/usr/share` path. Since we have used a non-default path, we have used the `MANPATH` environment variable to let the `man` command know the path of our documentation. Let's start by checking whether we can access the man pages by invoking the `man` command, as follows:

```
man chapter6_ex01_calculator
```

The `chapter6_ex01_calculator` name is automatically inferred by Doxygen from the `chapter6::ex01::calculator` class name. You should be able to see the man page output we covered in the previous section.

Up to now, we have learned a great deal about generating and packaging documentation. Up next, we will be learning about generating dependency graphs of CMake targets.

Creating dependency graphs of CMake targets

In the previous sections, we have been covering the documentation and graphing of the software code, but in a large project, we may also need to document and visualize the CMake code as well. The relations between CMake targets may be complex, and this may render keeping track of all the dependencies hard. Fortunately, CMake can again help with this by providing a graph showing all dependencies between targets. By calling `cmake --graphviz=my-project.dot /path/to/build/dir`, CMake will create files in the DOT language that contain how targets depend on each other.

The DOT language is a description language for graphs and can be interpreted by a multitude of programs, the most famous one being the freely available Graphviz. DOT files can be converted to images or even **Portable Document Format** (**PDF**) files using the `dot` command-line utility from Graphviz, like this: `dot -Tpng filename.dot -o out.png`.

Running these commands for the example project of this chapter will produce an output similar to this:

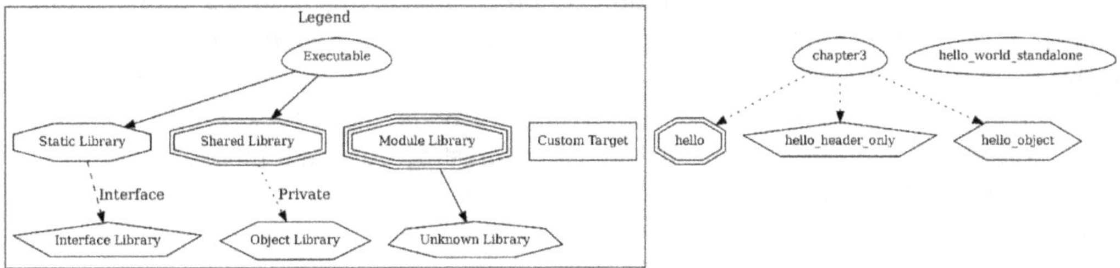

Figure 6.7 – The project structure of chapter3 visualized with the DOT language

Behavior and options can be controlled by the variables provided in `CMakeGraphVizOptions`. When creating DOT graphs, CMake will look for a file called `CMakeGraphVizOptions.cmake` in the `PROJECT_SOURCE_DIR` and `PROJECT_BINARY_DIR` directories and, if found, will use the values provided within. An example of such a config file might look like this:

```
set(GRAPHVIZ_GRAPH_NAME "CMake Best Practices")
set(GRAPHVIZ_GENERATE_PER_TARGET FALSE)
set(GRAPHVIZ_GENERATE_DEPENDERS FALSE)
```

By default, CMake creates dependency graphs for all targets. Setting `GRAPHVIZ_GENERATE_PER_TARGET` and `GRAPHVIZ_GENERATE_DEPENDERS` to `FALSE` will reduce the number of files generated. A full set of available options can be found in the CMake documentation at `https://cmake.org/cmake/help/latest/module/CMakeGraphVizOptions.html`.

Summary

In this chapter, we briefly introduced Doxygen and learned how to generate documentation from code, as well as how to package generated documentation for deployment. It is crucial to have these skills under your belt when it comes to any software project. Generating documentation from code greatly reduces the effort for technical documentation and has virtually no maintenance cost. From the perspective of a software professional, automating deterministic stuff and generating inferable information in different representations is most desired. This approach creates space and time for other engineering tasks that require more human problem-solving skills. Automating tasks reduces the maintenance cost, makes the product more stable, and reduces the overall need for human resources. It is a way of converting pure human effort to spent electricity by enabling a machine to do the same job. Machines are amazingly better at performing deterministic jobs than humans. They are never sick, rarely broken, and are easily scaled and never tired. Automation is a way of harnessing this untamed power.

The main goal of this book is not to teach you how to do things, but to teach you how to make a machine work for a particular task. This approach indeed requires learning first, but keep in mind that if you are doing a costly operation that can be done by a machine manually more than once, you are wasting your precious time. Invest in automating things—it is a profitable investment that pays for itself quickly.

In the next chapter, we will be learning how we can improve our code quality by integrating unit testing, code sanitizers, static code analysis, micro-benchmarking, and code coverage tools into our CMake projects, and of course, we will be automating all of this as well.

See you in the next chapter!

Questions

After completing this chapter, you should be able to answer the following questions:

1. What is Doxygen?
2. What is the easiest way to integrate Doxygen into a CMake project?
3. Can Doxygen draw diagrams and graphs? If so, how we can enable this behavior?
4. Which Doxygen tags should be used to embed a PlantUML diagram into Doxygen documentation?
5. Which configuration steps should be taken to enable Doxygen to use PlantUML?
6. Given that man/page output is present under the `build/` folder, how can we make this documentation installable?

7
Seamlessly Integrating Code Quality Tools with CMake

So far, we have focused on building and installing projects, as well as generating documentation and handling external dependencies. Another major task when writing quality software is testing and ensuring that the code quality is at the desired level by various other means. To achieve a high code quality, just writing unit tests and executing them once in a while is no longer enough. If you want to produce high-quality software, having proper test tools that integrate easily with your build system is not a luxury but a necessity. Only if building and testing work effortlessly together can the programmers focus on writing good tests instead of focusing on getting those tests to run. Methods such as test-driven development bring huge value to the software quality.

But, it is not just writing plain tests that increase the quality. Writing good tests is one thing; checking the effectiveness of the tests with coverage reports and ensuring general code quality with static code analysis is another.

While tests, coverage, and static code analysis help determine whether the code functions as intended, one problem is often that some of the tools only work with a specific compiler or require special compiler settings. In order to benefit from the tools, it might be necessary to compile the same source in different ways with different compilers. But, luckily, this is exactly what CMake is extremely good at, which is why CMake can help to boost the quality of the code by making these quality tools accessible.

The good thing about a lot of the technology for ensuring high code quality is that it can often be automated. With today's easy availability of CI/CD systems, creating a high degree of automated checks for good quality software is quite easy, especially since, with CMake, these things can often be configured and executed right where you define how the software is built. In this chapter, you will learn how to define and orchestrate tests using CMake, as well as how to create code coverage reports to see which parts of the software are tested at all. We will go over how to integrate various code sanitizers and static code analyzers to check code quality already while compiling. We will show you various ways to include all the tools, and how to create a dedicated build type for running static code quality tools.

This chapter covers the following main topics:

- Defining, discovering, and running tests
- Generating code coverage reports
- Sanitizing your code
- Static code analysis using CMake
- Creating custom build types for quality tools

Technical requirements

As with the previous chapters, the examples are tested with CMake 3.21 and run on any of the following compilers:

- GCC9 or newer
- Clang 12 or newer
- MSVC 19 or newer

Some of the examples for code coverage, sanitizers, and static code analysis require GCC or Clang to run and will not work with MSVC. For running Clang on Windows, have a look at *Chapter 8, Executing Custom Tasks with CMake*, where toolchain files are introduced. Some examples need the Catch2 unit testing suite installed to compile. Some examples pull dependencies from various online locations, so an internet connection is required as well.

In addition to a working compiler, the following software is used for the examples:

- Gcov and Gcovr for the examples on code coverage
- Cppcheck, Cpplint, and include-what-you-use for the examples on static code analyzers

All examples and source code are available on the GitHub repository of this book at `https://github.com/PacktPublishing/CMake-Best-Practices`.

If any of the software is missing, the corresponding examples will be excluded from the build.

Defining, discovering, and running tests

Testing is the staple diet for any software engineer who takes pride in quality software. The number of frameworks for writing unit tests in the various languages is huge and, especially for C++, CMake includes modules to work with most of the more popular ones.

At very abstract levels, all unit testing frameworks do the following:

- Allow the formulating and grouping of test cases.
- Contain some form of assertion to check for various test conditions.
- Discover and run the test cases, either altogether or a selection of them.
- Produce the test result in a variety of formats, such as plain text, JSON, XML, and possibly more.

With the CTest utility, CMake includes a built-in way to execute almost any test. Any CMake project that has set `enable_testing()` and added at least one test with `add_test()` has testing support enabled. Any call to `enable_testing()` will enable test discovery in the current directory and any directory below, so it is often a good idea to set it in the top-level `CMakeLists.txt`, before any calls to `add_subdirectory`. The CTest module of CMake automatically sets `enable_testing` if used with `include(CTest)`, unless the `BUILD_TESTING` option was set to `OFF`.

It is good practice to disable building and running the tests depending on the `BUILD_TESTING` option. A common pattern here is to put all parts of a project that concern testing into its own subfolder and only include the subfolder if `BUILD_TESTING` is set to `ON`.

The CTest module should generally be included only in the top-level `CMakeLists.txt` of a project. Since CMake version 3.21, the `PROJECT_IS_TOP_LEVEL` variable can be used to test if the current `CMakeLists.txt` is the top level. This variable will be `true` for the top-level directory of a project and top-level directories of projects added with `ExternalProject`. For directories added with `add_subdirectory` or `FetchContent`, the value is `false`. As such, CTest should be included like this:

```
project(CMakeBestPractice)
...
if(PROJECT_IS_TOP_LEVEL)
   include(CTest)
endif()
```

Unit tests are, in essence, small programs that run a list of assertions inside, and if any of the assertions fail, they return a non-zero return value. There are many frameworks and libraries that help with organizing tests and writing assertions, but from the outside, checking assertions and returning a corresponding value is the core functionality.

Tests can be added to any `CMakeLists.txt` with the `add_test` function:

```
add_test(NAME <name> COMMAND <command> [<arg>...]
         [CONFIGURATIONS <config>...]
         [WORKING_DIRECTORY <dir>]
         [COMMAND_EXPAND_LISTS])
```

`COMMAND` can be the name of an executable target defined in the project or a full path to an arbitrary executable. Any arguments needed for the test are also included. Using target names is the preferred way, as CMake will then substitute the path to the executable automatically. The `CONFIGURATION` option is used to tell CMake for which build configurations the test is valid. For most test cases, this is irrelevant, but for micro-benchmarking, for instance, this can be quite useful. `WORKING_DIRECTORY` should be an absolute path. By default, tests are executed in `CMAKE_CURRENT_BINARY_DIR`. `COMMAND_EXPAND_LISTS` ensures that any lists passed as part of the `COMMAND` option are expanded.

A simple project including a test might look like this:

```
cmake_minimum_required(VERSION 3.21)
project("simple_test" VERSION 1.0)
enable_testing()
add_executable(simple_test)
target_sources(simple_test PRIVATE src/main.cpp)
add_test(NAME example_test COMMAND simple_test)
```

In the example, an executable target called `simple_test` is used as a test called `example_test`.

CTest will consume the information about the tests and execute them. The tests are executed by running the `ctest` command standalone or as a special target as part of the build step of CMake. Either of the two following commands will execute the tests:

```
ctest --test-dir <build_dir>
cmake --build <build_dir> --target test
```

Invoking CTest as a target of the build has the advantage that CMake will check first whether all the needed targets are built and on the newest version, but CTest can do that as well, like this:

```
ctest --build-and-test <source_dir> <build_dir>
```

The output of ctest might look something like this:

```
Test project /workspaces/CMake-Best-Practices/build
    Start 1: example_test
1/3 Test #1: example_test .....................***Failed
0.00 sec
    Start 2: pass_fail_test
2/3 Test #2: pass_fail_test ..................   Passed
0.00 sec
    Start 3: timeout_test
3/3 Test #3: timeout_test ....................   Passed
0.50 sec

67% tests passed, 1 tests failed out of 3

Total Test time (real) =   0.51 sec

```

```
The following tests FAILED:
          1 - example_test (Failed)
Errors while running CTest
Output from these tests are in: /workspaces/CMake-Best-
  Practices/build/Testing/Temporary/LastTest.log
Use "--rerun-failed --output-on-failure" to re-run the failed
  cases verbosely.
```

Generally, the test suppresses all output to `stdout`. By passing the `-V` or `--verbose` command-line argument, the output is always printed. However, usually, you're only interested in the output of the failed tests. So, the `--output-on-failure` argument is often the better alternative. This way, only failed tests produce output. For very verbose tests, the output can be limited in size with the `--test-output-size-passed <size>` and `--test-output-size-failed <size>` options, where the size is the number of bytes.

Having one or more calls to `add_test` in the build tree will cause CMake to write out an input file for CTest in `CMAKE_CURRENT_BINARY_DIR`. The input files for CTest are not necessarily located at the top level of the project, but where they are defined. To list all tests but not execute them, the `-N` option for CTest is used.

A very useful feature of CTest is that it caches the states of the tests between runs. This allows you to only run tests that failed in the last run. For this, running `ctest --rerun-failed` will just run the tests that failed in the last run.

Sometimes, you do not want to execute the full test set, for instance, if a single failing test is to be fixed. The `-E` and `-R` command-line options take **regular expressions** (**regexes**) that are matched against test names. The `-E` option excludes tests matching the pattern, and the `-R` option selects tests to be included. The options can be combined. The following command would run all tests that begin with `FeatureX` but exclude the test called `FeatureX_Test_1`:

```
ctest -R ^FeatureX -E FeatureX_Test_1
```

Another way to selectively execute tests is to label them using the `LABELS` properties for tests and then select the labels to run with the `-L` option of CTest. A test can have multiple labels assigned separated by a semicolon, as shown in the following example:

```
add_test(NAME labeled_test_1 COMMAND someTest)
set_tests_properties(labeled_test PROPERTIES LABELS "example")
```

```
add_test(NAME labeled_test_2 COMMAND anotherTest)
set_tests_properties(labeled_test_2 PROPERTIES LABELS "will_
  fail" )

add_test(NAME labeled_test_3 COMMAND YetAnotherText)
set_tests_properties(labeled_test_3 PROPERTIES LABELS
  "example;will_fail")
```

The `-L` command line option takes a regex to filter for the labels:

```
ctest -L example
```

This will only execute `labeled_test_1` and `labeled_test_3`, as they both have the `example` label assigned, but not `labeled_test_2` or any other tests that have no label assigned.

By formulating the regex accordingly, multiple labels can be combined:

```
ctest -L "example|will_fail"
```

This would execute all the tests from the example, but no other tests that have no label assigned.

Using labels would be particularly useful to mark tests that are designed to fail or similar, or to mark tests that are only relevant in certain execution contexts.

The last alternative to the regex or label-based test selection is to use the `-I` option, which takes the assigned test numbers. The argument for the `-I` option is somewhat complicated:

```
ctest -I [Start,End,Stride,test#,test#,...|Test file]
```

With `Start`, `End`, and `Stride`, a range for the tests to be executed can be specified. The three numbers are for the range combined with explicit test numbers, `test#`. Alternatively, a file containing the argument can be passed.

The following call would execute all odd tests from `1` to `10`:

```
ctest -I 1,10,2
```

So, tests `1`, `3`, `5`, `7`, and `9` would be executed. The following command would execute only the tests and `8`:

```
ctest -I ,0,,6,7
```

Note that in this call, `End` is set to `0` so no test range is executed. To combine the range and explicit test numbers, the following command will execute all odd tests from `1` to `10`, and additionally test `6` and `8`:

```
ctest -I 1,10,2,6,8
```

The cumbersome handling of the `-I` option and the fact that adding new tests might reassign the numbers are two reasons why it is rarely used in practice. Usually, filtering either by labels or test names is preferred.

Another common pitfall when writing tests is that they are not independent enough. So, test `2` might accidentally depend on a previous execution of test `1`. To harden against this accidental dependency, CTest has the ability to randomize test execution order with the `--schedule-random` command-line argument. This will ensure that tests are executed in an arbitrary order.

Automatically discovering tests

Defining tests with `add_test` is one way to expose them to CTest. One drawback is that this will register the whole executable as a single test. In most cases, however, a single executable will contain many unit tests and not just one, so when one of the tests inside the executable fails, it might be hard to figure out which test exactly failed.

Consider a C++ file containing the following test code, and let's assume that the Fibonacci function contains a bug, so `Fibonacci(0)` will not return `1` as it should, but returns something else:

```
TEST_CASE("Fibonacci(0) returns 1"){ REQUIRE(Fibonacci(0) ==
  1);}
TEST_CASE("Fibonacci(1) returns 1"){ REQUIRE(Fibonacci(1) ==
  1); }
TEST_CASE("Fibonacci(2) returns 2"){ REQUIRE(Fibonacci(2) ==
  2); }
TEST_CASE("Fibonacci(5) returns 8"){ REQUIRE(Fibonacci(5) ==
  8); }
```

If all these tests are compiled into the same executable called `Fibonacci`, then adding them with `add_test` will only indicate that the executable failed, but not which of the scenarios seen in the previous code block.

The result of the test will look something like this:

```
Test project /workspaces/CMake-Best-Practices/build
    Start 5: Fibonacci
1/1 Test #5: Fibonacci ........................***Failed
  0.00 sec
0% tests passed, 1 tests failed out of 1
Total Test time (real) =   0.01 sec
The following tests FAILED:
          5 - Fibonacci (Failed)
```

That is hardly helpful to figure out which of the test cases failed. Luckily, with Catch2 and GoogleTest, there is a way to expose the internal tests to CTest so they are executed as regular tests. For GoogleTest, the module to do so is provided by CMake itself; Catch2 provides this functionality in its own CMake integration. Discovering the tests with Catch2 is done with `catch_discover_tests`, while for GoogleTest, `gtest_discover_tests` is used. The following example will expose tests written in the Catch2 framework to CTest:

```
find_package(Catch2)
include(Catch)
add_executable(Fibonacci)
catch_discover_tests(Fibonacci)
```

Note that, in order to have the function available, the `Catch` module has to be included. For GoogleTest it works very similarly:

```
include(GoogleTest)
add_executable(Fibonacci)
gtest_discover_tests(Fibonacci)
```

When using the discovery functions, each test case defined in a test executable will be treated as its own test by CTest. If tests are exposed like this, the result of a call to CTest might look as follows:

```
    Start 5: Fibonacci(0) returns 1
1/4 Test #5: Fibonacci(0) returns 1 .........***Failed    0.00
sec
    Start 6: Fibonacci(1) returns 1
2/4 Test #6: Fibonacci(1) returns 1 .........   Passed    0.00
sec
```

```
    Start 7: Fibonacci(2) returns 2
3/4 Test #7: Fibonacci(2) returns 2 .........   Passed    0.00
sec
    Start 8: Fibonacci(5) returns 8
4/4 Test #8: Fibonacci(5) returns 8 .........   Passed    0.00
sec

75% tests passed, 1 tests failed out of 4
Total Test time (real) =   0.02 sec
The following tests FAILED:
          5 - Fibonacci(0) returns 1 (Failed)
```

Now, we see exactly which of the defined test cases failed. In this case, the `Fibonacci(0) returns 1` test case did not behave as expected. This comes in especially handy when using an editor or IDE with integrated testing functionality. The discovery functions both work by running the specified executable with an option to only print out the test names to register them internally with CTest, so there is a slight overhead added to each build step. Discovering tests more granularly also has the advantage that the execution of them can be better parallelized by CMake, as described in the *Running tests in parallel and managing test resources* section of this chapter.

Both `gtest_discover_tests` and `catch_discover_tests` can take various options, such as adding a prefix or suffix to the test names or a list of properties to add to the generated tests. The full documentation for the functions can be found here:

- Catch2: `https://github.com/catchorg/Catch2/blob/devel/docs/cmake-integration.md`
- GoogleTest: `https://cmake.org/cmake/help/v3.21/module/GoogleTest.html`

Catch2 and GoogleTest are just two of the many testing frameworks out there; there might be more test suites that bring this functionality with them that might be unknown to the authors. Now, let's move on from finding tests and have a closer look at how to control test behavior.

Advanced ways to determine test success or failure

By default, CTest determines whether a test failed or passed based on the return value of the command. `0` means all tests were successful, anything other than `0` is interpreted as a failure.

Sometimes, the return value is not enough to determine whether a test passes or fails. If you need to check program output for a certain string, the FAIL_REGULAR_EXPRESSION and PASS_REGULAR_EXPRESSION test properties can be used, as shown in the following example:

```
set_tests_properties(some_test PROPERTIES
                  FAIL_REGULAR_EXPRESSION "[W|w]arning|[E|e]rror"
                  PASS_REGULAR_EXPRESSION "[S|s]uccess")
```

These properties would cause the some_test test to fail if the output contains either "Warning" or "Error". If the "Success" string is found, the test is considered passed. If PASS_REGULAR_EXPRESSION is set, the test is considered passed only if the string is present. In both cases, the return value will be ignored. If a certain return value of a test needs to be ignored, it can be passed with the SKIP_RETURN_CODE option.

Sometimes, a test is expected to fail. In those, setting WILL_FAIL to true will cause the test result to be inverted:

```
add_test(NAME SomeFailingTerst COMMAND SomeFailingTest)
set_tests_properties(SomeFailingTest PROPERTIES WILL_FAIL True)
```

This is often better than disabling the test because the test will still be executed on each test run, and if the test unexpectedly starts to pass again, the developer is made aware of it. A special case of test failures is when tests fail to return or take too much time to complete. For this case, CTest provides the means of adding timeouts of tests and even retrying tests in the case of failure.

Handling timeouts and repeating tests

Sometimes, we're not just interested in the success or failure of a test, but also in how long it takes to complete. The TIMEOUT test property takes a number of seconds to determine a maximum runtime for a test. If the test exceeds the time, it is terminated and considered failed. The following command would limit the test execution of the test to 10 seconds:

```
set_tests_properties(timeout_test PROPERTIES TIMEOUT 10)
```

The TIMEOUT property often comes in handy for tests that run the risk of falling into infinite loops or hanging forever for whatever reason.

Alternatively, CTest accepts the `--timeout` argument to set a global timeout that is applied to all tests that have no `TIMEOUT` property specified. For those tests that have `TIMEOUT` defined, the timeout defined in `CmakeLists.txt` takes precedence over the timeout passed over the command line.

To avoid long test execution, the CTest command line accepts the `--stop-time` argument, which takes the real time of the day as a time limit for the complete set of tests. The following command would set a default timeout of 30 seconds for each test, and the tests would have to be completed before 23:59:

```
ctest --timeout 30 --stop-time 23:59
```

Sometimes, it can be expected for a test to experience occasional timeouts due to factors outside of our control. Very common cases are tests that need some form of network communication or a resource that has some kind of bandwidth limitation. Sometimes, the only way to get the test to run is to try it again. For this, the `--repeat after-timeout:n` command-line argument can be passed to CTest, where *n* is a number.

The `--repeat` argument actually has three options:

- `after-timeout`: This retries the test a number of times if a timeout occurred. Generally, the `--timeout` option should be passed to CTest whenever repeating after timeouts.
- `until-pass`: This reruns a test until it passes or until the number of retries is reached. Setting this as a general rule in a CI environment is a bad idea, as tests should generally always pass.
- `until-fail`: Tests are rerun a number of times or until they fail. This is often used if a test fails occasionally to find out how frequently this happens. The `--repeat-until-fail` argument works exactly like `--repeat:until-fail:n`.

As mentioned, the reason for failing tests might be the unavailability of resources that the test depends on. One common case for external resources being unavailable is that they are flooded with requests from tests. The *Running tests in parallel and managing test resources* section describes a few options to avoid such complications. Another common cause for timeouts when accessing external resources is that the resource is not yet available when the tests run. In the next section, we will see how to write test fixtures that can be used to ensure that resources are started before a test run.

Writing test fixtures

Tests should, in general, be independent of each other. There are cases where tests might depend on a precondition that is not controlled by the test itself. For instance, a test might require a server to be running in order to test a client. These dependencies can be expressed in CMake by defining them as test fixtures using the `FIXTURE_SETUP`, `FIXTURE_CLEANUP`, and `FIXTURE_REQUIRED` test properties. All three properties take a list of strings to identify the fixture. A test might indicate that it needs a particular fixture by defining the `FIXTURE_REQUIRED` property. This will ensure that the test named as `fixture` completes successfully before it is executed. Similarly, a test might declare it in `FIXTURE_CLEANUP` to indicate that it must be run after the completion of the test requiring the fixture. The fixtures defined in the cleanup part are always run, regardless of whether a test succeeded or failed. Consider the following example:

```
add_test(NAME start_server COMMAND echo_server --start)
set_tests_properties(start_server PROPERTIES FIXTURES_SETUP
  server)
add_test(NAME stop_server COMMAND echo_server --stop)
set_tests_properties(stop_server PROPERTIES FIXTURES_CLEANUP
  server)


add_test(NAME client_test COMMAND echo_client)
set_tests_properties(client_test PROPERTIES FIXTURES_REQUIRED
  server)
```

In this example, a program called `echo_server` is used as a fixture so that another program called `echo_client` can use it. The execution of `echo_server` with the `--start` and `--stop` arguments is formulated as tests with the names `start_server` and `stop_server`. The `start_server` test is marked as the setup of the fixture with the name `server`. The `stop_server` test is set up likewise but is marked as the cleanup routine of the fixture. In the end, the actual test called `client_test` is set up and it gets passed the `server` fixture as a required precondition.

If, now, the `client_test` test is run using CTest, the fixtures are automatically invoked with it. The fixture tests show up as regular tests in the output for CTest, as shown in the following example output:

```
ctest -R client
Test project CMake-Best-Practices:
    Start  9: start_server
```

```
1/3 Test  #9: start_server .............   Passed    0.00 sec
    Start 11: client_test
2/3 Test #11: client_test...............   Passed    0.00 sec
    Start 10: stop_server
3/3 Test #10: stop_server ..............   Passed    0.00 sec
```

Note that CTest was invoked with a regex filter that only matches the client test, but that CTest starts the fixture anyway. To not overwhelm test fixtures when tests are executed in parallel, they can be defined as resources, as shown in the next section.

Running tests in parallel and managing test resources

If a project has many tests, executing them in parallel will speed up the tests. By default, CTest runs the tests serially; by passing the `-j` option to the call to CTest, the tests can be run in parallel. Alternatively, the number of parallel threads can be defined in the `CTEST_PARALLEL_LEVEL` environment variable. By default, CTest assumes that each test will run on a single CPU. If a test requires multiple processors to run successfully, setting the `PROCESSORS` property for the test can be set to define the number of processors required:

```
add_test(NAME concurrency_test COMMAND concurrency_tests)
set_tests_properties(concurrency_test PROPERTIES PROCESSORS 2)
```

This will tell CTest that the `concurrency_test` test requires two CPUs to run. When running tests parallel with `-j 8`, `concurrency_test` will occupy two of the eight available "slots" for parallel execution. If, in this case, the `PROCESSORS` property is set to `8`, this would mean that no other test can run parallel to `concurrency_test`. When setting a value for `PROCESSORS` that is higher than the available number of parallel slots or CPUs that are available on the system, the test will be run as soon as the full pool is available. Sometimes, there are tests that do not just require a specific amount of processors, but that need to run exclusively without any other test running. To achieve this, the `RUN_SERIAL` property can be set to `true` for a test. This might have a serious impact on the overall test performance, so use this with caution. A more granular way to control this is by using the `RESOURCE_LOCK` property, which contains a list of strings. The strings have no particular meaning, except that CTest prevents two tests from running in parallel if they list the same strings. In this way, partial serialization can be achieved without halting the whole test execution. It also is a nice way to specify whether tests need a particular unique resource, such as a certain file, a database, or similar. Consider the following example:

```
set_tests_properties(database_test_1 database_test_2 database_
  test_3 PROPERTIES RESOURCE_LOCK database)
```

```
set_tests_properties(some_other_test PROPERTIES RESOURCE_LOCK
  fileX)
set_tests_properties(yet_another_test PROPERTIES RESOURCE_LOCK
  "database;fileX ")
```

In this example, the `database_test_1`, `database_test_2`, and `database_test_3` tests would be prevented from running in parallel. The `some_other_test` test will not be affected by the database tests, but `yet_another_test` will not run together with any of the database tests and `some_other_test`.

> **Fixtures as Resources**
>
> While not technically required, if `RESOURCE_LOCK` is used together with `FIXTURE_SETUP`, `FIXTURE_CLEANUP`, and `FIXTURE_REQUIRED`, it is good practice to use the same identifiers for the same resources.

Managing parallelism of tests with `RESOURCE_LOCK` is very handy when tests need exclusive access to some resource. In most cases, it is entirely sufficient to manage the parallelism. Since CMake 3.16, this can be controlled on an even more granular level with the `RESOURCE_GROUPS` property. Resource groups allow not just the specification of *what* resources are used, but *how much* of a resource is used. Common scenarios are defining the amount of memory a particular greedy operation might need or avoiding overrunning the connection limit of a certain service. Resource groups often come into play when working on projects that use the GPU for general-purpose computing, to define how many slots of a GPU each test needs.

Resource groups are quite a step up in complexity compared to simple resource locks. To use them, CTest has to do the following things:

- Know which resources a test needs to run. This is defined by setting the test properties in the project.
- Know which resources a system has available. This is done from outside the project when running the tests.
- Pass the information on which resources to use for the test. This is done by using environment variables.

Like resource locks, resource groups are arbitrary strings used to identify resources. The definition of the actual resource tied to the label is left to the user. The resource groups are defined as `name:value` pairs, which are separated by commas if there are multiple groups. A test can define what resources to use with the `RESOURCE_GROUPS` property, like this:

```
set_property(TEST SomeTest PROPERTY RESOURCE_GROUPS
    cpus:2,mem_mb:500
    servers:1,clients:1
    servers:1,clients:2
    4,servers:1,clients:1
 )
```

In the preceding example, `SomeTest` states that it uses two CPUs and 500 MB of memory. It uses a total of six instances of a client-server pair, each pair having a number of servers and clients assigned. The first pair consists of one server instance and one client instance, the second pair requires one server but two client instances.

The last line, `4, servers:1,clients:1`, is shorthand to tell CTest to use four instances of the same pair, consisting of one `servers` resource and one `clients` resource. This means this test will not run unless a total of six servers and seven clients are available, in addition to the required CPUs and memory.

The available system resources are specified in a JSON file that is passed to CTest, either by the `ctest --resource-spec-file` command-line parameter or by setting the `CTEST_RESOURCE_SPEC_FILE` variable when calling CMake. Setting the variable should be done by using `cmake -D` and not in `CMakeLists.txt`, as specifying the system resources should be done from outside the project.

A sample resource specification file for the preceding example could look like this:

```
{
    "version": {
        "major": 1,
        "minor": 0
    },
    "local": [
        {
            "mem_mb": [
                {
                    "id": "memory_pool_0",
```

```
                    "slots": 4096
                }
            ],
            "cpus" :
            [
                {
                    "id": "cpu_0",
                    "slots": 8
                }
            ],
            "servers": [
                {
                    "id": "0",
                    "slots": 4
                },
                {
                    "id": "1",
                    "slots": 4
                }
            ],
            "clients": [
                {
                    "id": "0",
                    "slots": 8
                },
                {
                    "id": "1",
                    "slots": 8
                }
            ]

        }
    ]
}
```

This file specifies a system with 4,096 MB of memory, eight CPUs, 2x4 server instances, and 2x8 client instances for a total of eight servers and 16 clients. If a resource request of a test cannot be satisfied with the available system resources, it fails to run with an error like this:

```
ctest -j $(nproc) --resource-spec-file ../resources.json

Test project /workspaces/CMake-Best-Practices/chapter_7
  /resource_group_example/build
    Start 2: resource_test_2
                    Start 3: resource_test_3
Insufficient resources for test resource_test_3:

  Test requested resources of type 'mem_mb' in the following
    amounts:
    8096 slots
  but only the following units were available:
    'memory_pool_0': 4096 slots

Resource spec file:

  ../resources.json
```

The current example would be able to run with this specification, as it needs a total of six servers and seven clients. CTest has no way of ensuring that the specified resources are actually available or not; this is the task of the user or the CI system. For instance, a resource file might specify that there are eight CPUs available, while the hardware actually only contains four cores.

The information about the assigned resource groups is passed to the test over environment variables. The `CTEST_RESOURCE_GROUP_COUNT` environment variable specifies the total number of resource groups assigned to a test. If it is not set, this means that CTest was invoked without an environment file. Tests should check this and act accordingly. If a test cannot run without the resources, it should either fail or indicate that it did not run by returning the respective return code or string defined in the `SKIP_RETURN_CODE` or `SKIP_REGULAR_EXPRESSION` property. The resource groups assigned to the test are passed with pairs of environment variables:

- `CTEST_RESOURCE_GROUP_<ID>`, which will contain the type of the resource groups. In the example from earlier, this will be either `"mem_mb"`, `"cpus"`, `"clients"`, or `"servers"`.

- `CTEST_RESOURCE_GROUP_<ID>_<TYPE>`, which will contain a pair of `id:slots` for the types.

It is up to the implementation of the test on how the resource groups are used and internally distributed.

Writing and running tests is obviously one of the major boosters of code quality. However, another interesting metric is often how much of your code is actually covered by the tests. Surveying and reporting code coverage can give interesting hints, not just about how widely software is tested, but also about where the gaps lie.

Generating code coverage reports

Being aware of how much of your code is covered by tests is a great benefit and often gives a very good impression about how well-tested a given software is. It can also give hints to developers about the execution paths and edge cases that are not covered by tests.

There are a few tools that help you get code coverage in C++. Arguably, the most popular is *Gcov* from GNU. It's been around for years and works well with GCC and Clang. Although it does not work with Microsoft's Visual Studio, using Clang to build and run the software provides a viable alternative for Windows. Alternatively, the tool `OpenCppCoverage` can be used to get coverage data on Windows to build with MSVC.

The coverage information generated by Gcov can be collected in summarized reports with the Gcovr or LCOV tools.

Generating coverage reports using Clang or GCC

In this section, we will look at how to create such reports with Gcovr. Generating code coverage reports roughly works in the following way:

1. The program and libraries to be tested are compiled with special flags so they expose the coverage information.
2. The program is run and the coverage information is stored in a file.
3. The coverage analyzers, such as Gcovr or LCOV, analyze the coverage files and generate reports.
4. Optionally, the reports are stored or further analyzed to display trends in coverage.

A common setting for code coverage is that you want to have information about how much of the code of a project is covered by the unit tests. In order to do this, the code has to be compiled with the necessary flags so the information is exposed.

The `<LANG>_COMPILER_FLAGS` cache variable should be passed to CMake over the command line. When using GCC or Clang, this might look like this:

```
cmake -S <sourceDir> -B <BuildDir> -DCMAKE_CXX_FLAGS=--coverage
```

An alternative is to define the respective preset, as explained in *Chapter 9, Creating Reproducible Build Environments*. When building for coverage, it is often a good idea to compile with the debug information enabled and to disable any optimization with the `-Og` flag. Additionally, specifying the `-fkeep-inline-functions` and `-fkeep-static-consts` compiler flags will prevent optimizing away static and inlined functions if they are never used. This will make sure that all possible execution branches are compiled into the code, otherwise, the coverage report might be misleading, especially for inlined functions.

Coverage reports work not just for single executables, but also for libraries. However, the libraries also have to be compiled with the coverage flags on.

As the compiler flags for coverage are set globally, the options will be passed on to projects added with `FetchContent` or `ExternalProject`, which might increase the compile time considerably.

Compiling sources using GCC or Clang with the coverage flag enabled will create `.gcno` files in the build directories for each object file and executable. These files contain the meta information for Gcov about which calls and execution paths are available in the respective compilation units. In order to find out which of these paths are used, the programs have to be run.

For the setting where we want to find out the code coverage of the tests, running CTest will generate the coverage results. Alternatively, running the executables directly will produce the same results. Running executables with coverage enabled will generate `.gcda` files in the build directories that contain the information about the calls in the respective object files.

Once these files are generated, running Gcovr on them will create the information about coverage. By default, Gcovr outputs the information to `stdout`, but it can also generate HTML pages, JSON files, or SonarQube reports.

One of the pitfalls is that Gcovr expects that all source and object files are in the same directory, which is not the case with CMake. So, we have to pass the respective directories to Gcov with the `-r` option, like this:

```
gcovr -r <SOURCE_DIR> <BINARY_DIR> -html
```

Such a call might produce an HTML file that looks like this:

GCC Code Coverage Report

			Exec	Total	Coverage
Directory:	./				
Date:	2022-05-06 15:24:12	Lines:	44	249	17.7 %
Legend:	low: < 75.0 % medium: >= 75.0 % high: >= 90.0 %	Branches:	47	333	14.1 %

File	Lines			Branches	
chapter07/custom_build_type/src/coverage_example/coverage_example.cpp		81.8 %	9 / 11	37.5 %	6 / 16
chapter07/custom_build_type/src/coverage_test.cpp		100.0 %	7 / 7	50.0 %	9 / 18
chapter07/pass_fail_criteria/src/main.cpp		71.4 %	5 / 7	50.0 %	2 / 4
chapter07/resource_group_example/src/main.cpp		100.0 %	3 / 3	50.0 %	1 / 2

Generated by: GCOVR (Version 4.2)

Figure 7.1 – An example output for a coverage run

Another alternative to Gcovr is LCOV, which works in a very similar way. In contrast to Gcovr, LCOV cannot directly produce HTML or XML output but will assemble any coverage information in an intermediate format, which can then be consumed by various converters. To produce HTML output, the `genhtml` tool is often used. To generate a report using LCOV, the commands could look like this:

```
lcov -c -d <BINARY_DIR> -o <OUTPUT_FILE>
genhtml -o <HTML_OUTPUT_PATH> <LCOV_OUTPUT>
```

A coverage report generated with LCOV might look like this:

LCOV - code coverage report

			Hit	Total	Coverage
Current view:	top level				
Test:	lcov.info	Lines:	2415	8138	29.7 %
Date:	2022-04-30 21:08:29	Functions:	3562	9304	38.3 %

Directory	Line Coverage			Functions	
/home/runner/work/CMake-Best-Practices/CMake-Best-Practices/build/_deps/catch2-src/single_include/catch2		30.5 %	1380 / 4532	35.3 %	463 / 1313
/home/runner/work/CMake-Best-Practices/CMake-Best-Practices/chapter07/fixture_example/src		100.0 %	7 / 7	100.0 %	2 / 2
/home/runner/work/CMake-Best-Practices/CMake-Best-Practices/chapter07/pass_fail_criteria/src		71.4 %	5 / 7	100.0 %	1 / 1
/home/runner/work/CMake-Best-Practices/CMake-Best-Practices/chapter07/resource_group_example/src		100.0 %	3 / 3	100.0 %	1 / 1
/home/runner/work/CMake-Best-Practices/CMake-Best-Practices/chapter07/simple_test/src		100.0 %	2 / 2	100.0 %	1 / 1
/home/runner/work/CMake-Best-Practices/CMake-Best-Practices/chapter07/test_discovery_example/src		100.0 %	6 / 6	100.0 %	5 / 5
/home/runner/work/CMake-Best-Practices/CMake-Best-Practices/chapter07/test_labels/src		100.0 %	2 / 2	100.0 %	1 / 1
/home/runner/work/CMake-Best-Practices/CMake-Best-Practices/chapter07/timeout_example/src		100.0 %	3 / 3	100.0 %	1 / 1
/usr/include/x86_64-linux-gnu/c++/9/bits		50.0 %	2 / 4	50.0 %	1 / 2
9		50.0 %	73 / 146	37.3 %	124 / 332
9/bits		26.4 %	888 / 3367	36.7 %	2469 / 6728
9/ext		74.6 %	44 / 59	53.8 %	493 / 917

Generated by: LCOV version 1.14

Figure 7.2 – An example coverage report generated with LCOV

Note that these calls will only create coverage reports for the last run. If you want to assemble them into a time series to see whether code coverage goes up or down, there are various CI tools, such as Codecov and Cobertura, available to do this. Such tools can generally parse the output from Gcovr or LCOV and assemble it into fancy graphics showing the coverage trends. The detailed documentation for Gcovr can be found at `https://gcovr.com/en/stable/`.

Creating coverage reports for MSVC

When building software using Microsoft Visual Studio, the OpenCppCoverage tool is an alternative to Gcov. It works by analyzing the program databases (`.pdb`) produced by the MSVC compiler rather than by compiling the source with different flags. The command to generate an HTML coverage report for a single executable might look like this:

```
OpenCppCoverage.exe --export_type html:coverage.html --
  MyProgram.exe arg1 arg2
```

Since this will only generate the coverage report for a single executable, `OpenCppCoverage` provides the possibility of reading the input from previous rounds and combining them into a report like this:

```
OpenCppCoverage.exe --export_type binary:program1.cov --
  program1.exe
OpenCppCoverage.exe --export_type binary:program2.cov --
  program2.exe
OpenCppCoverage.exe --input_coverage=program1.cov --input_
coverage= program2.cov --export_type html:coverage.html
```

This will combine the output of the first two runs into a common report. To consume the coverage information, the `export_type` option has to be `binary`.

A common use for coverage reports is to find out how much of the code is covered by the tests defined in a project. In this case, using CTest as the test driver will be convenient. As CTest will run the actual tests as subprocesses, the `--cover_children` option has to be passed to `OpenCppCoverage`. To avoid generating coverage reports for the system libraries used, adding a module and a source filter might be needed. The command could look something like this:

```
OpenCppCoverage.exe  --cover_children --modules <build_dir> --
  sources <source_dir> -- ctest.exe --build-config Debug
```

A slight downside to this approach will be that the coverage report will include a coverage report for CTest itself. The generated HTML report might look like this:

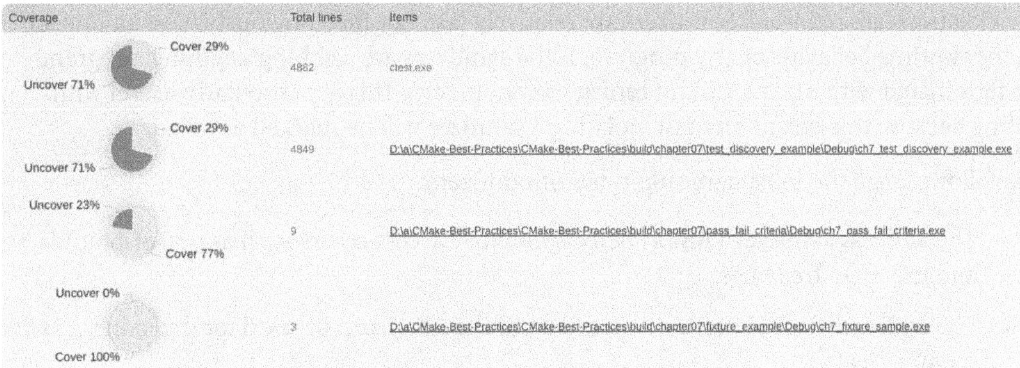

Figure 7.3 – A coverage report generated with OpenCppCoverage

If you are using Visual Studio, an alternative to the command line is to use a plugin for Visual Studio. The plugin can be found on the Visual Studio marketplace:

`https://marketplace.visualstudio.com/items?itemName=OpenCppCoverage.OpenCppCoveragePlugin`

For the full documentation, consult the GitHub page of OpenCppCoverage:

`https://github.com/OpenCppCoverage/OpenCppCoverage`

Knowing how much of your code is covered by the supplied tests is a piece of very valuable information regarding code quality. In fact, in a lot of regulated industries such as medical technology, aviation, or the car industry, providing code coverage reports might be required by the regulatory bodies. However, only knowing how much code is executed is obviously not enough; the quality of the underlying code is of even more importance. Some compilers provide useful tools to detect common errors in your code with the help of so-called sanitizers. In the next section, you will learn how to use and apply sanitizers using CMake.

Sanitizing your code

Today's compilers are often more than just programs to convert text to binary. They are complex software suites that have built-in functionality to ensure code quality. The focus on how much compilers are aware of code quality issues has drastically increased, especially with the advent of LLVM and Clang. These quality tools are commonly called **sanitizers** and are enabled by passing certain flags to the compiler and linker.

Code sanitizers are a way to bring additional quality checks into the code by using the compiler to decorate the binary code with annotations and hooks to detect various runtime issues. When the code is executed, the annotations are checked and confirmed if any violations are reported. Sanitizers are relatively fast, but they obviously have an impact on the runtime behavior of any program. If the sanitizers are catching anything, programs are terminated with `abort()` and return with non-zero. This is particularly useful with testing because this means any test violating a sanitizer will be marked as failed.

The following are the most common types of sanitizers:

- The **address sanitizer** (**ASan**) detects memory access errors such as out-of-bounds and use-after-free bugs.
- The **leak sanitizer** (**LSan**), which is part of the ASan, can be used for detecting memory leaks.
- In GCC and Clang, there are a few specialized versions of the general ASan, such as the **kernel address sanitizer** (**KASAN**) for detecting memory errors in the Linux kernel.
- On some platforms, the ASans can even be run with hardware assistance.
- The **memory sanitizer** (**MSan**) detects uninitialized memory reads.
- The **thread sanitizer** (**TSan**) will report data races. Because of the way the TSan works, it cannot be run together with the ASan and LSan.
- The **undefined behavior sanitizer** (**UBSan**) detects and reports cases where the code results in undefined behavior. Using variables before initialization or ambiguity regarding operator precedence are common examples.

The Clang suite leads the field with the availability of sanitizers, with GCC being a close second. Microsoft is a bit slower in adapting the features but they have started to include sanitizers in their compiler. At the time of writing this book, MSVC version 16.9, which comes with Visual Studio 19, was the first version that contained support for the ASan. For details on what the respective sanitizers do and how to configure them in detail, the documentation of the various compilers is of great help.

The sanitizers are enabled by passing various compiler flags that cause the compiler to add extra debugging information to the binaries. When the binaries are executed, the sanitizer code will perform its check and print out any errors to `stderr`. As the code needs to be executed for the sanitizers to find any potential bugs, having a high code coverage is essential to improve the reliability of the sanitizers.

To enable the ASan in GCC or Clang, the -fsanitize=<sanitizer> compiler flag has to be passed. For MSVC, the corresponding option is /fsanitize=<sanitizer>.

The compiler flags are passed into CMake with the CMAKE_CXX_FLAGS cache variable. So, calling CMake from the command line with a sanitizer enabled would look like this:

```
cmake -S <sourceDir> -B <BuildDir> -DCMAKE_CXX_FLAGS=-
  fsanitize=<sanitizer>
```

When using CMake presets, the cache variables to contain the compiler flags can also be defined there. Presets are covered in depth in *Chapter 9, Creating Reproducible Build Environments*. Setting the sanitizer option globally will also affect any projects included using FetchContent or ExternalProject after the flags are set, so act with caution there. For the ASan, use -fsanitizer=address on GCC and Clang, and /fsanitizer=address on MSVC. The MSan is enabled with -fsanitize=memory, the LSan with -fsanitize=leak, the TSan with -fsanitize=thread, and the UBSan is enabled with -fsanitize=undefined for GCC and Clang at the time of writing this book only. To get a more concise output for ASan, LSan, and MSan, tell the compiler to explicitly keep the frame pointer. This is done by setting -fno-omit-framepointer in GCC and Clang. MSVC only supports this for x86 builds with the /Oy- option.

> **Note**
>
> Setting the CMAKE_CXX_FLAGS variable to enable sanitizers in CMakeLists.txt itself is discouraged because the sanitizers are neither built nor have any usage requirements to use any of the targets defined by a project. Additionally, setting the CMAKE_CXX_FLAGS variable in CMakeLists.txt might conflict with what the user might be passing from the command line.

Sanitizers are a very powerful tool to increase code quality. Together with unit tests and the coverage report, they provide three of the four major concepts to ensure code quality. The fourth option for automatically ensuring code quality is using static code analyzers.

Static code analysis using CMake

Unit tests, sanitizers, and coverage reports all depend on the code being actually run to detect possible errors. Static code analysis analyzes the code without running it. The good thing about that is that all code that is compiled can be analyzed, not just the parts that are covered by tests. This, of course, also means that different kinds of glitches can be found. A downside of static code analysis is that it can take a very long time to run the tests.

CMake supports several tools for static code analysis that are enabled either by setting a property or a global variable. All of the tools, except *link what you use*, are external programs that need to be installed and found in the path of the system. *Link what you use* uses the linker of the system, so no further installation is necessary. The tools supported by CMake are the following:

- **Clang-Tidy** is a C++ linter tool covering a wide array of errors, including style violations and interface misuse. It is enabled with the <LANG>_CLANG_TIDY property or the CMAKE_<LANG>_CLANG_TIDY variable.
- **Cppcheck** is another static code analysis tool for C++. It is enabled with the <LANG>_CPPCHECK property or the CMAKE_<LANG>_CPPCHECK variable.
- **cpplint** is a style-checker for C++. It is enabled with the <LANG>_CPPLINT property or the CMAKE_<LANG>_CPPLINT variable. cpplint was originally developed at Google and because of this, it has the Google C++ style hardcoded within.
- **include what you use** (**iwyu**) is a Python program that parses the C++ source files and determines which of the included files are really required. It is enabled with the <LANG>_INCLUDE_WHAT_YOU_USE property or the CMAKE_<LANG>_INCLUDE_WHAT_YOU_USE variable.
- **link what you use** (**lwyu**) is a built-in feature of CMake that prints out a warning if an executable is linking libraries that are not used. It is enabled with the LINK_WHAT_YOU_USE property or the CMAKE_LINK_WHAT_YOU_USE variable. Note that this is not dependent on the language chosen.

For all tools, <LANG> is either C or CXX. The properties contain a semicolon-separated list containing the respective executable and command-line arguments. As of CMake 3.21, automatic execution of the static code analyzers is only supported for the Ninja and Makefile generators. Visual Studio handles the static code analyzers over settings of the IDE, which CMake cannot control. **lwyu** is a special case because it uses special flags for the ldd or ld linker, and not a special tool. So, the LINK_WHAT_YOU_USE property is just a Boolean value and not a command line. It also means that **lwyu** is only supported on ELF platforms.

Like the coverage reports and the sanitizers earlier in this chapter, the static code analysis tools are enabled by passing the command in the respective variable to CMake over the command line or by using a preset. If the variable is set, then the static code analyzers will be executed automatically when compiling the source files. Enabling clang-tidy for a build could look like this:

```
cmake -S <sourceDir> -B <buildDir>-DCMAKE_CXX_CLANG_
  TIDY="clang-tidy;-checks=*;-header-filter=<sourceDir>/*"
```

The command and the arguments are formatted as a semicolon-separated list. In the preceding example, all checks for clang-tidy are enabled with `-checks=*`, and a filter is added to only apply clang-tidy to the `include` files of the current project with `-header-filter=<sourceDir/*>`.

The same patterns work when using **Cppcheck**, **Cpplint** and **iwyu**, as the following examples show:

```
cmake -S <sourceDir> -B <buildDir> -DCMAKE_CXX_
   CPPCYHECK="cppcheck;--enable=warning;--inconclusive;--force;
     --inline-support"
cmake -S <sourceDir> -B <buildDir> -DCMAKE_CXX_CPPLINT="cpplint"
cmake -S <sourceDir> -B <buildDir> -CMAKE_CXX_INCLUDE_WHAT_YOU_
   USE="iwyu;-Xiwyu;any;-Xiwyu;iwyu;-Xiwyu;args;--verbose=5"
```

The static code analyzers will be run when compiling the files in a project. The output of any finding will be printed out with any normal compiler warnings or errors. By default, all non-critical findings of the analyzers will not cause the build to fail. For high-quality software where a zero-tolerance against warnings exists, the appropriate flags can be passed to Cppcheck and Clang-Tidy:

- For Clang-Tidy, passing `--warnings-as-errors=*` will cause compilation to fail on any issue found
- For Cppcheck, passing the `--error-exitcode=1` parameter will cause Cppcheck to exit with `1` instead of `0` if an issue is found, and the build will fail.

  iwyu and **cpplint**, unfortunately, lack similar flags.

A very nice feature of Clang-Tidy is that it can automatically apply fixes to the source files. This can be done by additionally passing the `--fix` and `--fix-error` flags to Clang-Tidy.

> **Attention When Building Incrementally**
>
> All the static code analyzers work only if a file is actually compiled. To ensure that the static code analyzers catch all errors, they have to be run on a clean build.

With the exception of **lwyu**, all the static code analyzers look at the source files to find any issues; on the other hand, **lwyu** will look at the binary files to find unused dependencies.

The **lwyu** analyzer is intended to help speed up builds and reduce the complexity of the dependencies tree. The command for **lwyu** is defined in `CMAKE_LINK_WHAT_YOU_USE_CHECK`. This variable is just a Boolean option, not a command for an external like the other tools. If set, it will pass the respective flags to the linker to output any unused direct dependencies. As of CMake version 3.21, this is defined as the `ldd -u -r` command. The usage of `ldd` means that this analyzer is only available for ELF platforms. **lwyu** can be enabled by passing a simple option, like this:

```
cmake -S <sourceDir> -B <buildDir> -DCMAKE_LINK_WHAT_YOU_
  USE=TRUE
```

The output of **lwyu** might be something like this:

```
[100%] Linking CXX executable ch7_lwyu_example
Warning: Unused direct dependencies:
        /lib/x86_64-linux-gnu/libssl.so.1.1
        /lib/x86_64-linux-gnu/libcrypto.so.1.1
```

In this example, it shows that `libssl.so` was linked but not used, even the ones indirectly linked by any dependencies.

The combination of the various static code analyzers together with **iwyu** and **lwyu** helps to keep code bases small and free from common code smells. So far in this chapter, we have looked at how tests are defined, sanitizers, and static code analysis, which deal with checking that the code functions correctly. One of the problems that we've seen now is, however, that if the various combinations have to be enabled for all the single targets, `CMakeLists.txt` can become a bit cluttered, especially for large projects. A clean alternative is to provide a custom build type that enables the compile-time code analysis globally.

Creating custom build types for quality tools

So far, we have only talked about build types such as `Debug`, `Release`, `RelWithDebInfo`, and `MinSizeRel`, which are provided by CMake by default. These build types can be extended with custom build types that pass global flags to all targets. For the code quality tools that rely on certain compiler flags, providing a custom build type can simplify `CMakeLists.txt` considerably, especially for large projects. Creating a custom build type is also much preferred to directly interfere with the global `CMAKE_<LANG>_FLAGS`.

> **Do Not Override CMAKE_<LANG>_FLAGS**
>
> Setting the global compiler option is preferred over the generic `CMAKE_<LANG>_FLAGS` in your `CMakeLists.txt`. These flags are intended to be set outside the project, either by passing them over the command line or by supplying them with a toolchain file. Modifying them inside a project has a high chance of interfering with the cases where they are set from the outside.

For multi-configuration generators such as MSVC or Ninja Multi-Config, the available build types are stored in the `CMAKE_CONFIGURATION_TYPES` cache variable. For single configuration generators such as Make or Ninja, the current build type is stored in the `CMAKE_BUILD_TYPE` variable. Custom build types should be defined on the top-level project.

A custom build type called `Coverage` could be added to `CMakeLists.txt` like this:

```
get_property(IS_MULTI_CONFIG GLOBAL PROPERTY GENERATOR_IS_
  MULTI)
if(IS_MULTI_CONFIG_GENERATOR)
  if(NOT "Coverage" IN_LIST CMAKE_CONFIGURATION_TYPES)
    list(APPEND CMAKE_CONFIGURATION_TYPES Coverage)
  endif()
else()

set(KNOWN_BUILD_TYPES Debug Release RelWithDebInfo Coverage)
set_property(CACHE CMAKE_BUILD_TYPE
           PROPERTY STRINGS ${KNOWN_BUILD_TYPES}
)
if(NOT CMAKE_BUILD_TYPE IN_LIST KNOWN_BUILD_TYPES)
    message(FATAL_ERROR "Unknown build type: ${CMAKE_BUILD_
      TYPE}")
endif()
```

Let's see what happens in the preceding example:

- First, it is determined whether the current `Generator` is a multi or single configuration generator. This is stored in the `GENERATOR_IS_MULTI_CONFIG` global property. Since the property cannot be used directly in an `if` statement, the property is retrieved and stored in the `IS_MULTI_CONFIG` variable.

- If the current generator is indeed a multi-configuration generator, the custom build configuration called `Coverage` is added to `CMAKE_CONFIGURATION_TYPES` and made available to the generator, but only if it does not yet exist.
- If the generator is a single configuration generator, a hint that the `Coverage` build exists is added by setting the `STRINGS` property of the `CMAKE_BUILD_TYPE` cache variable. This will create a drop-down menu with the valid options in the CMake GUI. For convenience, the supported build types are stored in the `KNOWN_BUILD_TYPES` variable
- As the current build type is usually supplied from the outside for single configuration generators, it is prudent to check for unknown build types and abort the configuration if an unknown build type was specified. Printing a message as `FATAL_ERROR` will cause CMake to stop the build.

With this, the `Coverage` build type is added to CMake, but the build type is not yet configured to add a custom compiler and linker flags to the build. To define the flags, two sets of cache variables are used:

- `CMAKE_<LANG>_FLAGS_<CONFIGURATION>`
- `CMAKE_<TARGET_TYPE>_LINKER_FLAGS_<CONFIGURATION>`

`<CONFIGURATION>` is the name of the custom build type, `<LANG>` is the programming language, and `<TARGET_TYPE>` for the linker flags are either executable or the various types of libraries. It can be useful to base the configuration for the custom build on existing build types to reuse any of the configuration options. The following example sets up the `Coverage` build type for a Clang or GCC compatible compiler based on the flags of the `Debug` build type:

```
set(CMAKE_C_FLAGS_COVERAGE
  "${CMAKE_C_FLAGS_DEBUG} --coverage" CACHE STRING ""
)
set(CMAKE_CXX_FLAGS_COVERAGE
  "${CMAKE_CXX_FLAGS_DEBUG} --coverage" CACHE STRING ""
)
set(CMAKE_EXE_LINKER_FLAGS_COVERAGE
  "${CMAKE_EXE_LINKER_FLAGS_DEBUG} --coverage" CACHE STRING ""
)
```

```
set(CMAKE_SHARED_LINKER_FLAGS_COVERAGE
    "${CMAKE_SHARED_LINKER_FLAGS_DEBUG} --coverage"
    CACHE STRING ""
)
```

The flags can also contain generator expressions to account for different compilers when setting the flags. Marking the flags as advanced will help prevent accidental changes to the variables by the user:

```
mark_as_advanced(CMAKE_C_FLAGS_COVERAGE
                 CMAKE_CXX_FLAGS_COVERAGE
                 CMAKE_EXE_LINKER_FLAGS_COVERAGE
                 CMAKE_SHARED_LINKER_FLAGS_COVERAGE
                 CMAKE_STATIC_LINKGER_FLAGS_COVERAGE
                 CMAKE_MODULE_LINKER_FLAGS_COVERAGE
)
```

Sometimes the filename for libraries should reflect that they are created with a special build type. Setting `CMAKE_<CONFIGURATION>_POSTFIX` for the custom build type will achieve that. This is already common practice for debug builds so the files can be distinguished from the release build when packaged together. Related to this is the `DEBUG_CONFIGURATIONS` global property, which contains the configurations that are considered non-optimized and are used for debugging. If the custom build is considered a non-release build, adding to the property, as follows, should be considered:

```
set_property(GLOBAL APPEND PROPERTY DEBUG_CONFIGURATIONS
  Coverage)
```

The `DEBUG_CONFIGURATION` property should be set on the top-level project before any calls to `target_link_libraries`. The `DEBUG_CONFIGURATIONS property` is currently only used by `target_link_libraries` where, for historical reasons, the libraries can be prefixed with `debug` or optimized to indicate that they should only be linked for the respective build configuration. Nowadays, this is rarely used, as generator expressions allow for more granular control of this.

With this, we conclude this chapter. We have covered the most common aspects of testing and quality tools, and hope that we can contribute to your journey towards excellent software quality.

Summary

Maintaining high software quality is a huge and complex task, and today, there is such a multitude of tools and techniques to test software that it can be hard to keep an overview. With the techniques and the tools described in this chapter, we hope to have given a brief overview of the most common tasks and tools used in modern C++ development. CTest and CMake can help orchestrate the various kinds of tests to get the most out of the tools. In this chapter, you've seen how tests can be defined and run, how to execute them in parallel, and how to manage test resources. We've gone over how to define test fixtures and how to define advanced ways to determine whether a test succeeded or failed based on its output.

We have illustrated how to set up code coverage reports using Gcov and how to define custom build types to pass the needed compiler flags. We've looked at how various tools for static code analysis can be included in CMake projects and how the sanitizers of various compilers can be used. And, in the end, we've briefly covered defining and running micro-benchmarks using the Catch2 framework.

In the next chapter, we will see how to use external programs to format your code and speed up your build.

Questions

Answer the following questions to test your knowledge of this chapter:

1. How are tests defined in CMake?
2. How can CTest be told to execute a specific test?
3. How can an unstable test be repeated until it either succeeds or fails?
4. How are tests run in parallel and in random order?
5. How can you prevent multiple tests from using a unique test resource at the same time?
6. How are static code analyzers enabled for targets?
7. How are custom build types defined?

8
Executing Custom Tasks with CMake

Building and shipping software can be a complex task and no tool can ever do all the different tasks that are needed to do so. At some point, you may want to execute a task that is not covered by the compiler or CMake's functionality. Common tasks include archiving build artifacts, creating hashes to verify downloads, or generating or customizing input files for the build. There are also lots of other specialized tasks that depend on the environment a certain software is built inside.

In this chapter, we will learn how to include such custom tasks in a CMake project and how to create custom build targets and custom commands. We will go over how to create and manage dependencies between targets and how to include or exclude them from the standard builds.

Including such external programs in the build steps of a project can help ensure that the code is kept consistent, even when many people contribute to it. As a CMake build is very easy to automate, using CMake to invoke the necessary commands makes it easy to apply these tools to various machines or a CI environment.

In this chapter, we will learn how custom tasks can be defined and how to control when they are executed. In particular, we will focus on managing the dependency between custom tasks and regular targets. As CMake is often used to provide build information across multiple platforms, you will also learn how to define common tasks in a way that they run everywhere where CMake runs.

This chapter will cover the following main topics:

- Using external programs with CMake
- Executing custom tasks at build time
- Executing custom tasks at configuration time
- Copying and modifying files
- Using CMake for platform-independent commands

So, let's begin!

Technical requirements

As with the previous chapters, the examples in this chapter have been tested with CMake 3.21 and can run on any of the following compilers:

- GCC 9 or newer
- Clang 12 or newer
- MSVC 19 or newer

All the examples and source code for this chapter are available in this book's GitHub repository at `https://github.com/PacktPublishing/CMake-Best-Practices`. If any of the software is missing, the corresponding examples will be excluded from the build.

Using external programs with CMake

CMake has a pretty broad functionality so that it can cover many tasks when building software. However, there are situations when developers will need to do something that is not covered. Common examples include running special tools that do some pre-or postprocessing of files for a target, using source code generators that produce input for the compiler, and compressing and archiving artifacts that are not handled with CPack. The list of such special tasks that must be accomplished during a build step is probably almost endless. CMake supports three ways of executing custom tasks:

- By defining a target that executes a command with `add_custom_target`
- By attaching a custom command to an existing target by using `add_custom_command` or by making a target depend on a file that's been generated by a custom command
- By using the `execute_process` function, which executes a command during the configuration step

Whenever possible, external programs should be called during the build step because the configuration step is far less controllable by the user and should generally run as fast as possible.

Let's learn how to define tasks that run at build time.

Executing custom tasks at build time

The most generic way to add a custom task is by creating a custom target that executes an external task as a sequence of commands. Custom targets are handled like any other library or executable target, with the difference that they do not invoke the compiler and linker; instead, they do something defined by the user. Custom targets are defined using the `add_custom_target` command:

```
add_custom_target(Name [ALL] [command1 [args1...]]
                  [COMMAND command2 [args2...] ...]
                  [DEPENDS depend depend depend ... ]
                  [BYPRODUCTS [files...]]
                  [WORKING_DIRECTORY dir]
                  [COMMENT comment]
                  [JOB_POOL job_pool]
```

```
                  [VERBATIM] [USES_TERMINAL]
                  [COMMAND_EXPAND_LISTS]
                  [SOURCES src1 [src2...]])
```

The core of the add_custom_target command is the list of commands that are passed with the COMMAND option. Although the first command can be passed without this option, it is good practice to always add the COMMAND option to any add_custom_target call. By default, custom targets are only executed if they're explicitly requested unless the ALL option is specified. Custom targets are always considered to be out of date, so the commands that are specified internally are always run, regardless of whether they produce the same result over and over again. With the DEPENDS keyword, custom targets can be made to depend on the files and outputs of custom commands that have been defined with the add_custom_command function or by other targets. To make the custom target depend on another target, use the add_dependencies function. The same works the other way round – any target may depend on a custom target. If the custom target creates files, these can be listed under the BYPRODUCTS option. Any file that's listed there will be marked with the GENERATED property, which is used by CMake to determine whether a build is out of date and to find out which files to clean. However, tasks that create files using add_custom_command might be better suited, as described later in this section.

By default, these commands are executed in the current binary directory, which is stored in the CMAKE_CURRENT_BINARY_DIRECTORY cache variable. If needed, this can be changed with the WORKING_DIRECTORY option. This option can either be an absolute path or a relative path if it is a relative path to the current binary directory.

The COMMENT option is used to specify a message that is printed right before the command runs, which can come in handy if a command runs silently. Unfortunately, not all generators show these messages, so using this to display critical information is somewhat unreliable.

The VERBATIM flag causes all the commands to be passed directly to the platform without further escaping or variable substitution by the underlying shell. CMake itself will still replace variables that are passed to the commands or the arguments. Whenever escaping might be an issue, passing the VERBATIM flag is recommended. It is also good practice to write custom tasks so that they are independent of the underlying platform. You can find more hints about how to create platform-independent commands later in this chapter in the *Using CMake for platform-independent commands* section.

The USES_TERMINAL option instructs CMake to give the command access to the terminal if possible. If the Ninja generator is used, this means that it runs in the terminal job pool. All the commands in this pool are executed serially.

The JOB_POOL option can be used when you're generating with Ninja to control the concurrency of the job. It is rarely used and cannot be used together with the USES_TERMINAL flag. You will rarely need to interfere with Ninja's job pool and handling it is not trivial. If you wish to learn more, further information can be found in the official documentation for CMake's JOB_POOLS property.

The SOURCES properties take a list of source files that are associated with the custom target. The property does not affect the source files but can help make the files visible in some IDEs. If a command relies on files such as scripts that are delivered together with the project, these should be added here.

The COMMAND_EXPAND_LISTS option tells CMake to expand lists before passing them to the command. This is sometimes necessary because, in CMake, lists are just strings separated by semicolons, which may lead to syntax errors. When you're passing the COMMAND_EXPAND_LISTS option, the semicolons are replaced with a suitable whitespace character, depending on the platform. The expansion includes lists that have been generated using the $<JOIN: generator expression.

The following is an example of a custom target that uses an external program called *CreateHash* to create a hash for the output of another target:

```
add_executable(SomeExe)
add_custom_target(CreateHash ALL COMMAND Somehasher
   $<TARGET_FILE:SomeExe>)
```

This example creates a custom target called *CreateHash* that invokes the external *SomeHasher* program with the binary file of the *SomeExe* target as an argument. Note that the binary file is retrieved using the $<TARGET_FILE:SomeExe> generator expression. This serves two purposes: it removes the need for the user to track the filename of the binary of the target and it adds an implicit dependency between the two targets. CMake will recognize these implicit dependencies and execute the targets in the correct order. If the target that produces the needed file hasn't been built yet, CMake will automatically build it. You can also use the $<TARGET_FILE: generator to directly execute an executable that's been created by another target. The following generator expressions cause implicit dependencies between targets:

- $<TARGET_FILE:target>: This contains the full path to the main binary file of the target, such as .exe, .so, or .dll.
- $<TARGET_LINKER_FILE: target>: This contains the full path to the file that's used for linking against the target. This is usually the library file itself, except on Windows, where it will be the .lib file associated with the DLL.

- `$<TARGET_SONAME_FILE: target>`: This contains the library file and its full name, including any number that's been set by the `SOVERSION` property, such as `.so.3`.
- `$<TARGET_PDB_FILE: target>`: This contains the full path to the generated program database file that's used for debugging. Creating custom targets is one way to execute external tasks during build time. Another way is to define custom commands. Custom commands can be used to add custom tasks to existing targets, including custom targets.

Adding custom tasks to existing targets

Sometimes, you may need to perform an additional, external task when building a target. In CMake, you can achieve this with `add_custom_command`, which has two signatures. One is used to hook commands into existing targets, while the other is used to generate files. We will cover this later in this section. The signature for adding commands to an existing target looks like this:

```
add_custom_command(TARGET <target>
                   PRE_BUILD | PRE_LINK | POST_BUILD
                   COMMAND command1 [ARGS] [args1...]
                   [COMMAND command2 [ARGS] [args2...] ...]
                   [BYPRODUCTS [files...]]
                   [WORKING_DIRECTORY dir]
                   [COMMENT comment]
                   [VERBATIM] [USES_TERMINAL]
                   [COMMAND_EXPAND_LISTS])
```

Most of the options work similarly to those in `add_custom_target`, as described earlier. The `TARGET` property can be any target that is defined in the current directory, which is a limitation of the command, although it is rarely an issue. Commands can be hooked into the build at the following times:

- `PRE_BUILD`: In Visual Studio, this command is executed before any other build steps are executed. When you're using other generators, it will run just before the `PRE_LINK` commands.
- `PRE_LINK`: This command will be run after the sources have been compiled, but before the executable or the archiver tool has been linked to static libraries.
- `POST_BUILD`: This runs the command after all the other build rules have been executed.

The most common way to execute custom steps is by using POST_BUILD; the other two are rarely used either because of limited support or because they can neither influence the linking nor the build.

Adding a custom command to an existing target is relatively straightforward. The following code adds a command that generates and stores the hash of a built file after each compilation:

```
add_executable(MyExecutable)


add_custom_command(TARGET MyExecutable
    POST_BUILD
   COMMAND hasher $<TARGET_FILE:ch8_custom_command_example>
     ${CMAKE_CURRENT_BINARY_DIR}/MyExecutable.sha256
COMMENT "Creating hash for MyExecutable"
)
```

In this example, a custom executable called `hasher` is being used to generate the hash of the output file of the `MyExectuable` target.

Often, the reason you may need to execute something before a build is to change files or to generate additional information. For this, the second signature is often the better choice. Let's take a closer look.

Generating files with custom tasks

Typically, we want custom tasks to produce a specific output file. This can be done by defining custom targets and setting the necessary dependencies between the targets, or by hooking into the build steps, as described earlier. Unfortunately, the PRE_BUILD hook is unreliable because only the Visual Studio generator supports it properly. So, a better way to do this is to create a custom command that creates the file by using the second signature of the `add_custom_command` function:

```
add_custom_command(OUTPUT output1 [output2 ...]
                   COMMAND command1 [ARGS] [args1...]
                   [COMMAND command2 [ARGS] [args2...] ...]
                   [MAIN_DEPENDENCY depend]
                   [DEPENDS [depends...]]
                   [BYPRODUCTS [files...]]
```

```
                    [IMPLICIT_DEPENDS <lang1> depend1
                                     [<lang2> depend2] ...]
                    [WORKING_DIRECTORY dir]
                    [COMMENT comment]
                    [DEPFILE depfile]
                    [JOB_POOL job_pool]
                    [VERBATIM] [APPEND] [USES_TERMINAL]
                    [COMMAND_EXPAND_LISTS])
```

This signature of `add_custom_command` defines a command that generates a file specified in `OUTPUT`. Most of the options of the command are very similar to `add_custom_target` and the signature for hooking custom tasks into build steps. The `DEPENDS` option can be used to manually specify a dependency to either files or targets. Note that in comparison, the `DEPENDS` option of custom targets can only point to files. If any of the dependencies are updated by a build or by CMake, the custom command is run again. The `MAIN_DEPENDENCY` option is closely related, which specifies the primary input file for the command. It works like the `DEPENDS` option does, except that it only takes one file. `MAIN_DEPENDENCY` is mainly used to tell Visual Studio where to add the custom command.

> **Note**
>
> If a source file is listed as `MAIN_DEPENDENCY`, then the custom command replaces the normal compilation of the file listed, which can lead to linker errors.

The other two dependency-related options, `IMPLICIT_DEPENDS` and `DEPFILE`, are rarely used because their support is limited to the Makefile generator. `IMPLICT_DEPENDS` tells CMake to use a C or C++ scanner to detect any compile-time dependencies of the files listed and create the dependencies from that. The other option, `DEPFILE`, can be used to point to a `.d` dependency file, which is generated by the Makefile project. The `.d` files originally stem from the GNU make project and can be powerful to use but are also complex and should not be manually managed for most projects. The following example illustrates how a custom command can be used to generate a source file before a regular target is run, based on another file that is being used for input:

```
add_custom_command(OUTPUT ${CMAKE_CURRENT_BINARY_DIR}/main.cpp
COMMAND sourceFileGenerator ${CMAKE_CURRENT_SOURCE_DIR}/
  message.txt
  ${CMAKE_CURRENT_BINARY_DIR}/main.cpp
```

```
COMMENT "Creating main.cpp frommessage.txt"
DEPENDS message.txt
VERBATIM
)

add_executable(
ch8_create_source_file_example
${CMAKE_CURRENT_BINARY_DIR}/main.cpp
)
```

Several things are happening in this example. First, the custom command defines the `main.cpp` file in the current binary directive as an `OUTPUT` file. Then, the command that generates this file is defined – here, this is using an imaginary program called `sourceFileGenerator` – which converts a message file into a `.cpp` file. The `DEPENDS` part states that this command should be rerun every time the `message.txt` file changes.

Later, the target for the executable is created. Since the executable is referencing the `main.cpp` file specified in the `OUTPUT` section of the custom command, CMake will implicitly add the necessary dependency between the command and the target. Using custom commands in this way is much more reliable and portable than using the `PRE_BUILD` directive as it works with all generators.

Sometimes, to create the desired output, more than one command is needed. If a previous command that produces the same output exists, commands can be chained by using the `APPEND` option. Custom commands that use `APPEND` may only define additional `COMMAND` and `DEPENDS` options; the other options are ignored. If two commands produce the same output file, CMake will print an error unless `APPEND` is specified. This is mainly useful if a command is only optionally executed. Consider the following example:

```
add_custom_command(OUTPUT archive.tar.gz
COMMAND cmake -E tar czf ${CMAKE_CURRENT_BINARY_DIR}/archive.
  tar.gz
  $<TARGET_FILE:MyTarget>
COMMENT "Creating Archive for MyTarget"
VERBATIM
)

add_custom_command(OUTPUT archive.tar.gz
```

```
COMMAND cmake -E tar czf ${CMAKE_CURRENT_BINARY_DIR}/archive.
   tar.gz
   ${CMAKE_CURRENT_SOURCE_DIR}/SomeFile.txt
APPEND
)
```

In this example, the output file of a target, `MyTarget`, has been added to a `tar.gz` archive; later, another file is added to the same archive. Note that the first command automatically depends on `MyTarget` because it uses the binary file that was created in the command. However, it will not automatically be executed by a build. The second custom command lists the same output file as the first command but adds the compressed file as a second output. By specifying `APPEND`, the second command is automatically executed whenever the first command is executed. If the `APPEND` keyword is missing, CMake will print out an error, similar to the following:

```
CMake Error at CMakeLists.txt:30 (add_custom_command):
  Attempt to add a custom rule to output
     /create_hash_example/build/hash_example.md5.rule
   which already has a custom rule.
```

As we mentioned previously, the custom commands in this example implicitly depend on `MyTarget` but they will not be executed automatically. To execute them, the recommended practice is to create a custom target that depends on the output file, which can be generated like so:

```
add_custom_target(create_archive ALL DEPENDS
    ${CMAKE_CURRENT_BINARY_DIR}/archive.tar.gz
)
```

Here, a custom target called `create_archive` has been created that is executed as part of the `All` build. Since it depends on the output of the custom commands, building the target will invoke the custom commands. The custom commands, in turn, depend on `MyTarget`, so building `create_archive` will also trigger a build of `MyTarget` if it is not already up to date.

Both the `add_custom_command` and `add_custom_target` custom tasks are executed during the build step of CMake. It is possible to add tasks at configuration time if necessary. We'll look at this in the next section.

Executing custom tasks at configuration time

To execute custom tasks at configuration time, you can use the `execute_process` function. Common needs for this are if the build requires additional information before a build, or if files need to be updated for any rerun of CMake. Another common case is when either the `CMakeLists.txt` file or other input files are generated during the configuration step, although this can also be achieved with the specialized `configure_file` command, as shown later in this chapter.

The `execute_process` function works very similarly to the `add_custom_target` and `add_custom_command` functions we saw earlier. However, one distinction is that `execute_process` can capture output to `stdout` and `stderr` in a variable or files. The signature of `execute_process` is as follows:

```
execute_process(COMMAND <cmd1> [<arguments>]
                [COMMAND <cmd2> [<arguments>]]...
                [WORKING_DIRECTORY <directory>]
                [TIMEOUT <seconds>]
                [RESULT_VARIABLE <variable>]
                [RESULTS_VARIABLE <variable>]
                [OUTPUT_VARIABLE <variable>]
                [ERROR_VARIABLE <variable>]
                [INPUT_FILE <file>]
                [OUTPUT_FILE <file>]
                [ERROR_FILE <file>]
                [OUTPUT_QUIET]
                [ERROR_QUIET]
                [COMMAND_ECHO <where>]
                [OUTPUT_STRIP_TRAILING_WHITESPACE]
                [ERROR_STRIP_TRAILING_WHITESPACE]
                [ENCODING <name>]
                [ECHO_OUTPUT_VARIABLE]
                [ECHO_ERROR_VARIABLE]
                [COMMAND_ERROR_IS_FATAL <ANY|LAST>])
```

The execute_process function takes a list of COMMAND properties to be executed in WORKING_DIRECTORY. The return codes of the last command to be executed can be stored in the variable defined with RESULT_VARIABLE. Alternatively, a semicolon-separated list of variables can be passed to RESULTS_VARIABLE. If you're using the list version, the commands will store the return codes of the commands in the same order as the variables that have been defined. If fewer variables have been defined than commands, any surplus return codes will be ignored. If a TIMEOUT value was defined and any of the child processes failed to return, the result variables will contain timeout. Since CMake version 3.19, the convenient COMMAND_ERROR_IS_FATAL option is available, which tells CMake to abort execution if any (or just the last) of the processes fails. This is much more convenient than retrieving all the return codes and then checking them individually after their execution. In the following example, if any of the commands return a non-zero value, the configuration step of CMake will fail with an error:

```
execute_process(
    COMMAND SomeExecutable
    COMMAND AnotherExecutable
    COMMAND_ERROR_IS_FATAL_ANY
)
```

Any output to stdout or stderr can be captured using the OUTPUT_VARIABLE or ERROR_VARIABLE variable, respectively. As an alternative, they can be redirected to files by using OUTPUT_FILE or ERROR_FILE or can be completely ignored by passing OUTPUT_QUIET or ERROR_QUIET. Capturing the output in both a variable and a file is not possible and will result in either of the two being empty. Which one is kept and which is discarded depends on the platform. If not, the OUTPUT_* option specifies that the output is sent to the CMake process itself.

If the output is captured in a variable but can still be displayed, ECHO_<STREAM>_VARIABLE can be added. CMake can also be told to output the command itself by passing STDOUT, STDERR, or NONE to the COMMAND_ECHO option. However, if the output is captured in files, this will have no effect. If the same variable or file is specified for both stdout and stderr, then the results will be merged. If necessary, the input stream of the first command can be controlled by passing a file to the INPUT_FILE option.

The output to variables can be controlled in a limited way by using the <STREAM>_STRIP_TRAILING_WHITESPACE option, which will trim any white spaces at the end of the output. When you're redirecting output to files, this has no effect. On Windows, the ENCODING option can be used to control the output. It takes the following values:

- NONE: Performs no reencoding. This will keep CMake's internal encoding, which is UTF-8.

- AUTO: Uses the current console's encoding. If this is not available, it uses ANSI.
- ANSI: Uses the ANSI code page for encoding.
- OEM: Uses the code page defined by the platform.
- UTF8 or UTF-8: Forced to use the UTF-8 encoding.

A common reason for using execute_process is gathering information that is needed for a build and then passing it to the project. Consider an example where we would like to compile the Git revision into an executable by passing it as a preprocessor definition. The downside of this approach is that for the custom tasks to be executed, CMake has to be invoked, not just the build system. So, using add_custom_command with an OUTPUT parameter would probably be the more realistic solution here, but for illustrative purposes, this example should serve well enough. The following is an example where the Git hash is read out at configuration time and passed as a compile definition to a target:

```
find_package(Git REQUIRED)
execute_process(COMMAND ${GIT_EXECUTABLE} "rev-parse" "--short"
  "HEAD"
OUTPUT_VARIABLE GIT_REVISION
OUTPUT_STRIP_TRAILING_WHITESPACE
COMMAND_ERROR_IS_FATAL ANY
WORKING_DIRECTORY ${CMAKE_CURRENT_SOURCE_DIR})


add_executable(SomeExe src/main.cpp)
target_compile_definitions(SomeExe PRIVATE VERSION=
  \"${GIT_REVISION}\")
```

In this example, the git command that's passed to execute_process is executed in the directory containing the CMakeLists.txt file, which is currently being executed. The resulting hash is stored in the GIT_REVISION variable and if the command fails for any reason, the configuration process is halted with an error.

Passing the information from execute_process into the compiler by using a preprocessor definition is far from optimal. A much nicer solution would be if we could generate a header file to be included that contains this information. CMake has another feature called configure_file that can be used for this purpose, as we will see in the next section.

Copying and modifying files

A relatively common task when building software is that some files must be copied to a specific location or modified before the build. In the *Executing custom tasks at configuration time* section, we saw an example where the Git revision was retrieved and passed to the compiler as a preprocessor definition. A much nicer way to do this would be to generate a header file containing the necessary information. While just echoing the code snippet and writing it into a file would be possible, it is dangerous as it may lead to platform-specific code. CMake's solution to this is the `configure_file` command, which can copy files from one location to another and modify their content while doing so. The signature of `configure_file` is as follows:

```
configure_file(<input> <output>
               NO_SOURCE_PERMISSIONS | USE_SOURCE_PERMISSIONS |
                FILE_PERMISSIONS <permissions>...]
                [COPYONLY] [ESCAPE_QUOTES] [@ONLY]
                [NEWLINE_STYLE [UNIX|DOS|WIN32|LF|CRLF] ])
```

The `configure_file` function will copy the `<input>` file to the `<output>` file. If necessary, the path to the output file will be created, and paths can be relative or absolute. If you're using relative paths, the input file will be searched from the current source directory but the path of the output file will be relative to the current build directory. If the output file cannot be written, the command will fail and the configuration will be halted. By default, the output file has the same permissions as the target file, although ownership may change if the current user is a different one than the one that the input file belongs to. If `NO_SOURCE_PERMISSION` is added, the permissions are not transferred and the output file gets the default `rw-r--r--` value. Alternatively, the permissions can be manually specified with the `FILE_PERMISSIONS` option, which takes a three-digit number as an argument. `USE_SOURCE_PERMISSION` is already the default and the option is only there to state the intent more explicitly.

As we mentioned earlier, `configure_file` will also replace parts of the input file when you're copying to the output path unless `COPYONLY` is passed. By default, `configure_file` will replace all the variables referenced as `${SOME_VARIABLE}` or `@SOME_VARIABLE@` with the value of any variable of the same name. If a variable is defined in `CMakeLists.txt`, when `configure_file` is called, the respective value is written into the output file. If a variable is not specified, the output file will contain an empty string in the respective place. Consider a `hello.txt.in` file that contains the following information:

```
Hello ${GUEST} from @GREETER@
```

In a `CMakeLists.txt` file, the `configure_file` function is used to configure the `hello.txt.in` file:

```
set(GUEST "World")
set(GREETER "The Universe")
configure_file(hello.txt.in hello.txt)
```

In this example, the resulting `hello.txt` file will contain `Hello World from The Universe`. If the `@ONLY` option is passed to `configure_file`, only `@GREETER@` would be replaced and the resulting content would be `Hello ${GUEST} from The Universe`. Using `@ONLY` is useful when you're transforming CMake files that may contain brace-enclosed variables that should not be replaced. `ESCAPE_QUOTES` will escape any quotes in the target file with a backslash. By default, `configure_file` will transform the newline character so that the target file matches the current platform. The default behavior can be changed by setting `NEWLINE_STYLE`. `UNIX` or `LF` will use `\n` for newlines, while `DOS`, `WIN32`, and `CRLF` will use `\r\n`. Setting the `NEWLINE_STYLE` and `COPYONLY` options together will cause an error. Note that setting `COPYONLY` will not affect the newline style.

Let's go back to the example where we want to compile the Git revision into an executable. Here, we would write a header file as input. It may contain a line that looks like this:

```
#define CMAKE_BEST_PRACTICES_VERSION "@GIT_REVISION@"
The CMakeLists.txt could look something like this:
execute_process(
    COMMAND ${GIT_EXECUTABLE} rev-parse --short HEAD
    OUTPUT_VARIABLE GIT_REVISION
    OUTPUT_STRIP_TRAILING_WHITESPACE
    COMMAND_ERROR_IS_FATAL ANY
```

```
    WORKING_DIRECTORY ${CMAKE_CURRENT_SOURCE_DIR})

configure_file(version.h.in ${CMAKE_CURRENT_SOURCE_DIR}/src
  /version.h @ONLY)
```

As shown in the example in the previous section, where the version information was passed as a compile definition, the Git revision is first retrieved with `execute_process`. Later, the file is copied using `configure_file`, and `@GIT_REVISION@` is replaced with the short hash of the current commit.

When you're working with preprocessor definitions, `configure_file` will replace any lines in the form of `#cmakedefine VAR ...` with either `#define VAR` or `/* undef VAR */`, depending on whether VAR contains a value that CMake interprets as `true` or `false`.

Consider a file called `version.in.h` that contains the following two lines:

```
#cmakedefine GIT_VERSION_ENABLE
#cmakedefine GIT_VERSION "@GIT_REVISION@"
```

The accompanying `CMakeLists.txt` file may look like this:

```
option(GIT_VERSION_ENABLE "Define revision in a header file"
  ON)
if(GIT_VERSION_ENABLE)
  execute_process(
    COMMAND ${GIT_EXECUTABLE} rev-parse --short HEAD
    OUTPUT_VARIABLE GIT_REVISION
    WORKING_DIRECTORY ${CMAKE_CURRENT_SOURCE_DIR}
)
endif()
configure_file(version.h.in ${CMAKE_CURRENT_SOURCE_DIR}/src/
  version.h @ONLY)
```

Once the configuration has been run, if `GIT_REVISION_ENABLE` is on, the resulting file will contain the following output:

```
#define GIT_VERSION_ENABLE
#define CMAKE_BEST_PRACTICES_VERSION "c030d83"
```

If GIT_REVISION_ENABLE is off, the resulting file will contain the following output:

```
/* #undef GIT_VERSION_ENABLE */
/* #undef GIT_REVISION */
```

All in all, the configure_file command is quite useful for preparing input for a build. Apart from generating source files, it is often used to generate CMake files that are then included in a CMakeLists.txt file. One of the strengths of this is that it allows for the platform-independent copying and modification of files, which is a major advantage when you're working cross-platform. Since configure_file and execute_process often go hand in hand, ensure that the commands that are executed are also platform-independent. In the next section, you will learn how CMake can be used to define platform-agnostic commands and scripts.

Using CMake for platform-independent commands

One of the keystones for the success of CMake is that it allows you to build the same software on a multitude of platforms. On the other hand, this means that CMakeLists.txt must be written in a way that does not assume a certain platform or compiler must be used. This can be challenging, especially when you're working with custom tasks. A big help here is that the cmake command-line utility supports the -E flag, which can be used to perform common tasks such as file operations and creating hashes. Most of the cmake -E commands are for file-related operations such as creating, copying, renaming, and deleting files, as well as creating directories. On systems that support filesystem links, CMake can also create symbolic links or hard links between files. Additionally, CMake can create file archives using the tar command and concatenate text files with the cat command. It can also be used to create various hashes for files.

There are also a few operations that provide information about the current system. The capabilities operation will print out CMake's capabilities, such as knowing which generators are supported and the version that CMake is currently running. The environment command will print a list of environment variables that have been set.

You can get a full reference to the command-line options by running cmake -E without any other arguments. The online documentation for CMake can be found at https://cmake.org/cmake/help/v3.21/manual/cmake.1.html#run-a-command-line-tool.

> **Platform-Agnostic File Operations**
>
> Whenever file operations must be performed by a custom task, use `cmake -E`.

With `cmake -E`, you can get pretty far in most cases. Sometimes, however, more complex operations need to be done. For this, CMake can run in script mode, which executes CMake files.

Executing CMake files as scripts

CMake's script mode is a very powerful feature when it comes to creating cross-platform scripts. It's powerful because it allows you to create scripts that are completely platform-agnostic. By invoking `cmake -P <script>.cmake`, the specified CMake file is executed. The script files may not contain any commands that define a build target. Arguments may be passed as variables with the `-D` flag, but this must be done before the `-P` option. Alternatively, the arguments may just be appended after the script name so that they can be retrieved with the `CMAKE_ARGV[n]` variables. The number of arguments is stored in the `CMAKE_ARGC` variable. The following script, which generates the hash of a file and stores it in another, demonstrates how to use positional arguments:

```
cmake_minimum_required(VERSION 3.21)
if(CMAKE_ARGC LESS 5)
    message(FATAL_ERROR "Usage: cmake -P CreateSha256.cmake
      file_to_hash target_file")
endif()
set(FILE_TO_HASH ${CMAKE_ARGV3})
set(TARGET_FILE ${CMAKE_ARGV4})
# Read the source file and generate the hash for it
file(SHA256 "${FILE_TO_HASH}" GENERATED_HASH)
# write the hash to a new file
file(WRITE "${TARGET_FILE}" "${GENERATED_HASH}")
```

This script can be invoked with `cmake -P CreateSha256.cmake <input file> <output_file>`. Note that the first three arguments are occupied with `cmake`, `-P`, and the name of the script (`CreateSha256.cmake`). Although not strictly required, script files should always contain a `cmake_minimum_required` statement at the beginning. An alternative way to define the script without positional arguments would be as follows:

```
cmake_minimum_required(VERSION 3.21)
if(NOT FILE_TO_HASH OR NOT TARGET_FILE)
```

```
    message(FATAL_ERROR "Usage: cmake -DFILE_TO_HASH=<intput_
        file> \
-DTARGET_FILE=<target file> -P CreateSha256.cmake")
endif()
# Read the source file and generate the hash for it
file(SHA256 "${FILE_TO_HASH}" GENERATED_HASH)
# write the hash to a new file
file(WRITE "${TARGET_FILE}" "${GENERATED_HASH}")
```

In this case, the script would have to be invoked with the variables passed explicitly, like so:

```
cmake -DFILE_TO_HASH=<input>
      -DTARGET_FILE=<target> -P CreateSha256.cmake
```

These two approaches can also be combined. A common pattern is to expect all simple mandatory arguments as positional arguments and any optional or more complex arguments as defined variables. Combining the script mode with `add_custom_command`, `add_custom_target`, or `execute_process` is a good way to create platform-independent build instructions. An example of generating hashes from the earlier sections could look like this:

```
add_custom_target(Create_hash_target ALL
COMMAND cmake -P ${CMAKE_CURRENT_SOURCE_DIR}/cmake/
  CreateSha256.cmake $<TARGET_FILE:SomeTarget>
   ${CMAKE_CURRENT_BINARY_DIR}/hash_example.sha256
)


add_custom_command(TARGET SomeTarget
POST_BUILD
COMMAND cmake -P ${CMAKE_CURRENT_SOURCE_DIR}/cmake
  /CreateSha256.cmake $<TARGET_FILE:SomeTarget>
    ${CMAKE_CURRENT_BINARY_DIR}/hash_example.sha256
)
```

Combining the script mode of CMake with the various ways to execute custom commands during the configuration or build phase of a project provides a lot of freedom when you're defining build processes, even for different platforms. However, one word of warning is that adding too much logic to a build process may make it harder to maintain than it should be. Whenever you need to write a script or add a custom command to a `CMakeLists.txt` file, it pays to take a quick break and consider whether this step belongs to the build process or whether it is something better left to the user when they're setting up the development environment.

Summary

In this chapter, you learned how to customize a build by executing external tasks and programs. We covered how to add custom build actions as targets, how to add them to existing targets, and how to execute them during the configuration step. We went over how commands can generate files and how CMake can copy and modify files with the `configure_file` command. Finally, we learned how the CMake command-line utility can be used to perform tasks in a platform-independent manner.

The ability to customize a CMake build is a very powerful tool but it also tends to make builds more brittle as the complexity of the builds often increases when any customized tasks are performed. Although sometimes not avoidable, relying on external programs other than a compiler and linker being installed may mean that a piece of software can be built on a platform where those programs haven't been installed or aren't available. This means special care must be taken to ensure that custom tasks do not assume anything about the system around CMake if possible. Lastly, executing custom tasks may carry a performance penalty for the build system, especially if they do heavy work on each build.

But if you are careful with custom build steps, they are a great way of increasing the cohesion of a build as many of the build-related tasks can be defined where the build definition is. This can make automating tasks such as creating hashes of the build artifacts or assembling all the documents in a common archive much easier.

In the next chapter, you will learn how to make the build environment portable between different systems. You will learn about how to use presets to define common ways to configure a CMake project, as well as how to wrap your build environment into a container and use `sysroots` to define toolchains and libraries as being portable between systems.

Questions

Answer the following questions to test your knowledge of this chapter:

1. What is the main difference between `add_custom_command` and `execute_process`?
2. What are the two signatures of `add_custom_command` used for?
3. What is the problem with the `PRE_BUILD`, `PRE_LINK`, and `POST_BUILD` options of `add_custom_command`?
4. What are the two ways of defining variables so that they can be substituted with `configure_file`?
5. How can the substitution behavior of `configure_file` be controlled?
6. What are the two flags for the CMake command-line tool to execute tasks?

9
Creating Reproducible Build Environments

Building software can be complex, especially when dependencies or special tools are involved. What compiles on one machine might not work on another because a crucial piece of software is missing. Relying on the correctness of the documentation of a software project to figure out all the build requirements is often not enough, and as a consequence, programmers spend a significant amount of time combing through various error messages to figure out why a build fails.

There are countless stories out there of people avoiding upgrading anything in a build or **continuous integration** (**CI**) environment because they fear that every change might break the ability to build the software. This goes as far as companies refusing to upgrade the compiler toolchains they are using for fear of no longer being able to ship products. Creating robust and portable information about build environments is an absolute game-changer. With presets, CMake provides the possibility to define common ways to configure a project. When combined with toolchain files, Docker containers, and sysroots, creating a build environment that can be recreated on different machines becomes much easier.

In this chapter, you will learn how to define CMake presets for configuring, building, and testing a CMake project and how to define and use a toolchain file. We will briefly go over using a container to build your software and learn how to use a system root toolchain file to create an isolated build environment. The main topics of this chapter will be as follows:

- Using CMake presets
- Using build containers with CMake
- Using sysroots to isolate build environments

So, let's buckle down and get started!

Technical requirements

As with the previous chapters, the examples are tested with CMake 3.21 and are run on any of the following compilers:

- GCC 9 or newer
- Clang 12 or newer
- MSVC 19 or newer

For the examples using build containers, Docker is needed.

All examples and source code are available on the GitHub repository for this book. For this chapter, the examples for the CMake presets and build container are in the root folder of the repository. If any of the software is missing, the corresponding examples will be excluded from the build. The repository is found here: `https://github.com/PacktPublishing/CMake-Best-Practices`.

Using CMake presets

While building software on a multitude of configurations, compilers, and platforms is CMake's greatest strength, it is also one of its greatest weaknesses as this often makes it hard for a programmer to figure out which build setups have actually been tested and are working for a given piece of software. Since version 3.19, CMake has a feature called presets that lets us handle these scenarios in a reliable and convenient way. Earlier, developers had to rely on documentation and fuzzy conventions to figure out the preferred configuration of a CMake project. Presets can specify the build directory, generators to use, target architecture, host toolchain, cache variables, and environment variables to use with a project. Since CMake 3.20, there are additional presets that affect the build and test phases as well.

For using presets, the top directory of a project must contain a file named either `CMakePresets.json` or `CMakeUserPresets.json`. If both files are present, they will be internally combined by parsing `CMakePresets.json` first and then `CMakeUserPresets.json`. Both files have the same format but serve slightly different use cases:

- `CMakePresets.json` should be provided by the project itself and handle project-specific things, such as running CI builds or knowing which toolchains to use for cross-compilation if they are provided with the project itself. As `CMakePresets.json` is project-specific, it should not refer to any files or paths outside the project structure. Since these presets are tied in closely with the project, it is usually also kept under version control.
- `CMakeUserPresets.json`, on the other hand, is usually defined by the developer for use on their own machine or build environment. `CMakeUserPresets.json` can be as specific as possible and contain paths outside the project or those that are unique to a particular system setup. As such, projects should not provide this file and also not put it under version control.

Presets are a great way of moving cache variables, compiler flags, and so on out of `CMakeLists.txt` files while still keeping the information available in a way that can be used with CMake and thus improve the portability of projects. If presets are available, they can be listed from the source directory by calling the following `cmake --list-presets`, which will produce an output similar to this:

```
Available configure presets:

  "ninja-debug"   - Ninja (Debug)
  "ninja-release" - Ninja (Release)
```

This will list the name of the preset in quotes and the `displayName` property if set. To use properties from the command line, the name in quotes is used.

The CMake GUI will show all available presets in a source directory like this:

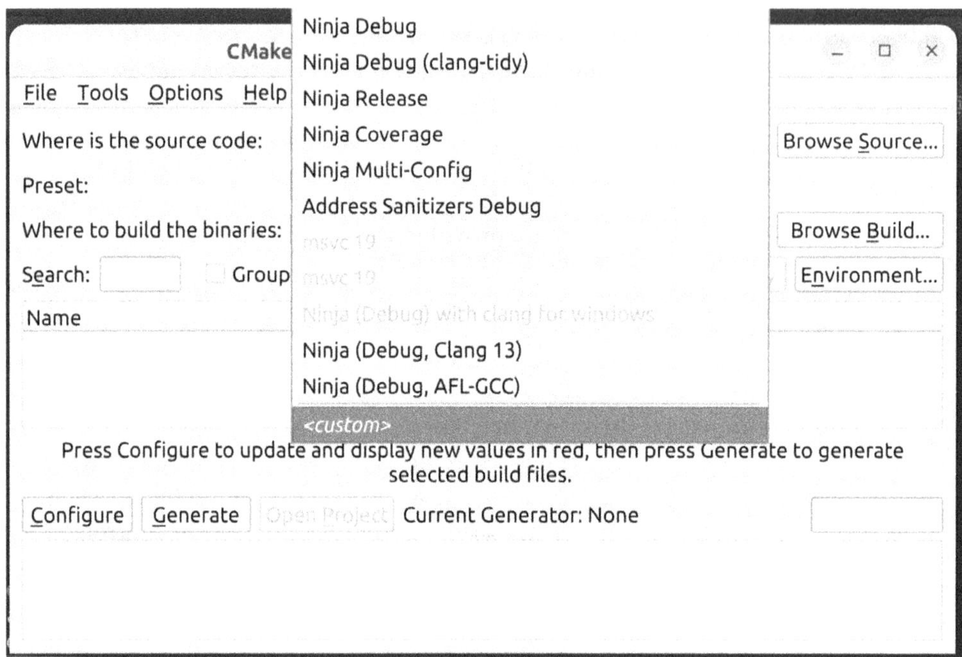

Figure 9.1 – Listing available presets in the CMake GUI

As of version 3.21 of CMake, the `ccmake` command-line configuration tool does not support presets. Configure presets can be selected from the top-level directory by calling the following:

```
cmake --preset=name
```

The overall structure of `CMakePresets.json` and `CMakeUserPresets.json` is like this:

```
{
   "version": 3,
   "cmakeMinimumRequired": {"major": 3,"minor": 21,"patch":
      0 },
   "configurePresets": [...],
   "buildPresets": [...],
   "testPresets": [...],
   "vendor": {
```

```
        "microsoft.com/VisualStudioSettings/CMake/1.9":
        {                    "intelliSenseMode": "windows-msvc-x64"
        } }
}
```

The `version` field specifies the JSON schema to use. Version 1 is the first release from CMake 3.19 and only supports `configurePresets`. Version 2 adds `buildPresets` and `testPresets` and is available from CMake 3.20 and the version 3 adds more options and is available from CMake 3.21.

The optional `cmakeMinimumRequired` field may be used to define the minimum version of CMake needed to build this project. As the minimum requirement is usually also stated in the `CMakeLists.txt` files, this is often omitted.

The three lists, `configurePresets`, `buildPresets`, and `testPresets`, each contain a list of configurations for configuring, building, and testing the project. The presets for building and testing require the presence of at least one configuration preset, as we will see later in this section.

The `vendor` field contains an optional map of vendor- or IDE-specific information. CMake does not interpret the contents of this field except to verify the JSON format. The keys for the map should be the vendor-specific domain separated by slashes. In the previous example, the key for the vender presets is `microsoft.com/VisualStudioSettings/CMake/1.9`. The values inside the vendor fields can be of any valid JSON format.

To use presets, at least one configure preset that defines an environment for CMake to configure the build system has to be present. They should at least specify the build path and the generator to use when configuring. Often, a configure preset also sets common cache variables such as `CMAKE_BUILD_TYPE` for single-configuration generators. A preset containing a configure preset to build a project with the Ninja generator in debug mode might look like this:

```
{
  "version": 3,
  "configurePresets": [
    {
      "name": "ninja",
      "displayName": "Ninja Debug",
      "description": "build in debug mode using Ninja
      generator",
      "generator": "Ninja",
      "binaryDir": "build",
```

```
            "cacheVariables": { "CMAKE_BUILD_TYPE": "Debug" }
        }
    ]
}
```

All presets must have a name that is unique within the preset block. As some GUI applications only show presets that have a `displayName` field assigned, setting this field is highly recommended.

> **Naming Conventions for Presets**
>
> It is good practice to name presets that are defined by the project in a `CMakePresets.json` file in a way that they do not clash with names the developer might define in `CMakeUserPresets.json`. A common convention is to prefix project-defined presets with `ci-` to mark them as used by the CI environment.

In versions 1 and 2 of the presets, the `binaryDir` and `generator` fields were mandatory; with version 3, they became optional. If either of the fields is not set, the behavior in this regard is the same as when CMake is used without presets. The command-line options for the CMake command will override the values specified in the presets where relevant. So, if `binaryDir` is set, it is automatically created when `cmake --preset=` is called, although the value of it would be overridden if the `-B` option is passed to CMake.

Cache variables can either be defined as `key:value` pairs, as shown in the preceding example, or as a JSON object, which allows specifying the variable type. A file path could be specified like this:

```
"cacheVariables": {
  "CMAKE_TOOLCHAIN_FILE": {
    "type": "FILEPATH",
    "value": "${sourceDir}/cmake/toolchain.cmake"
  }
}
```

If used in the `key:value` form, the type is treated as `STRING` unless it is `true` or `false` (without quotes), in which case, it is interpreted as `BOOL`. Note `${sourceDir}` in the example, which is a macro that is expanded when the preset is used.

Here are a few well-known macros:

- ${sourceDir}: This points to the project source directory and ${sourceParentDir} points to the parent directory of the source directory. The directory name without the path of the source directory can be obtained with ${sourceDirName}. For example, if ${sourceDir} is /home/sandy/MyProject, ${sourceDirName} would be MyProject and ${sourceParentDir} would be /home/sandy/.
- ${generator}: This contains the generator as specified by the current preset used. For build and test presets, this contains the generator of the configure preset used.
- ${hostSystemName}: The system name of the host operating system, which is the same as the CMAKE_HOST_SYSTEM variable. The value is either the result of uname -s or Linux, Windows, or Darwin (for macOS).
- $env{<variable-name>}: This contains the environment variable with the name <variable-name>. If the variable is defined in the preset with the environment field, this value is used instead of the value from the parent or system environment. Using $penv{<variable-name>} works similarly, but the value is always taken from the parent environment and not from the environment field, even if it is defined. This allows prepending or appending values to existing environment variables. Appending or prepending variables is not possible with $env{...} because it does not allow circular references. Note that while in a Windows environment, variables are case-insensitive; variables used in the presets are still case-sensitive. Because of this, it is recommended to keep the casing of environment variables consistent.
- $vendor{<macro-name>}: This is an extension point for vendors of IDEs to insert their own macros. Since CMake is not able to interpret these macros, presets using $vendor{...} macros will be ignored.
- ${dollar}: This is a placeholder for the literal dollar sign, $.

Modifying the environment for a preset works similarly to setting cache variables: by setting the environment field, which contains a map of key:value pairs. Environment variables are always set, even when the value is empty or null. Environment variables may reference each other as long as they do not contain circular references. Consider the following example:

```
{
  "version": 3,
  "configurePresets": [
```

```
        {
          "name": "ci-ninja",
          "generator": "Ninja",
          "binaryDir": "build",
          "environment": {
            "PATH": "${sourceDir}/scripts:$penv{PATH}",
            "LOCAL_PATH": "$env{PATH}",
            "EMPTY" : null
        }
    ]
}
```

In this example, the `PATH` environment variable is modified by prepending a path from inside the project structure. Using the `$penv{PATH}` macro ensures that the value is taken from outside the preset. The `LOCAL_PATH` variable then references the modified `PATH` environment variable by using the `$env{PATH}` macro. This reference is fine as long as the `PATH` environment variable does not contain `$env{LOCAL_PATH}`, which would create a circular reference. The `EMPTY` environment variable is unset by passing `null`. Note that `null` is not quoted. Unless a build preset or a test preset is used, the environment is *not* forwarded to the respective steps. If a build preset or a test preset is used, but the environment from the configure preset should not be applied, this can be explicitly stated when the `inheritConfigureEnvironment` field is set to `false`.

Inheriting from presets

Presets may inherit from other presets of the same type with the `inherits` field, which may contain either a single preset or a list of presets. When inheriting fields from the parent, a preset can be overridden or additional fields added. This is useful to avoid duplicating code for common building blocks. In combination with the `hidden` field, this can make a `CMakePreset.json` file smaller. Consider the following example:

```
{
  "version": 3,
  "configurePresets": [
    {
      "name": "ci-ninja",
      "generator": "Ninja",
      "hidden": true,
```

```
            "binaryDir": "build"
        },
        {
            "name": "ci-ninja-debug",
            "inherits": "ci-ninja",
            "cacheVariables": {
                "CMAKE_BUILD_TYPE": "Debug"
            }
        },
        {
             "name": "ci-ninja-release",
             "inherits": "ci-ninja",
             "cacheVariables": {
                 "CMAKE_BUILD_TYPE": "Release"
            }
        }
    ]
}
```

In the example, `the ci-ninja-debug` and `ci-ninja-release` presets both inherit from the hidden `ci-ninja build` preset and additionally set the `CMAKE_BUILD_TYPE` cache variable to the respective configuration. Hidden presets can still be used but will not show up when `cmake --list-presets` is invoked. Presets defined in `CMakeUserPreset.json` may inherit from `CMakePreset.json` but not the other way around.

In the preceding example, the preset inherits from a single parent, but presets can also inherit from multiple parents. The following example shows how `CMakeUserPreset.json` working with `CMakePreset.json` from the previous example might look:

```
{
    "version": 3,
    "configurePresets": [
    {
      "name": "gcc-11",
      "hidden": true,
      "binaryDir": "build",
      "cacheVariables": {
```

```
        "CMAKE_C_COMPILER": "gcc-11",
        "CMAKE_CXX_COMPILER": "g++-11"
      }
    },
    {
      "name": "ninja-debug-gcc",
      "inherits": ["ci-ninja-debug","gcc-11"]
    },
  ]
}
```

Here, the user supplies a preset that explicitly selects GCC 11 as a compiler named `gcc-11`. Later, the `ninja-debug-gcc` preset inherits the values from the `ninja-debug` preset that is defined in `CMakePreset.json` supplied by the project and combines it with the user-supplied `gcc-11` preset. If two parent presets define different values for the same field, the value from the one first appearing in the inherits list takes precedence.

Conditions for presets

Sometimes presets only make sense under certain conditions, such as for a certain build platform. For instance, a configure preset that uses the Visual Studio generator would only be useful in a Windows environment. For these cases, presets can be disabled if the conditions are not met with the `condition` option. Any conditions defined in a parent preset are inherited. Conditions can be constants, string comparisons, or a check of whether a list contains a value. They are available from version 3 of the presets. The following configure preset would only be enabled if working on Windows:

```
{
  "name": "ci-msvc-19",
  "generator": "Visual Studio 16 2019",
  "binaryDir": "build",
  "condition": {
    "type": "equals",
    "lhs": "${hostSystemName}",
    "rhs": "Windows"
  }
}
```

In the preceding example, the build preset is enabled if the name of the host system is retrieved using the `${hostSystemName}` macro and then compared to the `Windows` string. If `${hostSystemName}` matches, then the preset is enabled, else it is disabled and trying to use it will result in an error. When comparing strings, casing matters: for case-insensitive tests, the `matches` or `notMatches` type, which takes a regex, could be used.

For more complex conditions, support nesting with Boolean logic with the `allOf`, `anyOf`, and `not` operators. For example, if a configure preset should only be enabled for Windows and Linux but not for macOS, the preset and the condition could look like this:

```
{
"name": "WindowsAndLinuxOnly",
"condition": {
      "type": "anyOf",
      "conditions": [
            {
                  "type": "equals",
                  "lhs": "${hostSystemName}",
                  "rhs": "Windows"
            },
            {
                  "type": "equals",
                  "lhs": "${hostSystemName}",
                  "rhs": "Linux"
            }
      ]
}
```

Each of the conditions could also contain further nested conditions if needed, although doing so will quickly increase the complexity of the presets.

So far, we have only seen configure presets in the examples, but as mentioned at the beginning of the chapter, there are also build presets and test presets. The syntax for build and test presets is very similar to configure presets and a lot of fields, such as `name`, `displayName`, and `inherit`, and conditions work the same as with configure presets.

Build presets have to specify a configure preset in the `configurePreset` field or inherit from another build preset that specifies the configure preset. The build directory is determined by the configure preset and the environment from the configure preset is inherited unless the `inheritConfigureEnvironment` field is set to `false`. Optionally, build presets can specify a list of targets to build. An example of a build preset could look like this:

```
{
    "version": 3,
    "configurePresets": [
        {
            "name": "ci-msvc-19",
            "displayName": "msvc 19",
            "description": "Configuring for msvc 19",
            "generator": "Visual Studio 16 2019",
            "binaryDir" : "build"
        }
    ],
    "buildPresets": [
        {
            "name": "ci-msvc-debug",
            "configurePreset": "ci-msvc-19",
            "configuration": "Debug"
        },
        {
            "name": "ci-msvc-release",
            "configurePreset": "ci-msvc-19",
            "configuration": "Release"
        },
        {
            "name": "ci-documentation",
            "configurePreset": "ci-msvc-19",
            "targets": [
                "api-doc",
                "doc"
```

```
                ]
            }
        ]
    }
```

In the preceding example, three build presets are defined. Two are used to specify the build configuration for Visual Studio and the third one lists the `api-doc` and `doc` targets as part of the documentation build. Invoking any of the `ci-msvc` build presets will build the `"all"` target, while `ci-documentation` will only build the listed targets. The list of available build presets can be retrieved with `cmake --build --list-presets`.

Test presets work very similar to build presets, except that they are used with CTest. Similarly, a call to `ctest --list-presets` on the project root will list the available test presets. Test presets can be used to select or exclude certain tests, specify fixture options, or control the output of the tests. Most of the options for tests described in *Chapter 7, Seamlessly Integrating Code Quality Tools with CMake*, can be controlled from the test presets. An example of test presets could look like this:

```
{
    "version": 3,
    "configurePresets": [
        {
            "name": "ci-ninja",
            ...
        }


    "testPresets": [
        {
            "name": "ci-feature-X",
            "configurePreset": "ci-ninja",
            "filter": {
                "include": {
                    "name": "feature-X"
                },
                "exclude": {
                    "label": "integration"
                }
```

```
            }
        }
    ]
}
```

In the previous example, a test preset is added that filters the tests for any that include `feature-X` but excludes any tests that were labeled as `integration`. This is equivalent to invoking the following command from the build directory:

```
ctest --tests-regex feature-X --label-exclude integration
```

CMake presets are arguably one of the few features that changed how CMake is used the most since the introduction of targets. They are a good compromise to deliver common configuration and build options together with the projects while still keeping the `CMakeLists.txt` file platform-agnostic. However, sometimes providing the necessary settings is not enough and you also want to share a build environment in which you are sure that the software compiles. One option to do this is by defining a build container that contains CMake and the necessary libraries.

Using build containers with CMake

Containerization brings the benefit that developers can control the build environment to some extent. Containerized build environments are also a tremendous help for setting up CI environments. There are quite a few container runtimes out there, with Docker being the most popular. It would exceed the scope of this book to look in depth at containerization, so we will use Docker for the examples in this book.

A build container contains a fully defined build system including CMake and any tools and libraries needed to build a certain software. By providing the container definition, for example, the Dockerfile, along with the project, or over a publicly accessible container registry, anyone can use the container to build the software. The huge advantage is that developers do not need to install and possibly pollute their host machine by installing additional libraries or tools except the software needed to run the containers. The downside is that building might take longer and not all IDEs and tools support working with containers in a convenient way. Notably, Visual Studio Code has very good support for working in containers. You can visit `https://code.visualstudio.com/docs/remote/containers` for more details.

At a very high level, the workflow for using a build container is the following:

1. Define the container and build it.
2. Mount a local copy of the source code into the build container.
3. Run any commands for building inside the container.

A very simple Docker definition for building a simple C++ application could look like this:

```
FROM alpine:3.15
RUN apk add --no-cache cmake ninja g++ bash make git
RUN <any command to install additional libraries etc.>
```

This will define a small container based on the Alpine Linux 3.15 and install `cmake`, `ninja`, `bash`, `make`, and `git`. Any real-life container will probably have additional tools and libraries installed inside to work conveniently; however, just to illustrate how building software with a container works, having such a minimal container is enough. The following Docker command builds the container image and tags it with the name `builder_minimal`:

```
docker build . -t builder_minimal
```

Once the container is a clone of the local, the source is mounted inside the container and all CMake commands are executed inside the container. Assuming the user is executing the Docker command from the `source` directory, the commands to configure a CMake build project might look like this:

```
docker run --user 1000:1000 --rm -v $(pwd):/workspace
  builder_minimal cmake -S /workspace -B /workspace/build
docker run --user 1000:1000 --rm -v $(pwd):/workspace
  builder_minimal cmake --build /workspace/build
```

This will start up the container we created and execute the CMake commands within. The local directory is mounted inside the container as `/workspace` with the `-v` option. Since our Docker containers use *root* as the default user, the user ID and group to use are passed with the `--user` option. On Unix-like operating systems, this should match the user ID of the host, so any files created can also be edited from outside the container. The `--rm` flag tells Docker to remove the image once it is done with it.

An alternative way to work with the container is to run it in interactive mode by passing the `-ti` flag to the `docker run` command:

```
docker run --user 1000:1000 --rm -ti -v $(pwd):/workspace
   builder_minimal
```

This will start a shell inside the container where the `build` command can be invoked without the need to restart the container every time.

There are several strategies on how editors or IDEs and build containers can work together. The most convenient way is, of course, if the IDE supports it natively or through a convenient extension like Visual Studio Code does. If this is not the case, packing a suitable editor inside the container and executing it from within can be a strategy to develop software conveniently. Another way is to run the editor on the host system and reconfigure it so it invokes CMake not directly but starts the container and executes CMake within.

What we have shown here is the bare minimum for working with containers as build environments but we hope it serves as a very first stepping stone to working with containers. As more and more IDEs start to support working with containerized build environments, using them will become much easier. Containers make build environments very portable between various machines and can help to ensure that all developers of a project are using the same build environment. It is also a good idea to put container definition files under version control so that necessary changes to the build environment are tracked together with the code.

Containers are a good and portable way of creating isolated build environments. But if this is not an option for any reason, another way to create an isolated and portable build environment is using sysroots.

Using sysroots to isolate build environments

In a nutshell, a **system root**, or just **sysroot**, is a directory that a build system considers to be the root directory from which to locate headers and libraries. In brief, they contain a stripped-down version of the root filesystem for the platform for which software is being compiled. They are often used when cross-compiling software for other platforms, as described in *Chapter 12, Cross-Platform Compiling and Custom Toolchains*. If containers for shipping whole build environments are not an option, sysroots can be an alternative to provide a defined build environment.

To use a sysroot with CMake, a toolchain file is needed. As the name suggests, these files define the tools to use to compile and link the software as well as where to find any libraries. In a normal build, CMake automatically detects the toolchain by introspecting the system. Toolchain files are passed to CMake with the `CMAKE_TOOLCHAIN_FILE` variable like this:

```
cmake -S <source_dir> -B <binary_dir> -DCMAKE_TOOLCHAIN_FILE=
  <path/to/toolchain.cmake>
```

Since version 3.21, CMake additionally supports the `--toolchain` option to pass toolchain files, which is equivalent to passing the `CMAKE_TOOLCHAIN_FILE` cache variable.

Alternatively, the toolchain file can be passed with a CMake preset as a cache variable. At a minimum, a toolchain file to use with a sysroot will define the `CMAKE_SYSROOT` variable to point to the sysroot and the `CMAKE_<LANG>_COMPILER` variable to point to a compiler that is compatible with the libraries in the sysroot. To avoid mixing dependencies from outside the sysroot with the files installed on the host system, the variables for controlling where the `find_` commands look for stuff are usually also set. A minimal toolchain file might look like this:

```
set(CMAKE_SYSTEM_NAME Linux)

set(CMAKE_SYSROOT /path/to/sysroot/)
set(CMAKE_STAGING_PREFIX path/to/staging/directory)

set(CMAKE_C_COMPILER /path/to/sysroot/usr/bin/gcc-10)
set(CMAKE_CXX_COMPILER /path/to/sysroot/usr/bin/g++-10)

set(CMAKE_FIND_ROOT_PATH_MODE_PROGRAM NEVER)
set(CMAKE_FIND_ROOT_PATH_MODE_LIBRARY ONLY)
set(CMAKE_FIND_ROOT_PATH_MODE_INCLUDE ONLY)
set(CMAKE_FIND_ROOT_PATH_MODE_PACKAGE BOTH)
```

Let's see what happens here in detail:

1. First, the system name of the target system is set, by setting the CMAKE_SYSTEM_NAME variable. This is the system for which the files are compiled inside the sysroot.
2. Then, the path to the sysroot itself is set by setting the CMAKE_SYSROOT variable. CMAKE_STAGING_PREFIX is optional and is used to specify a location on the host machine to install any artifacts of the project. Specifying a staging prefix helps keep the sysroot and the host filesystem clean, as without it, any installation of the artifacts will happen on the host filesystem.
3. Next, the compilers are set to the compilers binaries delivered with the sysroot by setting the CMAKE_C_COMPILER and CMAKE_CXX_COMPILER variables.
4. Lastly, the search behavior of any find_ command in CMake is set. The CMAKE_FIND_ROOT_PATH_MODE_* variables take any of the values of ONLY, NEVER, and BOTH. If set to ONLY, CMake will only search the type of file inside the sysroot; if set to NEVER, searches will only consider the host file structure. If set to BOTH, then the host system path and the sysroot path will be searched. Note that CMAKE_STAGING_PREFIX is considered a system path, so in order to search the sysroot and the staging directory, BOTH must be selected. In the example, this is configured in a way that all header files and libraries are restricted to the sysroot, while any call for find_program will look only on the host system and find_package will look in both places.

Setting the CMAKE_SYSROOT variable will not automatically set the place where build artifacts are installed. For situations where the resulting binaries are compatible with the host system, this might be the intended behavior. In a lot of cases, such as when cross-compiling, this is not what is wanted, so setting CMAKE_STAGING_PREFIX is often recommended. Setting the staging directory has two effects: first, it will cause any artifacts to be installed in the staging directory, and second, the staging directory will be added to the search prefix for the find_ commands. One caveat is that the staging directory will be added to CMAKE_SYSTEM_PREFIX_PATH, which has the downside that the CMAKE_FIND_ROOT_PATH_MODE_XXX variables from the preceding example have to be set to BOTH so the packages, libraries, and programs installed in the staging area are found.

CMAKE_STAGING_PREFIX and CMAKE_INSTALL_PREFIX

If both CMAKE_STAGING_PREFIX and CMAKE_INSTALL_PREFIX are set, the staging prefix will take precedence. So, as a rule of thumb, whenever the toolchain is compatible with the host system, the staging might be omitted, else it tends to be defined.

One downside of sysroots compared to containers is that they cannot be started and be used to execute commands within just like that. So, if the toolchain and the sysroot are not compatible with the host platform, any files produced will not be executable without either moving to the target platform or using an emulator.

Summary

One of the main strengths of CMake is its versatility to build software using a variety of toolchains for a large number of platforms. The downside of this is that it sometimes can be hard for developers to find a working configuration for software. But by supplying CMake presets, containers, and sysroots, it often gets easier to get started with a CMake project.

In this chapter, we looked in detail into how to define CMake presets to define working configuration setups, along with creating build and test definitions. Then, we briefly covered how to create a Docker container and how to invoke CMake commands within, before closing the chapter with a brief look into sysroots and toolchain files. More about toolchains and sysroots will be covered in *Chapter 12, Cross-Platform Compiling and Custom Toolchains*.

In the next chapter, you will learn how to work with big, distributed projects as super builds. There, you will learn how to handle different versions and how to assemble projects from multiple repositories in a manageable way.

Questions

Answer the following questions to test your knowledge of this chapter:

1. What is the difference between `CMakePresets.json` and `CMakeUserPresets.json`?
2. How are presets used on the command line?
3. Which three types of presets exist and how do they depend on each other?
4. What is the minimum a configure preset should define?
5. When inheriting from multiple presets, which one takes precedence if a value is specified multiple times?
6. What strategies for working with build containers are common?
7. What do toolchain files to be used with sysroots usually define?

10
Handling Big Projects and Distributed Repositories in a Superbuild

As we should have learned by now, every big project comes with its own set of dependencies. The easiest way of dealing with these dependencies is by using a package manager, such as **Conan** or **vcpkg**. But using a package manager might not always be possible or feasible, due to company policies, project requirements, or lack of resources. Thus, the project authors might consult the traditional, old-style ways to deal with the dependencies. The usual way of dealing with these dependencies may include shipping all dependencies embedded into the repository's build code. Alternatively, project authors may decide to let the end user deal with the dependencies from scratch. Neither of these ways is clean and has its own drawbacks. What if I told you there is a middle ground? Welcome to the *super-build* approach.

A super-build is a method that can be used for decoupling the logic required for satisfying dependencies from project code, similar to how package managers work. In fact, we can call this method *the poor man's package manager*. Separating the dependency logic from the project code allows us to have a more flexible and maintainable project structure. In this chapter, we will learn how to achieve this in detail.

To understand the skills shared in this chapter, we'll cover the following main topics:

- The requirements and prerequisites for a super-build
- Building across multiple code repositories
- Ensuring version consistency in a super-build

Let's begin with the technical requirements.

Technical requirements

Before you dive further into this chapter, you should have a good grasp of the content covered in *Chapter 5, Integrating Third-Party Libraries and Dependency Management*. This chapter will follow the teaching by example approach. Thus, it is recommended to obtain this chapter's example content from `https://github.com/PacktPublishing/CMake-Best-Practices/tree/main/chapter_10`. All of the examples assume that you will be using the container provided by the `https://github.com/PacktPublishing/CMake-Best-Practices` project.

Let's begin learning about super-builds by inspecting the prerequisites and requirements.

Requirements and prerequisites for a super-build

Super-builds can be structured as a grand build that builds multiple projects or as an in-project submodule that deals with dependencies. Therefore, having the means of acquiring repositories is a must. Fortunately, CMake has stable and established ways of doing so. To name a few, `ExternalProject` and `FetchContent` are the most popular CMake modules for dealing with external dependencies. We will be using the `FetchContent` CMake module in our examples, since it is cleaner and easier to deal with. Please note that using the means provided by CMake is not a strict requirement but a convenience. A super-build can also be structured by using version control system utilities, such as `git submodule` or `git subtree`. Since CMake is this book's focal point and Git support for `FetchContent` is quite decent, we prefer to utilize it.

That's it for now. Let's continue with learning about building a project that spans multiple code repositories.

Building across multiple code repositories

Software projects, either directly or indirectly, span multiple code repositories. Dealing with local project code is the easiest, but software projects are rarely standalone. Things may get complicated really fast without a proper dependency management strategy. The first recommendation of this chapter is to *use a package manager if you can*. Package managers greatly reduce the effort spent on dependency management. If you are not able to use a package manager, you may need to roll your very own mini project-specific package manager, which is called a **super-build**.

Super-builds are mostly used for making a project self-sufficient dependency-wise, which means the project is able to satisfy its very own dependencies without the intervention of the user. Having such an ability is very convenient for all consumers. To demonstrate this technique, we will start with an example of such a scenario. Let's begin.

The recommended way – FetchContent

We will be following *Chapter 10, Example 01* for this part. Let's start by inspecting *Chapter 10, Example 01's* `CMakeLists.txt` file, as usual. The first seven lines are left out for simplicity:

```
if(CH10_EX01_USE_SUPERBUILD)
  include(superbuild.cmake)
else()
  find_package(GTest 1.10.0 REQUIRED)
  find_package(benchmark 1.6.1 REQUIRED)
endif()

add_executable(ch10_ex01_tests)
target_sources(ch10_ex01_tests PRIVATE src/tests.cpp)
target_link_libraries(ch10_ex01_tests PRIVATE GTest::Main)

add_executable(ch10_ex01_benchmarks)
target_sources(ch10_ex01_benchmarks PRIVATE src
  /benchmarks.cpp)
target_link_libraries(ch10_ex01_benchmarks PRIVATE
  benchmark::benchmark)
```

As we can see, it is a simple `CMakeLists.txt` file that defines two targets, named `ch10_ex01_tests` and `ch10_ex01_benchmarks`. These targets depend on `Google Test` and `Google Benchmark` libraries, respectively. These libraries are found and defined by either the super-build or the `find_package(...)` calls, depending on the `CH10_EX01_USE_SUPERBUILD` variable. The `find_package(...)` path is the way we have followed until now. Let's inspect the super-build file, `superbuild.cmake`, together:

```
include(FetchContent)
FetchContent_Declare(benchmark
    GIT_REPOSITORY https://github.com/google/benchmark.git
    GIT_TAG        v1.6.1
)
FetchContent_Declare(GTest
    GIT_REPOSITORY https://github.com/google/googletest.git
    GIT_TAG        release-1.10.0
)
FetchContent_MakeAvailable(GTest benchmark)
add_library(GTest::Main ALIAS gtest_main)
```

In the first line, the `FetchContent` CMake module is included, since we are going to utilize it for the dependencies. In the following six lines, the `FetchContent_Declare` function is used to declare two external targets, `benchmark` and `GTest`, which are instructed to be fetched via `Git`. Consequently, the `FetchContent_MakeAvailable(...)` function is called to make the declared targets available. Lastly, an `add_library(...)` call is made for defining an alias target named `GTest::Main` for the `gtest_main` target. This is done to keep the compatibility between `find_package(...)` and super-build target names. There is no alias target defined for `benchmark`, since its `find_package(...)` and super-build target names are already compatible.

Let's configure and build the example by invoking the following commands:

```
cd chapter_10/ex01_external_deps
cmake -S ./ -B build -DCH10_EX01_USE_SUPERBUILD:BOOL=ON
cmake --build build/ --parallel $(nproc)
```

In the first two lines, we are going into the `example_10/` folder and configuring the project. Note that we are setting the `CH10_EX01_USE_SUPERBUILD` variable to `ON` in order to enable the super-build code. In the last line, we are building the project with *N* parallel jobs, where *N* is the result of the `nproc` command.

Thanks to the alternate `find_package(...)` path, the build will also work fine without enabling the super-build, given that `google test >= 1.10.0` and `google benchmark >= 1.6.1` are available in the environment. This will allow package maintainers to change dependency versions without patching the project. Small customization points such as this are important for portability and reproducibility.

Up next, we'll be taking a look at a super-build example that uses the `ExternalProject` module instead of the `FetchContent` module.

The legacy way – ExternalProject_Add

Before `FetchContent` was a thing, most people implemented the super-build approach by utilizing the `ExternalProject_Add` CMake function. This function is provided by the `ExternalProject` CMake module. In this section, we will see a super-build example with `ExternalProject_Add` to see how it differs from using the `FetchContent` module.

Let's take a look at the `CMakeLists.txt` file in *Chapter 10, Example 02* together (comments and the project directive are omitted):

```
# ...
include(superbuild.cmake)
add_executable(ch10_ex02_tests)
target_sources(ch10_ex02_tests PRIVATE src/tests.cpp)
target_link_libraries(ch10_ex02_tests PRIVATE catch2)
```

Again, the project is a unit-test project containing a single C++ source file, but this time it is with `Catch2` instead of Google Test. The `CMakeLists.txt` file includes the `superbuild.cmake` file directly, defines an executable target, and links the `Catch2` library to the target. You may have noticed that this example does not use `FindPackage(...)` to discover the `Catch2` library. The reason for this is that, unlike `FetchContent`, `ExternalProject` fetches and builds the external dependencies at build time. Since the content of the Catch2 library is not available at configuration time, we are unable to use `FindPackage(...)` here. `FindPackage(...)` runs at configure time and requires the package files to be present. Let's take a look at `superbuild.cmake` as well:

```
include(ExternalProject)
ExternalProject_Add(catch2_download
    GIT_REPOSITORY https://github.com/catchorg/Catch2.git
    GIT_TAG v2.13.9
```

```
    INSTALL_COMMAND ""
    # For disabling the warning that treated as an error
    CMAKE_ARGS -DCMAKE_CXX_FLAGS="-Wno-error=pragmas"
)
SET(CATCH2_INCLUDE_DIR ${CMAKE_CURRENT_BINARY_DIR}
  /catch2_download-prefix/src/catch2_download/single_include)
file(MAKE_DIRECTORY ${CATCH2_INCLUDE_DIR})
add_library(catch2 IMPORTED INTERFACE GLOBAL)
add_dependencies(catch2 catch2_download)
set_target_properties(catch2 PROPERTIES "INTERFACE_INCLUDE_
  DIRECTORIES" "${CATCH2_INCLUDE_DIR}")
```

The `superbuild.cmake` module includes the `ExternalProject` CMake module. The code calls the `ExternalProject_Add` function to declare a target named `catch2_download` with the `GIT_REPOSITORY`, `GIT_TAG`, `INSTALL_COMMAND`, and `CMAKE_ARGS` arguments. As you may recall from previous chapters, the `ExternalProject_Add` function can fetch dependencies from different sources. Our example is trying to fetch the dependency via using Git. The `GIT_REPOSITORY` and `GIT_TAG` arguments are for specifying the target Git repository URL and the tag to be checked out after the `git clone`, respectively. Since Catch2 is a CMake project, the amount of parameters we need to supply to the `ExternalProject_Add` function is minimal. The `ExternalProject_Add` function knows how to configure, build, and install a CMake project by default, so no `CONFIGURE_COMMAND` or `BUILD_COMMAND` arguments are needed. The empty `INSTALL_COMMAND` argument is for disabling and installing the dependency after the build. The last argument, `CMAKE_ARGS`, is for passing CMake arguments to the external project's configure step. We use this to suppress a GCC warning (treated as an error) about legacy pragmas in the Catch2 compilation.

The `ExternalProject_Add` command fetches the desired library into a prefix path and builds it. So, to use the fetched content, we have to import it to the project first. Since we cannot utilize `FindPackage(...)` to let CMake deal with the importing of the library, we have some manual work to do. One of them is to define the `include` directory of the Catch2 target. Since Catch2 is a header-only library, defining an interface target with the headers will be sufficient. We are declaring the `CATCH2_INCLUDE_DIR` variable to set the directory that will contain the Catch2 headers. We are using this variable to set the `INTERFACE_INCLUDE_DIRECTORIES` property of the imported target created in this example. Up next, the file `(MAKE_DIRECTORY ${CATCH2_INCLUDE_DIR})` CMake command is called to create the include directory. The reason for that is, because of how `ExternalProject_Add` works, the Catch2 content is not present until the build step executes. Setting a target's `INTERFACE_INCLUDE_DIRECTORIES` requires the given directories to be present, so we're doing a small hack as a workaround. In the last three lines, we declare an `IMPORTED INTERFACE` library for Catch2, making this library dependent on the `catch2_download` target and setting `INTERFACE_INCLUDE_DIRECTORIES` of the imported library.

Let's try to configure and build our example to check whether it works or not:

```
cd chapter_10/ex02_external_deps_with_extproject
cmake -S ./ -B build
cmake --build build/ --parallel $(nproc)
```

If everything goes as expected, you should be seeing output similar to this:

```
[ 10%] Creating directories for 'catch2_download'
[ 20%] Performing download step (git clone) for
  'catch2_download'
Cloning into 'catch2_download'...
HEAD is now at 62fd6605 v2.13.9
[ 30%] Performing update step for 'catch2_download'
[ 40%] No patch step for 'catch2_download'
[ 50%] Performing configure step for 'catch2_download'
/* ... */
[ 60%] Performing build step for 'catch2_download'
/* ... */
[ 70%] No install step for 'catch2_download'
[ 80%] Completed 'catch2_download'
[ 80%] Built target catch2_download
/* ... */
[100%] Built target ch10_ex02_tests
```

Alright, it seems we have built our test executable successfully. Let's run it to check whether it works by running the `./build/ch10_ex02_tests` executable:

```
===========================================================
All tests passed (4 assertions in 1 test case)
```

Up next, we'll see a simple QT application using the QT framework from a super-build.

Bonus – using the Qt 6 framework with a super-build

Until now, we have dealt with libraries that have rather small footprints. Let's try something more complex, such as using a big framework such as the Qt framework in a super-build. For this part, we are going to follow the *Chapter 10, Example 03* example.

> **Important note**
>
> If you are going to try this example outside of the provided Docker container, you may have to install some additional dependencies required by the Qt runtime. The required packages for the Debian-like systems are as follows: `libgl1-mesa-dev libglu1-mesa-dev '^libxcb.*-dev' libx11-xcb-dev libglu1-mesa-dev libxrender-dev libxi-dev libxkbcommon-dev libxkbcommon-x11-dev`.

The example contains a single source file, `main.cpp`, that outputs a simple Qt window application with a message. The implementation is as follows:

```
#include <qapplication.h>
#include <qpushbutton.h>
int main( int argc, char **argv )
{
    QApplication a( argc, argv );
    QPushButton hello( "Hello from CMake Best Practices!",
      0 );
    hello.resize( 250, 30 );
    hello.show();
    return a.exec();
}
```

Our goal is to be able to compile this Qt application without requiring the user to install the Qt framework themselves. The super-build should automatically install the Qt 6 framework, and the application should be able to use that. Let's take a look at the `CMakeLists.txt` file of the example, as usual:

```
if(CH10_EX03_USE_SUPERBUILD)
  include(superbuild.cmake)
else()
    set(CMAKE_AUTOMOC ON)
    set(CMAKE_AUTORCC ON)
    set(CMAKE_AUTOUIC ON)
```

```
    find_package(Qt6 COMPONENTS Core Widgets REQUIRED)
endif()

add_executable(ch10_ex03_simple_qt_app main.cpp)
target_compile_features(ch10_ex03_simple_qt_app PRIVATE
  cxx_std_11)
target_link_libraries(ch10_ex03_simple_qt_app Qt6::Core
  Qt6::Widgets)
```

Like the first example, the `CMakeLists.txt` file includes the `superbuild.cmake` file, depending on an `option` flag. If the user is opted in to use a super-build for the example, it will include the super-build module. Otherwise, the dependency will try to be located in the system via `find_package(...)`. In the last three lines, an executable target is defined, C++ standards for the target are set, and the defined target is linked to the `QT6::Core` and `QT6::Widgets` targets. These targets are either defined by the super-build or the `find_package(...)` call, depending on whether the user is opted in to use the super-build or not. Let's continue by taking a look at the `superbuild.cmake` file:

```
include(FetchContent)
message(STATUS "Chapter 10, example 03 superbuild enabled.
  Will try to satisfy dependencies for the example.")
set(FETCHCONTENT_QUIET FALSE) # Enable message output for
  FetchContent commands
set(QT_BUILD_SUBMODULES "qtbase" CACHE STRING "Submodules
  to build")
set(QT_WILL_BUILD_TOOLS on)
set(QT_FEATURE_sql off)
set(QT_FEATURE_network off)
set(QT_FEATURE_dbus off)
set(QT_FEATURE_opengl off)
set(QT_FEATURE_testlib off)
set(QT_BUILD_STANDALONE_TESTS off)
set(QT_BUILD_EXAMPLES off)
set(QT_BUILD_TESTS off)

FetchContent_Declare(qt6
    GIT_REPOSITORY https://github.com/qt/qt5.git
    GIT_TAG        v6.3.0
```

```
    GIT_SHALLOW TRUE
    GIT_PROGRESS TRUE # Since the clone process is lengthy,
      show progress of download
    GIT_SUBMODULES qtbase # The only QT submodule we need
)

FetchContent_MakeAvailable(qt6)
```

The `superbuild.cmake` file uses the `FetchContent` module to fetch the Qt dependency. Since the fetching and preparing process for the Qt may be lengthy, some of the unused Qt framework features are disabled. The `FetchContent` message output is enabled for better tracking of the progress. Let's try to configure and compile the example by running the following commands:

```
cd chapter_10/ex03_simple_qt_app/
cmake -S ./ -B build -DCH10_EX03_USE_SUPERBUILD:BOOL=ON
cmake --build build/ --parallel $(nproc)
```

If everything goes as expected, you should see similar output, as shown here:

```
/*...*/
[ 11%] Creating directories for 'qt6-populate'
[ 22%] Performing download step (git clone) for
  'qt6-populate'
Cloning into 'qt6-src'...
/*...*/
[100%] Completed 'qt6-populate'
[100%] Built target qt6-populate
/*...*/
-- Configuring done
-- Generating done
/*...*/
[  0%] Generating ../../mkspecs/modules
  /qt_lib_widgets_private.pri
[  0%] Generating ../../mkspecs/modules
```

```
   /qt_lib_gui_private.pri
[  0%] Generating ../../mkspecs/modules/qt_lib_core_private.pri
/* ... */
[ 98%] Linking CXX executable ch10_ex03_simple_qt_app
[ 98%] Built target ch10_ex03_simple_qt_app
/*...*/
```

If all goes well, you have succeeded in compiling the example. Let's check whether it works by running the produced executable with the following command:

```
./build/ch10_ex03_simple_qt_app
```

If everything is alright, a small GUI window should pop up. The window should look similar to the following figure:

Figure 10.1 – The simple Qt application window

With this sorted out, we have concluded the part that learns about how to use super-build in a CMake project. Next, we will be taking a look at ensuring version consistency in a super-build.

Ensuring version consistency in a super-build

Version A consistency is an important aspect of all software projects. As you should have learned by now, nothing is set in stone in the software world. Software evolves and changes over time. Such changes often need to be acknowledged in advance, either by running a series of tests against the new version or by making changes to the consuming code itself. Ideally, changes in upstream code should not have an effect on reproducing existing builds, until we want them to do so. A project's `x.y` version should always be built with the `z.q` dependency version if the software verification and testing have been done against this combination. The reason for this is that the smallest changes in an upstream dependency may affect the behavior of your software, even though there are no API or ABI changes. Your software will not have well-defined behavior if version consistency is not provided. Thus, having a way to provide version consistency is very crucial.

Ensuring version consistency in a super-build depends on the way that the super-build is organized. For repositories fetched over a version control system, it is relatively easy. Instead of cloning a project and using it as is, check out to a specific branch or tag. If there are no such anchor points, check out to a specific commit instead. This will future-proof your super-build. But even this may not be enough. Tags may get overridden, branches may be force-pushed, and history may be rewritten. To mitigate this risk, you may prefer to fork the project and use that as upstream instead. This way, you will have total control over the content that upstream has. But bear in mind that this method comes with a maintenance burden.

The moral of the story is, do not track an upstream blindly. Always keep an eye on recent changes. For third-party dependencies that are consumed as archive files, always check for their `hash` digest. This way, you will ensure that you are indeed using the intended revision for your project, and if there is a change, you will have to manually acknowledge it.

Summary

In this chapter, we have briefly introduced the concept of a super-build and how we can utilize super-builds for dependency management. Super-builds are a non-intrusive and powerful way of managing dependencies where there is a lack of package managers.

In the next chapter, we will be learning about the concept of fuzzing in software and how to automate fuzzing tasks with CMake.

See you soon in the next chapter!

Questions

After completing this chapter, you should be able to answer the following questions:

1. What is a super-build?
2. In which primary scenario can we use super-builds?
3. What can be done to achieve version consistency in a super-build?

11
Automated Fuzzing with CMake

Software testing is crucial for ensuring that a software system behaves as expected. Unsurprisingly, the need for scenario-based testing strategies is still at its peak. But the scenario-based approach often does not suffice. The real world is wild. Things out there are not sterile as in a test environment. Many different variables contribute to a software's behavior. Environments, users, hardware, operating systems, or third-party libraries can affect software unexpectedly. As software becomes more extensive, it becomes painfully hard to come up with a testing suite that covers all possible aspects of it. Therefore, the need for a technique that *performs the unexpected* arises. Thankfully, fuzzing comes to the rescue.

In this chapter, we will learn the basics of fuzzing and learn how to use this technique to test our software to make it more reliable. Furthermore, we will learn how we can integrate fuzzing tools into CMake to easily define fuzzing targets.

To understand the skills shared in this chapter, we'll cover the following main topics:

- A brief look at fuzzing in CMake projects
- Integrating `AFL/libfuzzer` into your CMake project

Let's begin with the technical requirements.

Technical requirements

Before you dive further into this chapter, it is recommended to take a look at *Chapter 7, Seamless Integration of Code-Quality Tools with CMake*. This chapter will follow the teaching by example approach. Thus, it is recommended to obtain this chapter's example content from `https://github.com/PacktPublishing/CMake-Best-Practices/tree/main/chapter_11`. All of the examples assume that you will be using the container provided by the `https://github.com/PacktPublishing/CMake-Best-Practices` project.

Note that libFuzzer is a part of the LLVM toolchain and requires LLVM to be installed and used as a toolchain. The bundled development environment has **llvm-13** pre-installed. AFL++ is a third-party tool that can be obtained from `https://github.com/AFLplusplus/AFLplusplus` to follow examples regarding `AFL++`. The installation guide is available at `https://github.com/AFLplusplus/AFLplusplus/blob/stable/docs/INSTALL.md`. Note that some distros have a pre-built version of AFL++ on their repositories. You may find it convenient to perform the installation via a package manager if that is the case.

Let's start by learning some basics about fuzzing.

A quick glance into fuzzing in CMake projects

Before further ado, let's learn a bit about fuzzing itself. Fuzzing, or fuzz testing, is a testing method that feeds random, unexpected data to a software system to see how a system behaves with certain input. The fuzzer reports the unexpected behaviors that it encounters. This allows us to discover critical bugs that are otherwise missed by other testing strategies and code reviews. Finding whether input causes a security issue or failure has proven to be hard. Surprisingly, fuzzing is pretty effective against this. It is known that the vast majority of critical security bugs such as remote code execution or privilege escalation can be discovered with ease when fuzzing is employed correctly. Therefore, it is important to understand the fuzzing technique to harness the power that comes with it.

Fuzzing can be done either manually by hand or automatically with the help of software. The second approach is more favorable, since it allows us to harness computing power for fuzzing. A corpus-based, coverage-guided fuzzing tool is a must-have in many software projects. Luckily for us, quite decent tooling is available for fuzzing C and C++ projects. Fuzzing can be basically divided into two sub-categories – *guided* and *black-box* fuzzing. Black-box fuzzing is the brute-force way that only relies on **System-Under-Test** (**SUT**)'s reactions to the fuzzing input, whereas guided fuzzing is guided automatically by either coverage analysis or the user itself. Guided fuzzing can be considered as white-box testing or gray-box testing, since guiding relies on feedback from the implementation. Guided fuzzing is a good way to discover the unknown edge-cases of software.

The generated input should not be random to make the fuzzing process effective. The input should be valid enough to pass from the fuzzed target's initial basic checks so that the fuzzer can delve into the delicate parts of the system. Therefore, a user may need to *kick-start* a fuzzer by providing a set of inputs that provides the most coverage to the SUT. That set of inputs is known as the corpus. The fuzzer mutates the initial corpus data to generate input variations similar to the original input but triggers different behavior. The input data that triggers an unexpected behavior or covers previously uncovered paths can be saved into the corpus by the fuzzing tool to extend the corpus data for later usage. A fuzzer may also not require any initial corpus data, meaning fuzzing will generate its input data from scratch. We will consider the corpus-based, guided approaches for this chapter.

Before diving any further into the topic, let's state the obvious thing first – *fuzzing is not a replacement for other testing strategies or types, such as unit testing or system testing.* Fuzzing aims to heuristically discover bugs and unwanted behaviors, so you should use fuzzing to only enhance your existing testing strategy. Note that fuzzing requires your system or unit to be testable. Hence, if you do not have any testing in place yet, it is recommended to invest and learn about the traditional testing strategies first.

Okay, we have learned quite a bit about what fuzzing is. Up next, we will learn how we can use the two prominent fuzzing software, AFL++ and libFuzzer, in C and C++ projects. Let's dive right in.

Integrating AFL++/libFuzzer into your project

In this section, we will take a look into the two prominent fuzzing tools you can encounter in the wild, namely libFuzzer and AFL++. Let's start by learning about libFuzzer.

Using libFuzzer in your CMake project

libFuzzer is a fuzzing library that is part of the LLVM project. It is a compiler-aided fuzzer that has powerful fuzzing techniques, It is the default go-to fuzzer if your project is already compilable with the LLVM toolchain, since using libFuzzer only requires an additional compiler/linker flag and defining a single function. We will start learning more about these details by digging into an example. Let's get our hands dirty.

To showcase fuzzing in practice, we will follow examples as usual. Let's begin with the example in `chapter_11/ex01_libfuzzer_static_lib` first. In this example, we have a hypothetical vulnerable static library target, named `message_printer`. This is a simple library that has only one interface function, named `print()`, which takes `const char*` and `length` as parameters. We are going to assume this library has a vulnerability that is triggered by an edge-case condition that we haven't been aware of. The implementation of the function looks like this:

```
void message_printer::print(const char *msg, std::uint32_t
  len) {
    constexpr char a [] = "Hello from CMake Best
      Practices!";
    if(len < sizeof(a)){
        return;
    }
    if(std::memcmp(a, msg, len) == 0){
        volatile char* ptr{nullptr};
        // attempt to write an invalid memory location
        // it is undefined behavior
        *ptr = 'a';
    }
    /* ... */
}
```

The `print()` function will cause an undefined behavior via dereferencing a `null` pointer when the `msg` input exactly contains the string in the `a` variable, which represents the undiscovered vulnerability in our function. The vulnerability in our example is pretty easy to spot, but real-world vulnerabilities and undefined behaviors will be way more subtle than this.

Okay, we have our target to fuzz in our hands. Up next, we will need a small driver application that will perform the fuzzing by feeding the fuzz data generated by `libFuzzer` to the `message_printer::print()` function. In order to do that, we must link our driver application against `message_printer` library and, of course, `libFuzzer`. Let's take a look into `CMakeLists.txt` for our example driver application first:

```
add_executable(ch11_ex01_libfuzzer_fuzz)
target_sources(ch11_ex01_libfuzzer_fuzz PRIVATE
  fuzz_library.cpp)
```

```
target_compile_features(ch11_ex01_libfuzzer_fuzz PRIVATE
  cxx_std_11)
target_link_libraries(ch11_ex01_libfuzzer_fuzz PRIVATE
  ch11_ex01_libfuzzer_static)
target_compile_options(ch11_ex01_libfuzzer_fuzz PRIVATE -
  fsanitize=fuzzer)
target_link_libraries(ch11_ex01_libfuzzer_fuzz PRIVATE -
  fsanitize=fuzzer)
```

We have defined an executable target named `ch11_ex01_libfuzzer_fuzz`, which links against our fuzzing target, `ch11_ex01_libfuzzer_static`. Different from usual, there is an additional `-fsanitize=fuzzer` argument passed to both the compiler and the linker. Passing this argument to the compiler and the linker enables `libFuzzer` for the LLVM toolchain. Unfortunately, libFuzzer is LLVM-specific and does not work with GCC at the moment. That's it for the CMake side of the fuzzing driver. Let's continue with the code next.

The fuzzing driver implementation is pretty simple. To get the generated fuzzing data from libFuzzer, we need to implement a special function that libFuzzer can recognize. This function is known as the fuzzing entry point. This function has a specific signature and looks like this:

```
extern "C" int LLVMFuzzerTestOneInput(const std::uint8_t
  *data, std::size_t size) {
    /* fuzzing entry-point impl. */
}
```

The libFuzzer generates fuzzing data as a `byte` span and calls `LLVMFuzzerTestOneInput` with the generated byte span. The byte span ranges from `&data[0]` to `&data[size-1]`. The span should be fed into the fuzzing target in the body of the function. This function is implemented in our driver application, `chapter_11/ex01_libfuzzer_static_lib/fuzz/fuzz_library.cpp`, as follows:

```
#include <cstdint>
#include <library/library.hpp>

extern "C" int LLVMFuzzerTestOneInput(const std::uint8_t
  *data,std::size_t size) {
  chapter11::ex01::message_printer printer;
```

```
  printer.print(reinterpret_cast<const char *>(data),
    size);
  return 0;
}
```

As we can see, the fuzzing function implementation is straightforward. We have constructed an instance of `message_printer` and called the `print()` function with the `data` and `size` generated by libFuzzer. The rest is automatically handled by libFuzzer. You may have noticed that the example application does not have a `main()` function, and that is intentional. The libFuzzer library provides a `main()` function that bootstraps the fuzzing framework and eventually calls the fuzzing entry point.

Let's see our driver application in action now. As always, let's configure and build our example with the following commands:

```
cd chapter_11/
cmake --preset="ninja-debug-clang-13" -S ./ -B build/
cmake --build build/
```

As you can see, we have explicitly specified the `ninja-debug-clang-13` preset, because we need the `clang` compiler for our libFuzzer example. After building our example, let's run our fuzzing driver application by executing the following command:

```
cd build/ex01_libfuzzer_static_lib/fuzz/
./ch11_ex01_libfuzzer_fuzz
After running for a second, the driver application should
output something like this:

./build/ex01_libfuzzer_static_lib/fuzz/ch11_ex01_libfuzzer_
fuzz
INFO: Running with entropic power schedule (0xFF, 100).
INFO: Seed: 1815993229
INFO: Loaded 1 modules   (1 inline 8-bit counters): 1
[0x485fa0, 0x485fa1),
INFO: Loaded 1 PC tables (1 PCs): 1 [0x4723d8,0x4723e8),
INFO: -max_len is not provided; libFuzzer will not generate
inputs larger than 4096 bytes
```

```
INFO: A corpus is not provided, starting from an empty
corpus
#2      INITED cov: 1 ft: 1 corp: 1/1b exec/s: 0 rss: 26Mb
UndefinedBehaviorSanitizer:DEADLYSIGNAL
==228856==ERROR: UndefinedBehaviorSanitizer: SEGV on
unknown address 0x000000000000 (pc 0x000000464194 bp
0x7ffdad9733f0 sp 0x7ffdad9733a0 T228856)
==228856==The signal is caused by a WRITE memory access.
==228856==Hint: address points to the zero page.
    #0 0x464194 in chapter11::ex01::message_printer::
    print(char const*, unsigned int) /home/mustafa
    /workspace/personal/CMake-Best-Practices/chapter_11
    /ex01_libfuzzer_static_lib/src/library.cpp:26:14
    #1 0x4640ea in LLVMFuzzerTestOneInput /home/mustafa
    /workspace/personal/CMake-Best-Practices/chapter_11
    /ex01_libfuzzer_static_lib/fuzz/fuzz_library.cpp:7:11
/* ... */
==228856==ABORTING
MS: 4 CopyPart-CMP-CrossOver-CMP- DE: "Hello from CMake
Best Practices!\x004\x97\xad\xfd\x7f\x00\x00\x00dH\x00"-
"Hello from CMake Best Practices!\x004\x97\xad\xfd\
x7f\x00\x00\x00dH\x00.\x00"-; base unit: adc83b19e793491b
1c6ea0fd8b46cd9f32e592fc
0x48,0x65,0x6c,0x6c,0x6f,0x20,0x66,0x72,0x6f,0x6d,0x20,0x43
,0x4d,0x61,0x6b,0x65,0x20,0x42,0x65,0x73,0x74,0x20,0x50,0x7
2,0x61,0x63,0x74,0x69,0x63,0x65,0x73,0x21,0x0,0x34,0x97,0xa
d,0xfd,0x7f,0x0,0x0,0x0,0x64,0x48,0x0,0x2e,0x0,
Hello from CMake Best Practices!\x004\x97\xad\xfd\x7f\x00
\x00\x00dH\x00.\x00
artifact_prefix='./'; Test unit written to ./crash-e95d3
e40c04fcaf86897df986d66f7345bb0d4b1
Base64: SGVsbG8gZnJvbSBDTWFrZSBCZXN0IFByYWN0a
WNlcyEANJet/X8AAABkSAAuAA==
```

As you can see, libFuzzer was able to reach the code path that triggers undefined behavior in an instant, even though we have not provided any prior corpus data. The reason for this is that LLVM helps libFuzzer to get more coverage on the target function. As you can imagine, finding the `Hello from CMake Best Practices!` string with randomly generated data is highly unlikely. It is analogous to cracking a 32-digit password, which means it is practically nearly impossible with today's hardware. With this approach, it is impossible for the fuzzer to reach the code path hidden behind the comparison. So, what LLVM with the `-fsanitize=fuzzer` flag does here is split the comparison into more manageable, multiple comparisons. For a 32-character-long string, LLVM emits {*N*} small comparisons. So, the resulting comparison is as if we have written code like this:

```
if (*msg[0] == a[0]) {
  if (*msg[1] == a[1]) {
    if (*msg[2] == a[2]) {
      if (*msg[3] == a[3]) {
        /*...*/
      }
    }
  }
}
```

Laying out the comparisons in this fashion allows libFuzzer to discover each character via code coverage data quickly. It is still similar to 32-digit password cracking, only this time the password validation system also tells whether you have correctly guessed each digit. This kind of feedback reduces the discovery time from infinity to an instant. Due to this behavior, I refer to libFuzzer as *the gray-box fuzzing library*. The technique is aware of the implementation; in fact, the technique consists of tailoring the emitted assembly code to make fuzzing more efficient. But I don't think this awareness makes the fuzzing biased. For this reason, I do not consider this approach a fully white-box strategy; also, it is clearly not a black-box strategy. Hence, the term *gray-box* suitably describes this approach.

Without a doubt, using a compiler-guided fuzzing toolkit has many benefits. Besides libFuzzer, the LLVM toolchain has many other good utilities up its sleeve, such as **Undefined Behavior Sanitizer** (**UBSan**), **Address Sanitizer** (**ASan**), and **Memory Sanitizer** (**MSan**). Consider investing in supporting LLVM compilation in your projects to benefit from all of these helpful utilities. It is a good idea to enable these tools for fuzzing targets to discover bugs that would otherwise go unnoticed.

With that said, we conclude the libFuzzer part of this chapter. For a real-life example, I recommend you to take a look at `https://google.github.io/clusterfuzz/setting-up-fuzzing/heartbleed-example/` to learn about how OpenSSL's famous Heartbleed vulnerability could have been discovered easily if libFuzzer was used. Also, consider reading LLVM's official libFuzzer documentation at `https://llvm.org/docs/LibFuzzer.html`. It contains useful information about libFuzzer's fuzzing concept and usage examples.

Up next, we will investigate another fuzzing tool, the infamous **American Fuzzy Lop** (**AFL**).

Using AFL++ in your CMake project

If you are stuck with GCC for whatever reason and you can't compile your code with LLVM, an alternative fuzzer to libFuzzer is AFL. Unfortunately, the original project is no longer maintained, but don't get downhearted. An even better, enhanced version of the project known as AFL++ is still maintained. In this section of the chapter, we will learn how to use AFL++ with the GCC toolchain.

We will follow an example similar to the previous libFuzzer example. But this time, we will utilize AFL++ and GCC instead of libFuzzer and LLVM. Before moving any further, ensure that a CMake configure preset using `afl-gcc` and `afl-g++` is available under the `CMakePresets.json` file. The example project already contains such a preset, as shown here:

```
"configurePresets": [
/*...*/
,{
   "name": "afl-gcc",
   "description": "AFL GCC compiler",
   "hidden": true,
   "cacheVariables": {
     "CMAKE_C_COMPILER": {
       "type": "STRING",
       "value": "/usr/bin/afl-gcc"
     },
     "CMAKE_CXX_COMPILER": {
       "type": "STRING",
       "value": "/usr/bin/afl-g++"
     }
   }
```

```
  },
  /*...*/
]
```

With that said, let's begin.

In this section, we will follow the `chapter11/ex02_afl_static_lib` example. In this example, we have a simple, URI decoder implementation that we want to fuzz with AFL++. The URI decoder is intentionally implemented without care, and as a result, it contains severe bugs. URI decoder contains a single method named `decode()`, which is implemented as follows:

```
char *uri_helper::decode(const char *str) {
  thread_local char result[100];
  for (auto *p = result; *str; ++str) {
    if (*str == '%') {
      const auto a = *++str;
      const auto b = *++str;
      *p++ = (a <= '9' ? a - '0' : a - 'a') * 16 + (b <=
        '9' ? b - '0' : b - 'a');
    } else if (*str == '+') {
      // replace + with space
      *p++ = ' ';
    } else {
      // copy as-is
      *p++ = *str;
    }
  }
  return result;
}
```

The intended effects of the function are to replace each two-digit URI encoded character with the corresponding ASCII code, replace the plus (+) sign with a space, and leave all other characters as-is. There are also unintended effects in this implementation, however. The first one is the function not checking whether the result variable has enough space to store the decoded character. This can cause a buffer overrun when the resulting output is greater than the result buffer's size. The second one is the function using the `NUL-terminator` as a `loop` condition but incrementing and dereferencing the loop variable, `str`, without any checks. This results in potential out-of-bounds writes and reads.

I am sure that the number of bugs in this function is proportional to the time spent reviewing it. So, let's begin fuzzing.

In order to fuzz our buggy implementation with AFL++, we need to implement a simple driver application, as we did for the previous libFuzzer example. The driver application consumes our `uri_helper` library and feeds the input from standard input to the `decode` function. The implementation of the driver application is as follows:

```
#include <library/uri_helper.hpp>
#include <iostream>

int main() {
 const auto input = std::string(
std::istreambuf_iterator<char>(std::cin),
std::istreambuf_iterator<char>()
);
chapter11::ex02::uri_helper uut{};
std::cout << uut.decode(input.c_str());
}
```

The preceding driver application implementation is enough to use it with AFL++. With that sorted out, there is one last thing that needs our attention, which is the initial corpus data for AFL++. Unlike libFuzzer, AFL++ needs some example inputs to get started. The example input should look like the typical input that the unit-under-test expects. Since our target is a URI decoder, a few URI-encoded strings should do the trick. For that purpose, we have provided a few URI encoded strings under the `chapter_11/ex02_afl_static_lib/corpus` folder. So, with everything set, let's compile the example by running the following commands:

```
cd chapter_11/
cmake --preset="ninja-debug-afl-gcc" -S ./ -B build/
cmake --build build/
```

Note that we are using `ninja-debug-afl-gcc` as a CMake configure preset. This preset uses `afl-gcc` and `afl-g++` as compilers. This allows AFL++ to perform the required instrumentation over the compiled code, similar to libFuzzer's `-fsanitize=fuzzer` flag. After the build completes, run the `afl-fuzz` command against the fuzzing driver, `ch11_ex02_afl_fuzz`, with the following command:

```
afl-fuzz -i ex02_afl_static_lib/corpus/ -o build/FINDINGS -
m none -- build/ex02_afl_static_lib/fuzz/ch11_ex02_afl_fuzz
```

> **Important note**
>
> If you see an error regarding core dump notifications, that is due to your system using an external utility for handling the core dump notifications. In order to resolve this issue, either apply the recommended action by `afl-fuzz` or prepend `AFL_I_DONT_CARE_ABOUT_MISSING_CRASHES=1` to the `afl-fuzz` command before executing. Note that the second option is only recommended for demonstration purposes and not recommended for daily use.

The command may look scary at first glance, but it is pretty simple. Let's dissect the command as parameters one by one. The `-i ex02_afl_static_lib/corpus/` parameter is used for specifying the corpus data, the example encoded URIs, as we have talked about before. The `-o build/FINDINGS` parameter is the path for saving interesting findings that AFL++ will make. All the input that causes weird behavior such as crashes and stalls will be saved here. Lastly, the `-m none` parameter is for specifying the memory limit for the run, where `none` stands for no limit. After running the command, you should be seeing the AFL++'s fuzzing progress report screen, as shown here:

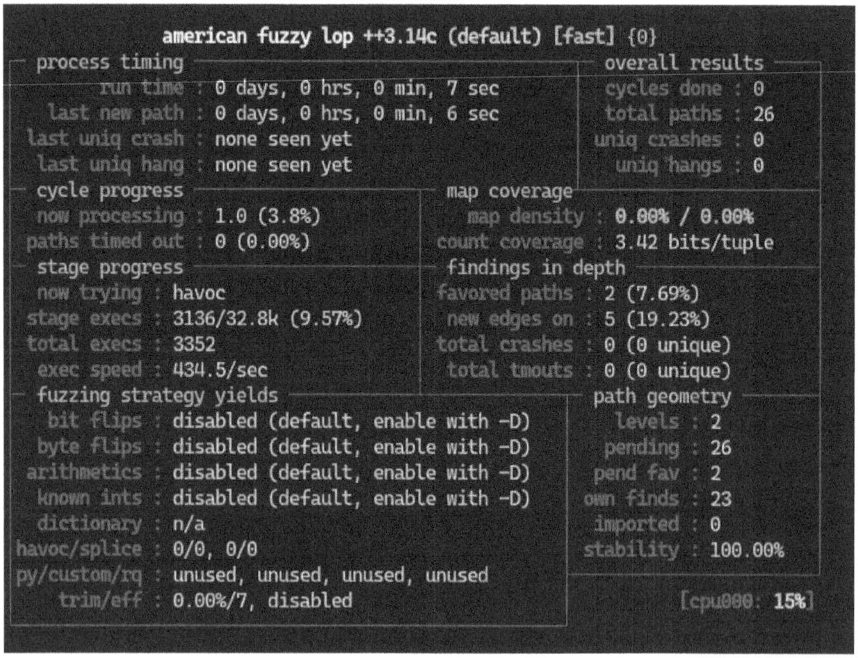

```
           american fuzzy lop ++3.14c (default) [fast] {0}
- process timing ------------------------------------ overall results ---
        run time : 0 days, 0 hrs, 0 min, 7 sec      cycles done : 0
   last new path : 0 days, 0 hrs, 0 min, 6 sec      total paths : 26
 last uniq crash : none seen yet                   uniq crashes : 0
  last uniq hang : none seen yet                     uniq hangs : 0
- cycle progress ------------------ map coverage -------------------------
  now processing : 1.0 (3.8%)          map density : 0.00% / 0.00%
 paths timed out : 0 (0.00%)        count coverage : 3.42 bits/tuple
- stage progress ------------------ findings in depth --------------------
  now trying : havoc                favored paths : 2 (7.69%)
 stage execs : 3136/32.8k (9.57%)    new edges on : 5 (19.23%)
 total execs : 3352                 total crashes : 0 (0 unique)
  exec speed : 434.5/sec             total tmouts : 0 (0 unique)
- fuzzing strategy yields ----------------------- path geometry ----------
   bit flips : disabled (default, enable with -D)     levels : 2
  byte flips : disabled (default, enable with -D)    pending : 26
 arithmetics : disabled (default, enable with -D)   pend fav : 2
  known ints : disabled (default, enable with -D)  own finds : 23
  dictionary : n/a                                  imported : 0
havoc/splice : 0/0, 0/0                            stability : 100.00%
py/custom/rq : unused, unused, unused, unused
    trim/eff : 0.00%/7, disabled                          [cpu000: 15%]
```

Figure 11.1 – The AFL++ fuzzing progress screen

There are interesting statistics presented on the screen. The most prominent ones are `uniq crashes` and `uniq hangs`, which are the statistics for the number of unique crashes and hangs on this fuzzing run. If these numbers are non-zero, this means the fuzzer may have found some bugs on the unit-under-test. The fuzzer will continuously generate new input, and statistics will be updated accordingly. You may need to wait from several minutes to weeks before anything interesting happens. After letting the tool work for a few minutes, it will begin to discover a few unique crashes, as shown in the following screenshot:

```
            american fuzzy lop ++3.14c (default) [fast] {0}
- process timing ------------------------------------ overall results ----
       run time : 0 days, 0 hrs, 5 min, 34 sec      cycles done : 5
  last new path : 0 days, 0 hrs, 3 min, 43 sec      total paths : 30
last uniq crash : 0 days, 0 hrs, 2 min, 45 sec     uniq crashes : 4
 last uniq hang : none seen yet                       uniq hangs : 0
- cycle progress --------------------- map coverage ----------------------
 now processing : 21.98 (70.0%)          map density : 0.00% / 0.00%
paths timed out : 0 (0.00%)           count coverage : 3.85 bits/tuple
- stage progress --------------------- findings in depth -----------------
 now trying : splice 9                favored paths : 5 (16.67%)
stage execs : 9/36 (25.00%)            new edges on : 5 (16.67%)
total execs : 156k                    total crashes : 1157 (4 unique)
 exec speed : 471.5/sec                total tmouts : 0 (0 unique)
- fuzzing strategy yields ----------------------------- path geometry ----
  bit flips : disabled (default, enable with -D)          levels : 3
 byte flips : disabled (default, enable with -D)         pending : 16
arithmetics : disabled (default, enable with -D)        pend fav : 0
 known ints : disabled (default, enable with -D)       own finds : 27
 dictionary : n/a                                       imported : 0
havoc/splice : 25/80.9k, 6/75.3k                       stability : 100.00%
py/custom/rq : unused, unused, unused, unused
    trim/eff : 3.78%/107, disabled                       [cpu000:  6%]
```

Figure 11.2 – The AFL++ fuzzing progress screen after discovering some crashes

Let's cancel the run by pressing *Ctrl* + *C* on the keyboard and take a look at the input that caused the reported crashes. The crash-causing input should be saved under the `chapter_11/build/FINDINGS/default/crashes` folder. For the illustrated run, there are four crash logs in the `crashes/` folder, as can be seen here:

```
18:03 $ ls
README.txt
id:000000,sig:11,src:000021+000026,time:90095,op:splice,rep:4
id:000001,sig:11,src:000021+000026,time:92289,op:splice,rep:4
id:000002,sig:11,src:000021+000026,time:92325,op:splice,rep:16
id:000003,sig:11,src:000001+000028,time:168241,op:splice,rep:2
```

Figure 11.3 – The crash-causing input discovered by AFL++

Let's take a look at the contents of the first file (starting with `id:000000`):

```
x=%���%B9D0JJJJJJJJJJJJJJJJJJJJJ������������������
��������D1%88%D0JJJJJJJJJ%I5�������������������������
�������%D0%B9D0JJJJJJJJJJJJJJJJJJJJJJJJJJIJJJJJ%BB%D1%XB
```

AFL++ has reported that our test driver application, `ch11_ex02_afl_fuzz`, has crashed with this input. Let's check that is indeed the case by running the driver application manually with the preceding string:

```
18:09 $ ./build/ex02_afl_static_lib/fuzz/ch11_ex02_afl_
fuzz  <<< x=%���%B9^FD0JJJJJJJJJJJJJJJJJJJJJJ������������
������������D1%88%D0JJJJJJJJJ%I5������������
��������������������%D0%B9^FD0JJJJJJJJJJJJJJJJJ
JJJJJIJJJJJJ%BB%D1%XB
Segmentation fault
```

The test driver crashes due to a segmentation fault, which means the program is accessing a memory location that is not allowed for the program to access, which is likely caused by a buffer overrun in the `result` array in the `decode()` method. It seems AFL++ has succeeded in discovering a bug in our URL decoder. From this point on, it is a pretty straightforward bug-fixing process. Run the program under the debugger, give the input, and diagnose.

Summary

In this chapter, we have learned about what fuzzing is and how we can utilize it in our CMake projects to detect bugs earlier and enhance the quality of software. Fuzzing allows us to harness computing power for discovering new and interesting test cases, which increases overall coverage and eventually leads to more stable software systems.

In the next chapter, we will be learning about cross-platform and cross-toolchain building with CMake.

See you soon in the next chapter!

Questions

After completing this chapter, you should be able to answer the following questions:

1. What is fuzzing?
2. What is corpus data in fuzzing?
3. How can you enable libFuzzer for a target in CMake?
4. Name one popular software bug that could have been found if fuzzing was employed.

Part 3: Mastering the Details

In this part, you will learn how to use CMake on cross-platform projects, reuse CMake code, optimize and maintain existing projects, and migrate non-CMake projects to CMake. You will also learn about ways of contributing to a CMake project and the recommended further reading materials.

This part contains the following chapters:

- *Chapter 12, Cross-Platform and Cross-Toolchain Building*
- *Chapter 13, Reusing CMake Code*
- *Chapter 14, Optimizing and Maintaining CMake Projects*
- *Chapter 15, Migrating to CMake*
- *Chapter 16, Contributing to CMake and Further Reading Material*

12
Cross-Platform Compiling and Custom Toolchains

One of the powerful features of CMake is its support for the cross-platform building of software. Simply said, this means that with CMake, a project from any platform can be built for any other platform, as long as the necessary tools are available on the system running CMake. When building software, we typically talk about compilers and linkers, and they are of course essential tools for building software, but on a closer look, there are often some other tools, libraries, and files involved when building software. Collectively, these are commonly known as toolchains in CMake.

So far in this book, all the examples were built for the same system CMake was running on. In these cases, CMake usually does a pretty good job of finding the correct toolchain to use. However, if the software is to be built for another platform, the toolchain usually has to be specified by the developer. Toolchain definitions might be relatively straightforward and just specify the target platform, or they might be as complex as specifying paths to individual tools needed to build the software or specific compiler flags to create binaries for a specific chipset.

In the context of cross-compiling, toolchains are often accompanied by **system roots** (**sysroots**), which were introduced in *Chapter 9*, *Creating Reproducible Build Environments*. Sysroots are directories that contain a stripped-down filesystem for the target platform to build. When building software, they are considered the root folder for finding the necessary libraries and files to compile and link the software for the intended target platform.

While cross-compiling might be intimidating at first, it is often not as hard as it seems when using CMake properly. In this chapter, we will look at how to use toolchain files and how to write them yourself. We will look in detail into which tools are involved at particular stages of building software. Finally, we will look at how to set up CMake so that it can run tests with an emulator.

We'll cover the following main topics in this chapter:

- Using existing cross-platform toolchain files
- Creating toolchain files
- Testing cross-compiled binaries
- Testing a toolchain for supported features

By the end of the chapter, you will be proficient in handling existing toolchains and in how to build and test software for different platforms using CMake. We will have a deeper look into how to test a compiler for a certain feature to determine whether it suits our purposes.

Technical requirements

As with the previous chapters, the examples are tested with CMake 3.21 and are run on either of the following compilers:

- The **GNU Compiler Collection 9** (**GCC 9**) or newer, including the cross-compiler for the **arm hard float** (**armhf**) architecture.
- Clang 12 or newer.
- **Microsoft Visual Studio C++ 19** (**MSVC 19**) or newer.
- For the Android examples, the **Android Native Development Kit** (**Android NDK**) 23b or newer is required.
- For the Apple-embedded examples, Xcode 12 or newer and the **iOS Software Development Kit 12.4** (**iOS SDK 12.4**) are recommended.

All examples and source code are available in the GitHub repository for this book. If any of the software is missing, the corresponding examples will be excluded from the build. The repository can be found here: `https://github.com/PacktPublishing/CMake-Best-Practices`.

Using existing cross-platform toolchain files

When building software for multiple platforms, the most straightforward way to do this is to compile software on the target system itself. The downside of that is that each developer has to have a running version of the target system to build the software. If these are desktop systems, that might work reasonably well, although moving between different installations for developing the software also makes the developer workflow quite tedious. Less powerful devices such as embedded systems might be very uncomfortable because of the lack of proper development tools or because compiling the software takes very long.

So, a much more convenient way from the developer's perspective is to use cross-compiling. This means the software engineer writes code and builds the software on their own machine, but the resulting binaries are for a different platform. The machine and platform on which the software is built are usually called the *host machine* and *host platform*, whereas the platform on which the software should run is called the *target platform*. For instance, the developer writes code on their *x64* desktop machine running Linux, but the resulting binaries are for embedded Linux on an *arm64* processor. So the host platform is *x64 Linux* and the target platform is *arm64 Linux*. To cross-compile software, the following two things are needed:

- A toolchain capable of producing binaries in the correct format
- Any dependencies of the project compiled for the target system

The toolchain is a set of tools such as a compiler, linker, and archiver to produce binaries that run on the host system but produce output for the target system. The dependencies are usually collected in a *sysroot* directory. Sysroots are directories containing a reduced version of a root filesystem where the needed libraries are stored. For cross-compilation, these directories serve as the root for searching any dependencies.

Some tools, such as the **Yocto Project** (**YP**) for building embedded Linux distributions, can build sysroots and toolchains for cross-compiling out of the box. This automation is often convenient, but when this is not available, creating a sysroot manually can be as simple as copying or mounting the filesystem from the target platform into a directory on the host machine. Sometimes, the combination of a toolchain and a sysroot is also referenced as SDKs. These SDKs might also contain further tools for debugging or definitions for running an emulator for the cross-platform build. CMake uses so-called toolchain files to configure the build tools and sysroots for cross-compiling. Toolchain files are ordinary CMake scripts that mostly set some cache variables to describe the target platform and the location of various components of the toolchain. Toolchain files can be passed to CMake by setting the `CMAKE_TOOLCHAIN_FILE` variable or, since CMake 3.21, with the `--toolchain` option, like this:

```
cmake -DCMAKE_TOOLCHAIN_FILE=toolchain.cmake -S <SourceDir> -B
  <BuildDir>
cmake  --toolchain toolchain.cmake -S <SourceDir> -B <BuildDir>
```

These calls are equivalent. If `CMAKE_TOOLCHAIN_FILE` is set as an environment variable, CMake will interpret this as well. If using CMake presets, the configure presets may configure a toolchain file with the `toolchainFile` option, like this:

```
{
    "name": "arm64-build-debug",
    "generator" : "ninja",
    "displayName": "Arm 64 Debug",
    "toolchainFile": "/path/to/toolchain.cmake",
    "cacheVariables": {
        "CMAKE_BUILD_TYPE": "Debug"
    }
},
```

The `toolchainFile` option supports macro expansion, as described in *Chapter 9, Creating Reproducible Build Environments*. If the path to the toolchain file is a relative path, CMake will first look relative to the build directory, and if the file is not found there, it will then search from the source directory. As `CMAKE_TOOLCHAIN_FILE` is a cache variable, it only needs to be specified for the first run of CMake; subsequent runs will use the cached value.

On the first run, CMake will perform some internal queries to establish which features a toolchain supports. This happens regardless of whether the toolchain is specified with a toolchain file or whether the default system toolchain is used. A more in-depth look at how these tests happen is done in the *Testing a toolchain for supported features* section. CMake will output the results of the tests for the various features and properties on the first run, like this:

```
-- The CXX compiler identification is GNU 9.3.0
-- The C compiler identification is GNU 9.3.0
-- Detecting CXX compiler ABI info
-- Detecting CXX compiler ABI info - done
-- Check for working CXX compiler: /usr/bin/arm-linux-
   gnueabihf-g++- 9 - skipped
-- Detecting CXX compile features
-- Detecting CXX compile features - done
-- Detecting C compiler ABI info
-- Detecting C compiler ABI info - done
-- Check for working C compiler: /usr/bin/arm-linux-gnueabi-
   gcc-9 - skipped
-- Detecting C compile features
-- Detecting C compile features - done
```

The detection of the features mostly happens on the first call to `project()` in a `CMakeLists.txt` file. However, for any later call to `project()` that enables a previously disabled language, further detection will be triggered. The same happens if `enable_language()` is used to enable an additional programming language in a `CMakeLists.txt` file.

As the features and test results of the toolchains are cached, changing the toolchain of a configured build directory cannot be done. CMake might detect that the toolchain has changed, but usually, the replacement of the cache variables is incomplete. Because of this, the build directory should be fully deleted before changing the toolchain.

Switching Toolchains After Configuring

Always clear the build directory completely before switching toolchains. Deleting only the `CMakeCache.txt` file will not be enough, as toolchain-related things might be cached in various locations.

CMake works under the paradigm that a project should use the same toolchain for everything. Because of this, using multiple toolchains is not directly supported. If this is really needed, the parts of a project that need a different toolchain have to be configured as sub-builds, as described in *Chapter 10, Handling Big Projects and Distributed Repositories in a Superbuild.*

Toolchains should be kept as small as possible and be completely decoupled from any project. Ideally, they are reusable for different projects. Often, toolchain files come bundled together with any SDK or sysroot used for cross-compiling. However, sometimes they need to be written manually.

Creating toolchain files

Toolchain files might seem intimidating at first, but on a closer look, they are often relatively trivial. The misconception that defining cross-compilation toolchains is hard stems from the fact that there are many overly complicated examples of toolchain files found on the internet. Many of them were written for early versions of CMake and thus implemented many additional tests and checks that are now part of CMake itself. CMake toolchain files basically do the following things:

- Define the target system and architecture.
- Provide paths to any tools needed to build the software for the defined platform. Often, these are just compilers.
- Set default flags for the compiler and linkers.
- Point to the sysroot and possibly any staging directory if cross-compiling.
- Set hints for the search order for any `find_` commands of CMake. Changing the search order is something the project might define, and it is debatable whether this belongs in the toolchain file or should be handled by the project. See *Chapter 5, Integrating Third-Party Libraries and Dependency Management*, for details on `find_` commands.

A sample toolchain doing all these things might look like this:

```
set(CMAKE_SYSTEM_NAME Linux)
set(CMAKE_SYSTEM_PROCESSOR arm)

set(CMAKE_C_COMPILER /usr/bin/arm-linux-gnueabi-gcc-9)
set(CMAKE_CXX_COMPILER /usr/bin/arm-linux-gnueabihf-g++-9)
```

```
set(CMAKE_C_FLAGS_INIT -pedantic)
set(CMAKE_CXX_FLAGS_INIT -pedantic)

set(CMAKE_SYSROOT /home/builder/raspi-sysroot/)
set(CMAKE_STAGING_PREFIX /home/builder/raspi-sysroot-staging/)

set(CMAKE_FIND_ROOT_PATH_MODE_PROGRAM NEVER)
set(CMAKE_FIND_ROOT_PATH_MODE_LIBRARY ONLY)
set(CMAKE_FIND_ROOT_PATH_MODE_INCLUDE ONLY)
set(CMAKE_FIND_ROOT_PATH_MODE_PACKAGE BOTH)
```

This example would define a toolchain targeting a build for a Linux operating system running on an **Advanced RISC Machine** (**ARM**) processor. The C and C++ compilers to use are a version of GCC and are installed in the `/usr/bin/` folders on the host system. Then, the compiler flags to print all warnings demanded by strict **International Standards Organization** (**ISO**) C and ISO C++ standardization with the `-pedantic` flag. Next, the sysroot for finding any needed libraries is set to `/home/builder/raspi-sysroot/`, and the staging directory for installing things when cross-compiling is set to `/home/builder/raspi-sysroot-staging/`. Finally, the search behavior for CMake is changed so that programs are searched only on the host system but libraries, include files, and packages are only searched in the sysroot. It is a controversial debate whether a toolchain file should influence the search behavior or not. Typically, only the project knows what it is trying to find, so making assumptions in the toolchain file might break that. However, only the toolchain knows which system root to use and what kinds of files are present inside it, so having the toolchain define this might be convenient. A good middle way is to use a CMake preset to define the toolchain and the search behavior instead of putting it either into the project or the toolchain file.

Defining the target system

The target system for cross-compiling is defined by the following three variables: `CMAKE_SYSTEM_NAME`, `CMAKE_SYSTEM_PROCESSOR`, and `CMAKE_SYSTEM_VERSION`. Corresponding to them are the `CMAKE_HOST_SYSTEM_NAME`, `CMAKE_HOST_SYSTEM_PROCESSOR`, and `CMAKE_HOST_SYSTEM_VERSION` variables, which describe the platform on which the build is performed.

The CMAKE_SYSTEM_NAME variable describes the target operating system for which the software is to be built. Setting this variable is important as this will cause CMake to set the CMAKE_CROSSCOMPILING variable to true. Typical values are Linux, Windows, Darwin, Android, or QNX, although you could use more specific platform names such as WindowsPhone, WindowsCE, WindowsStore, and more. For bare-metal embedded devices, the CMAKE_SYSTEM_NAME variable is set to Generic. Unfortunately, at the time of writing this book, there exists no official list of supported systems in the CMake documentation. However, if really needed, the files in the /Modules/Platform folder in the local CMake installation can be inspected.

The CMAKE_SYSTEM_PROCESSOR variable is used to describe the hardware architecture of the platform. If not specified, the value of the CMAKE_HOST_SYSTEM_PROCESSOR variable will be assumed. The target processor architecture should also be set when cross-compiling for a 32-bit platform from a 64-bit platform, even if the processors are of the same type. For Android and Apple platforms, the processor is often not specified. When cross-compiling for Apple targets, the actual device is defined by the SDK that is used, which is specified by the CMAKE_OSX_SYSROOT variable. When cross-compiling for Android, specialized variables such as CMAKE_ANDROID_ARCH_ABI, CMAKE_ANDROID_ARM_MODE, and—optionally—CMAKE_ANDROID_ARM_NEON are used to control the target architecture. Building for Android will be covered in more depth in the *Cross-compiling for Android* section.

The last variable for defining a target system is CMAKE_SYSTEM_VERSION. The content of it depends on the system being built. For WindowsCE, WindowsStore, and WindowsPhone, it will be used to define which version of the Windows SDK to use. On Linux, it is often omitted or might contain the kernel revision of the target system if this is relevant.

With the CMAKE_SYSTEM_NAME, CMAKE_SYSTEM_PROCESSOR, and CMAKE_SYSTEM_VERSION variables, target platforms are usually fully specified. However, some generators such as Visual Studio support their native platforms directly. For these, the architecture can be set with the -A command-line option of CMake, like this:

```
cmake -G "Visual Studio 2019" -A Win32 -T host=x64
```

When using a preset, the architecture setting may be used in a configure preset for the same effect. Once the target system is defined, the tools to actually build the software are defined.

Some compilers such as Clang and **QNX is Not Unix GCC** (**QNX GCC**) are inherently cross-compilers and take the platform as arguments. To pass the argument to those compilers, the `CMAKE_<LANG>_COMPILER_TARGET` variable is used. For Clang, the value is a target triple such as `arm-linux-gnueabihf`, and for QNX GCC, the compiler name and target have a value such as `gcc_ntoarmv7le`. The supported triples for Clang are described in its official documentation at `https://clang.llvm.org/docs/CrossCompilation.html`.

For available options for QNX, the QNX documentation found at `https://www.qnx.com/developers/docs/` should be consulted.

So, a toolchain file that uses Clang might look something like this:

```
set(CMAKE_SYSTEM_NAME Linux)
set(CMAKE_SYSTEM_PROCESSOR arm)

set(CMAKE_C_COMPILER /usr/bin/clang)
set(CMAKE_C_COMPILER_TARGET arm-linux-gnueabihf)

set(CMAKE_CXX_COMPILER /usr/bin/clang++)
set(CMAKE_CXX_COMPILER_TARGET arm-linux-gnueabihf)
```

In this example, Clang is used to compile C and C++ code for a Linux system running on an ARM processor with hardware floating-point support. Defining the target system often has a direct influence on the build tools that are to be used. In the next section, we will look into how the compiler and the related tools are selected for cross-compilation.

Selecting the build tools

When building software, the compiler is the tool that comes to mind, and in most cases, it is sufficient to set the compiler in a toolchain file. The path to the compiler is set by the `CMAKE_<LANG>_COMPILER` cache variable, which may be set in the toolchain file or passed manually to CMake. If the path is absolute it will be used directly; else, the same search order as when using `find_program()` will be used, which is one of the reasons why changing the search behavior in a toolchain file has to be treated with care. If neither the toolchain file nor the user does not specify the compiler, CMake will try to choose one automatically based on the target platform and generator specified. Additionally, the compiler can be set over an environment variable that is named after `<LANG>`. So, `C` will set the C compiler, `CXX` the C++ compiler, `ASM` the assembler, and so on.

Some generators such as Visual Studio may support their own toolset definition that works differently. They can be set with the `-T` command-line option. The following command would tell CMake to generate code for Visual Studio to produce binaries for a 32-bit system but to use 64-bit compilers to do so:

```
cmake -G "Visual Studio 2019" -A Win32 -T host=x64
```

The values might also be set with the `CMAKE_GENERATOR_TOOLSET` variable from a toolchain file. This should not be set inside a project as it obviously breaks the idea of CMake project files being agnostic to the generator and platform used.

For Visual Studio users, it is possible to have multiple competing instances of the same Visual Studio version installed by having the preview and the official release of the same version installed. If this is the case, the `CMAKE_GENERATOR_INSTANCE` variable may be set to the absolute installation path of Visual Studio in the toolchain file.

By specifying the compiler to use, CMake will choose the default flags for the compiler and linker and make them available in the project by setting `CMAKE_<LANG>_FLAGS` and `CMAKE_<LANG>_FLAGS_<CONFIG>`, where `<LANG>` stands for the respective programming language and `<CONFIG>` for the build configuration such as debug or release, and so on. The default linker flags are set by the `CMAKE_<TARGETTYPE>_LINKER_FLAGS` and `CMAKE_<TARGETTYPE>_LINKER_FLAGS_<CONFIG>` variables, where `<TARGETTYPE>` is either `EXE`, `STATIC`, `SHARED`, or `MODULE`.

To add custom flags to the default flags, there exists a variable with `_INIT` appended for each of the variables—for example, `CMAKE_<LANG>_FLAGS_INIT`. When working with toolchain files, `_INIT` variables are used to set any necessary flags. A toolchain file compiling with GCC for a 32-bit target from a 64-bit host would look like this:

```
set(CMAKE_SYSTEM_NAME Linux)
set(CMAKE_SYSTEM_PROCESSOR i686)

set(CMAKE_C_COMPILER  gcc)
set(CMAKE_CXX_COMPILER g++)

set(CMAKE_C_FLAGS_INIT -m32)
set(CMAKE_CXX_FLAGS_INIT -m32)
```

```
set(CMAKE_EXE_LINKER_FLAGS_INIT -m32)
set(CMAKE_SHARED_LINKER_FLAGS_INIT -m32)
set(CMAKE_STATIC_LINKER_FLAGS_INIT -m32)
set(CMAKE_MODULE_LINKER_FLAGS_INIT -m32)
```

For simple projects, setting up the target system and the toolchain might already be enough to start creating binaries, but for more complex projects, they might need access to libraries and headers of the target system. For such cases, sysroots can be specified in the toolchain files.

Setting the sysroot

When cross-compiling, all linked dependencies obviously have to match the target platform as well, and a common way of dealing with this is to create a sysroot, which is the root filesystem of the target system in a folder. While sysroots may contain full systems, they are often stripped down to just provide what is needed. Sysroots are described in detail in *Chapter 9, Creating Reproducible Build Environments.*

Setting the sysroot is done by setting `CMAKE_SYSROOT` to the path of the sysroot. If this is set, CMake will by default look in the sysroot first for libraries and header files unless specified differently, as described in *Chapter 5, Integrating Third-Party Libraries and Dependency Management*. In most cases, CMake will also automatically set the necessary compiler and linker flags to make the tools work with the sysroot.

In cases where the build artifacts should not be installed directly in the sysroot, the `CMAKE_STAGING_PREFIX` variable can be set to provide an alternative installation path. This is often the case when the sysroot should be kept clean or when it is mounted as read-only. Note that the `CMAKE_STAGING_PREFIX` setting will not add this directory to `CMAKE_SYSTEM PREFIX_PATH`, so things installed in the staging directory will only be found with `find_package()` if the `CMAKE_FIND_ROOT_PATH_MODE_PACKAGE` variable in the toolchain is set to `BOTH` or `NEVER`.

Defining the target system and setting the toolchain configuration, the sysroot, and staging directory are often all that are needed for cross-compiling. Two exceptions are cross-compiling for Android and Apple's iOS, tvOS, or watchOS.

Cross-compiling for Android

In the past, the compatibility between Android's NDK and various CMake versions was sometimes a bit of an unhappy relationship, as new versions of the NDK suddenly no longer worked the same way with CMake as previous releases did. However, this condition has now been vastly improved, as the r23 version of the Android NDK will now use CMake's internal support for handling the Android NDK if CMake is version 3.21 or higher, so using these versions is recommended. The official documentation of the Android NDK's integration with CMake can be found here: `https://developer.android.com/ndk/guides/cmake`.

As of r23, the NDK provides its own CMake toolchain file located at `<NDK_ROOT>/build/cmake/android.toolchain.cmake`, which may be used like any regular toolchain file. The NDK also includes all the necessary tools for a Clang-based toolchain, so usually, no further tools have to be defined. To control the target platform, the following CMake variables should be passed over the command line or by a CMake preset:

- `ANDROID_ABI`: Specifies the target **Application Binary Interface** (**ABI**) to be used. Valid values are `armeabi-v7a`, `arm64-v8a`, `x86`, and `x86_64`. This should always be set when cross-compiling for Android.
- `ANDROID_ARM_NEON`: Enables NEON support for `armeabi-v7a`. This variable has no effect on other ABI versions. When using an NDK above version r21, NEON support is enabled by default, and it is rare that this needs to be disabled.
- `ANDROID_ARM_MODE`: Specifies whether to generate ARM or Thumb instructions for `armeabi-v7a`. Valid values are either `thumb` or `arm`. This variable has no effect on other ABI versions.
- `ANDROID_LD`: Selects whether the default linker or the experimental `lld` from `llvm` should be used. Valid values are `default` or `lld`, but this variable is usually omitted for production builds because of the experimental state of `lld`.
- `ANDROID_PLATFORM`: Specifies the minimum **Application Programming Interface** (**API**) level supported by the application in the format `$API_LEVEL`, `android-$API_LEVEL`, or `android-$API_LETTER`, where `$API_LEVEL` is a number and `$API_LETTER` is the version code of the platform. `ANDROID_NATIVE_API_LEVEL` is a variable alias for this. While it is not strictly necessary to set the API level, this is usually done.
- `ANDROID_STL`: Specifies which **Standard Template Library** (**STL**) to use for this application. This can either be `c++_static` (which is the default), `c++_shared`, `none`, or `system`. Either `c++_shared` or `c++_static` is needed for modern C++ support. The `system` library only provides `new` and `delete` and C++ wrappers for the C library headers, while `none` provides no STL support at all.

A call to CMake to configure an Android build could look like this:

```
cmake -S . -B build --toolchain <NDK_DIR>/build/cmake/android
   .toolchain.cmake -DANDROID_ABI=armeabi-v7a -DANDROID_
      PLATFORM=23
```

This call would specify a build that requires API level 23 or higher, which corresponds to Android 6.0 or higher for a 32-bit ARM **Central Processing Unit** (**CPU**).

An alternative to using the toolchain provided by the NDK is to point CMake to the location of the Android NDK, which is the recommended way with NDKs newer than version r23. The configuration of the target platform then happens with the respective CMake variables. By setting the `CMAKE_SYSTEM_NAME` variable to `android` and the `CMAKE_ANDROID_NDK` variable to the location of the Android NDK, CMake is told to use the NDK. This can either happen over the command line or in a toolchain file. Alternatively, if the `ANDROID_NDK_ROOT` or `ANDROID_NDK` *environment variables* are set, they will be used as the value for `CMAKE_ANDROID_NDK`.

When using the NDK in this way, the configuration is defined over the `CMAKE_` equivalent of the variables used when invoking the toolchain file of the NDK directly, as follows:

- `CMAKE_ANDROID_API` or `CMAKE_SYSTEM_VERSION` is used to specify the minimum API level to build for.
- `CMAKE_ANDROID_ARCH_ABI` is used to specify which ABI mode to use.
- `CMAKE_ANDROID_STL_TYPE` specifies which STL to be used.

A sample toolchain file to configure CMake with the Android NDK might look like this:

```
set(CMAKE_SYSTEM_NAME Android)
set(CMAKE_SYSTEM_VERSION 21)
set(CMAKE_ANDROID_ARCH_ABI arm64-v8a)
set(CMAKE_ANDROID_NDK /path/to/the/android-ndk-r23b)
set(CMAKE_ANDROID_STL_TYPE c++_static)
```

When cross-compiling for Android using Visual Studio generators, CMake requires either *NVIDIA Nsight Tegra Visual Studio Edition* or *Visual Studio tools for Android*, which uses the Android NDK. When using Visual Studio to build Android binaries, the built-in support for CMake's Android NDK may be used by setting the `CMAKE_ANDROID_NDK` variable to the location of the NDK.

Cross-compiling native code for Android has become much easier with the recent versions of the NDK and the CMake versions from 3.20 on. Another special case for cross-compiling is when targeting Apple's iOS, tvOS, or watchOS.

Cross-compiling for iOS, tvOS, or watchOS

The recommended way to cross-compile for Apple's iPhone, Apple TV, or Apple Watches is to use the Xcode generator. Apple is quite restrictive in what to use for building apps for these devices, so using macOS or a **Virtual Machine** (**VM**) running macOS is needed. While using Makefiles or Ninja files is possible, they require much more in-depth knowledge of the Apple ecosystem to be configured correctly.

To cross-compile for any of these devices, the Apple device SDK is needed, and the `CMAKE_SYSTEM_NAME` variable is set to `iOS`, `tvOS`, or `watchOS`, as in the following example:

```
cmake -S <SourceDir> -B <BuildDir> -G Xcode -DCMAKE_SYSTEM_
  NAME=iOS
```

For relatively modern SDKs and a CMake version of 3.14 or higher, this is usually all that is needed. By default, the latest device SDK available on the system is used, but different SDKs can be selected by setting the `CMAKE_OSX_SYSROOT` variable to the path of the SDK if really needed. The minimum target platform version can be specified with the `CMAKE_OSX_DEPLOYMENT_TARGET` variable.

When cross-compiling for iPhone, Apple TV, or Apple Watches, the target can either be the real devices or the device simulator that comes with different SDKs. However, Xcode has built-in support to switch this during the build part, so CMake does not have to be run twice. CMake uses the `xcodebuild` command-line tool internally if the Xcode generator is selected, which supports the `-sdk` option to select the desired SDK. When building through CMake, this option can be passed like so:

```
cmake -build <BuildDir> -- -sdk <sdk>
```

This will pass the `-sdk` option with the specified value to `xcodebuild`. The allowed values are `iphoneos` or `iphonesimulator` for iOS, `appletvos` or `appletvsimulator` for Apple TV devices, and `watchos` or `watchsimulator` for Apple Watches.

Apple-embedded platforms require the mandatory signing of certain build artifacts. For the Xcode generator, the development team **Identifier** (**ID**), which is usually a short string of around 10 characters, can be specified with the `CMAKE_XCODE_ATTRIBUTE_DEVELOPMENT_TEAM` cache variable.

When building for Apple-embedded devices, the simulators come in handy for testing the code without the need to deploy to the devices each time. In this case, the testing is best done through Xcode or `xcodebuild` itself, but for other platforms, cross-compiled code may be tested through CMake and CTest directly.

Testing cross-compiled binaries

Being able to effortlessly cross-compile binaries for different architectures adds much convenience to the developer workflows of the people involved, but often, these workflows do not stop at building the binaries but also include running tests. If the software also compiles on the host toolchain and the tests are generic enough, running tests on the host might be the easiest way to test the software, although it might cost some time when switching the toolchains and rebuilding frequently. If this is not possible or too time-consuming, one alternative is of course to run any tests on the real target hardware, but depending on the availability of the hardware and the effort of setting up the tests on the hardware, this might also be rather cumbersome. So, often, a practicable middle way is to run tests inside an emulator for the target platform if this is available.

To define an emulator for running tests, the `CROSSCOMPILING_EMULATOR` target property is used. It can either be set for individual targets or globally by setting the `CMAKE_CROSSCOMPILING_EMULATOR` cache variable, which contains a semicolon-separated list of the command and arguments to run the emulator. If set globally, the command will be prefixed to all commands specified in `add_test()`, `add_custom_command()`, and `add_custom_target()`, and it will be used to run any executables generated by `try_run()` commands. This means that all custom commands that are used for a build must be accessible and runnable inside the emulator as well. The `CROSSCOMPILING_EMULATOR` property does not necessarily have to be an actual emulator—it can be any arbitrary program, such as a script that copies the binaries to a target machine and executes it there.

Setting `CMAKE_CROSSCOMPILING_EMULATOR` should either happen over a toolchain file, the command line, or a configured prefix. An example toolchain file used for cross-compiling C++ code for ARM that uses the popular open source emulator *QEMU* to run the tests could look like this:

```
set(CMAKE_SYSTEM_NAME Linux)
set(CMAKE_SYSTEM_PROCESSOR arm)

set(CMAKE_SYSROOT /path/to/arm/sysroot/)

```

```
set(CMAKE_CXX_COMPILER /usr/bin/clang++)
set(CMAKE_CXX_COMPILER_TARGET arm-linux-gnueabihf)

set(CMAKE_CROSSCOMPILING_EMULATOR "qemu-arm;-L;${CMAKE_
   SYSROOT}")
```

In addition to setting the information of the target system and the toolchain for cross-compiling the last line in the example, it sets the emulator command to `qemu-arm -L/path/to/arm/sysroot`. Let's assume a `CMakeLists.txt` file contains a test defined like this:

```
add_test(NAME exampleTest COMMAND exampleExe)
```

When CTest is run instead of running `exampleExe` directly, the test command is transformed into the following:

```
qemu-arm "-L" "/path/to/arm/sysroot/" "/path/to/build-dir/
   exampleExe"
```

Running tests in an emulator can speed up developer workflows quite a bit, as it might eliminate the need to switch between the host toolchain and the target toolchain and does not require the build artifacts to be moved to the target hardware for each superficial test. Using emulators like this also comes in handy for **Continuous Integration** (**CI**) builds where it might be hard to build on the real target hardware.

A trick regarding `CMAKE_CROSSCOMPILING_EMULATOR` is that it also can be used to temporarily wrap tests in a diagnostic utility such as *Valgrind* or similar diagnostic tools. As running the specified emulator executable is not dependent on the `CMAKE_CROSSCOMPILING` variable, which indicates whether a project is cross-compiled or not, a common pitfall with this workaround is that setting the `CMAKE_CROSSCOMPILING_EMULATOR` variable will impact the `try_run()` command often used to test for supported features of the toolchain or any dependencies, and as a diagnostic utility might cause a compiler test to fail, it might be necessary to run it on an already cached build where any results of `try_run()` are already cached. Because of this, using the `CMAKE_CROSSCOMPILING_EMULATOR` variable to run diagnostic utilities should not be done permanently but rather in a specific development situation, when hunting defects.

In this section, we mentioned CMake's `try_run()` command, which, along with the closely related `try_compile()` command, is used to check for the availability of certain features in the compiler or the toolchain. In the next section, we will have a closer look at the two commands and into feature testing toolchains.

Testing a toolchain for supported features

When CMake is run for the first time on a project tree, it performs various tests for compiler and language features. Any call to `project()` or `enable_language()` will trigger testing again, but the results may already be cached from a previous run. Caching is also the reason why switching toolchains on an existing build is not recommended.

As we will see in this section, CMake can check for quite a few features out of the box. Most of the checks will internally use the `try_compile()` command to perform these tests. This command essentially builds a small binary with the toolchain that is either detected or supplied by the user. All relevant global variables such as `CMAKE_<LANG>_FLAGS` will be forwarded to `try_compile()`.

Closely related to it is the `try_run()` command, which internally calls `try_compile()` and, if that succeeds, will try to run the program. For regular compiler checks, `try_run()` is not used, and any calls to it are usually defined in the project.

To write custom checks, rather than invoking `try_compile()` and `try_run()` directly, it is recommended to use the `CheckSourceCompiles` or `CheckSourceRuns` modules and the respective functions' `check_source_compiles()` and `check_source_runs()` commands, which have been available since CMake 3.19. For most cases, they will suffice to produce the needed information without the need for more complicated handling of `try_compile()` or `try_run()`. The signatures of the two commands are very similar, as we can see here:

```
check_source_compiles(<lang> <code> <resultVar>
    [FAIL_REGEX <regex1> [<regex2>...]]  [SRC_EXT <extension>])
check_source_runs(<lang> <code> <resultVar>
[SRC_EXT <extension>])
```

The `<lang>` parameter specifies one of the languages supported by CMake, such as `C` or `CXX` for C++. `<code>` is the code as a string to be linked as an executable, so it must contain a `main()` function. The result of the compilation will be stored as a Boolean value in the `<resultVar>` cache variable. If `FAIL_REGEX` is provided for `check_source_compiles`, the output of the compilation will be checked against the expressions supplied. The code will be saved in a temporary file with the extension matching the language selected; if the file should have a different extension than the default, it can be specified by the `SRC_EXT` option.

There are also language-specific versions of the modules called `Check<LANG>SourceCompiles` and `Check<LANG>SourceRuns` that provide the respective command, as illustrated in the following example:

```
include(CheckCSourceCompiles)
check_c_source_compiles(code resultVar
  [FAIL_REGEX regexes...]
)

include(CheckCXXSourceCompiles)
check_cxx_source_compiles(code resultVar
  [FAIL_REGEX regexes...]
)
```

Let's assume that there is a C++ project that could either use the atomics functionality of the standard library or, if this is not supported, fall back to a different implementation. A compiler check for this could look like this:

```
include(CheckSourceCompiles)

check_source_compiles(CXX "
#include <atomic>
int main(){
    std::atomic<unsigned int> x;
    x.fetch_add(1);
    x.fetch_sub(1);
}" HAS_STD_ATOMIC)
```

After including the module, the `check_source_compiles()` function is called with a small program that uses the functionality to be checked. If the code compiles successfully, `HAS_STD_ATOMIC` will be set to `true`; else, it will be set to `false`. The test is executed during the configuration of the project and will print a status message like this:

```
[cmake] -- Performing Test HAS_STD_ATOMIC
[cmake] -- Performing Test HAS_STD_ATOMIC - Success
```

The result will be cached so that any subsequent run of CMake will not perform the test again. In a lot of cases, checking whether a program compiles already gives enough information about a certain feature of the toolchain, but sometimes the underlying program has to be run to obtain the needed information. For this, `check_source_runs()` works analogously to `check_source_compiles()`. One caveat of `check_source_runs()` is this: if `CMAKE_CROSSCOMPILING` is set but no emulator command is set, then the test will only compile the test but not run it unless `CMAKE_CROSSCOMPILING_EMULATOR` is set.

There are a number of variables of the form `CMAKE_REQUIRED_` to control how the checks compile the code. Note that these variables lack the language-specific part, which needs special care if working on projects that run tests for different languages. An explanation of some of these variables is provided here:

- `CMAKE_REQUIRED_FLAGS` is used to pass additional flags to the compiler after any flags specified in the `CMAKE_<LANG>_FLAGS` or `CMAKE_<LANG>_FLAGS_<CONFIG>` variables.
- `CMAKE_REQUIRED_DEFINITIONS` specifies a number of compiler definitions of the form `-DFOO=bar`.
- `CMAKE_REQUIRED_INCLUDES` specifies a list of directories to search for additional headers.
- `CMAKE_REQUIRED_LIBRARIES` specifies a list of libraries to add when linking programs. These can be the filenames of the libraries or imported CMake targets.
- `CMAKE_REQUIRED_LINK_OPTIONS` is a list of additional linker flags.
- `CMAKE_REQUIRED_QUIET` may be set to `true` to suppress any status messages from the checks.

In situations where the checks need to be isolated from each other, the `CMakePushCheckState` module provides the `cmake_push_check_state()`, `cmake_pop_check_state()`, and `cmake_reset_check_state()` functions for storing the configuration, restoring a previous configuration, and resetting the configuration, as illustrated in the following example:

```
include(CMakePushCheckState)
cmake_push_check_state()
# Push the state and clean it to start with a clean check state
cmake_reset_check_state()
```

```
include(CheckCompilerFlag)
check_compiler_flag(CXX -Wall WALL_FLAG_SUPPORTED)

if(WALL_FLAG_SUPPORTED)
    set(CMAKE_REQUIRED_FLAGS -Wall)

    # Preserve -Wall and add more things for extra checks
    cmake_push_check_state()
        set(CMAKE_REQUIRED_INCLUDES ${CMAKE_CURRENT_SOURCE_DIR}
          /include)
        include(CheckSymbolExists)

        check_symbol_exists(hello "hello.hpp" HAVE_HELLO_
          SYMBOL)

    cmake_pop_check_state()

endif()
# restore all CMAKE_REQUIRED_VARIABLEs to original state
cmake_pop_check_state()
```

Underlying the commands for checking compilation or running the test program are the more complicated `try_compile()` and `try_run()` commands. While available for use, they are mainly intended for internal use, and thus we refer you to the official documentation of the commands instead of explaining them here.

Checking compiler features by compiling and running programs is a very versatile approach to checking for toolchain features. Some checks are so common that CMake provides dedicated modules and functions for them.

Common checks for toolchain and language features

For some of the most common feature checks, such as checking whether a compiler flag is supported or whether a header file exists, CMake provides its own modules for convenience. Since CMake 3.19, the general modules that take the language as an argument exist, but the corresponding `Check<LANG>...` language-specific modules may still be used.

A very basic test to check whether a compiler for a certain language is available is the `CheckLanguage` module. It can be used to check whether a compiler for a certain language is available if the `CMAKE_<LANG>_COMPILER` variable is not set. An example to check whether Fortran is available could look like this:

```
Include(CheckLanguage)
check_language(Fortran)if(CMAKE_Fortran_COMPILER)
enable_language(Fortran)else()  message(STATUS "No Fortran
  support")endif()
```

If the check succeeds, the corresponding `CMAKE_<LANG>_COMPILER` variable is set. If the variable was set before the check, it has no effect.

`CheckCompilerFlag` provides the `check_compiler_flag()` function to check whether the current compiler supports a certain flag. Internally, a very simple program will be compiled, and the output will be parsed for a diagnostic message. The check assumes that any compiler flags already present in `CMAKE_<LANG>_FLAGS` will run successfully through; else, the `check_compiler_flag()` function will always fail. The following example checks whether the C++ compiler supports the `-Wall` flag:

```
include(CheckCompilerFlag)
check_compiler_flag(CXX -Wall WALL_FLAG_SUPPORTED)
```

If the `-Wall` flag is supported, the `WALL_FLAG_SUPPORTED` cache variable will be `true`; else, it will be `false`.

The corresponding module to check linker flags is called `CheckLinkerFlag` and works similarly to the check for compiler flags, but the linker flag will not be passed directly to the linker. As the linker will typically be invoked through the compiler, passing additional flags to the linker may use a prefix such as `-Wl` or `-Xlinker` to tell the compiler to pass the flag through. As this flag is compiler-specific, CMake provides the `LINKER:` prefix to automatically substitute the command. For example, to pass a flag to generate statistics about execution time and memory consumption to the linker, the following command would be used:

```
include(CheckLinkerFlag)
check_linker_flag(CXX LINKER:-stats LINKER_STATS_FLAG_
  SUPPORTED)
```

If the linker supports the `-stats` flag, the `LINKER_STATS_FLAG_SUPPORTED` variable will be `true`.

Other useful modules to check various things are the `CheckLibraryExists`, `CheckIncludeFile`, and `CheckIncludeFileCXX` modules for checking whether a certain library or include file exists in certain locations.

CMake offers even more detailed checks that might be very specific to the project; for example, the `CheckSymbolExists` and `CheckSymbolExistsCXX` modules check whether a certain symbol exists either as a preprocessor definition, a variable, or a function. `CheckStructHasMember` will check whether a struct has a certain member, while `CheckTypeSize` can check the size of non-user types and the definition of C and C++ function prototypes with `CheckPrototypeDefinition`.

As we have seen, CMake offers quite a lot of checks, and the list of available checks will probably grow as CMake evolves further. While checks are useful in certain situations, we should be careful not to carry the number of tests too far. The number and complexity of the checks will have quite an impact on the speed of the configuration step while sometimes not providing too much benefit. Having a lot of checks in a project could also be a hint toward the unnecessary complexity of the project.

Summary

Broad support for cross-compiling is one of the striking features of CMake. In this chapter, we looked into how to define a toolchain file for cross-compiling and how to use sysroots to use libraries for a different target platform. A special case of cross-compiling is Android and Apple mobile devices, which rely on their specific SDKs. With a brief excursion into using emulators or simulators for testing for other platforms, you will have all the essential information to start building quality software for various target platforms.

The last part of the chapter concerned itself with the advanced topic of testing toolchains for certain features. While most projects will not have to concern themselves with these details, they are nevertheless useful to know.

The next chapter will be about making CMake code reusable across multiple projects without the need to rewrite all the things again and again.

Questions

Answer the following questions to test your knowledge of this chapter:

1. How are toolchain files passed to CMake?
2. What is usually defined in a toolchain file for cross-compiling?
3. What is a staging directory in the context of a sysroot?

4. How can an emulator be passed to CMake for testing?
5. What triggers the detection of compiler features?
6. How can the configuration context for compiler checks be stored and restored?
7. What is the effect of the `CMAKE_CROSSCOMPILING` variable on compiler checks?
8. Why should the build directory be fully cleared when switching toolchains and not just the cache deleted?

13 Reusing CMake Code

Writing build system code for a project is no easy task. Project maintainers and developers are spending a lot of effort on writing CMake code for configuring compiler flags, project build variants, third-party libraries, and tool integrations. Writing CMake code for project-agnostic details from scratch may start to incur a significant burden when dealing with multiple CMake projects. Most of the CMake code written for a project to configure the aforementioned details could be reused between the projects. With that in mind, it is for our benefit to develop a strategy to make our CMake code reuse-friendly. The straightforward way to approach this problem is to treat CMake code as regular code and apply some of the most basic coding principles: the **Don't Repeat Yourself** (**DRY**) principle and the **Single Responsibility Principle** (**SRP**).

CMake code can be easily reused if structured with reusability in mind. Achieving essential reusability is pretty straightforward: separate CMake code into modules and functions. You may have realized that the way to make CMake code reusable is no different from making software code reusable. Remember, CMake itself is a scripting language, after all. So, it is natural to treat CMake code as regular code and apply software design principles while dealing with it. As with any functional scripting language, CMake has the following basic abilities for reusability:

- Ability to include other CMake files
- Functions/macros
- Portability

In this chapter, we will learn the ways of writing CMake code for a project with reusability in mind and reusing CMake code in CMake projects. We will also discuss the ways of versioning and sharing common CMake code between projects.

To understand the skills shared in this chapter, we'll cover the following main topics:

- What is a CMake module?
- Fundamental building blocks of modules – functions and macros
- Writing your first very own CMake module

Let's begin with the technical requirements.

Technical requirements

Before you dive further into this chapter, revisit *Chapter 1, Kickstarting CMake* to brush up the skills we learned in the chapter. This chapter follows the teaching over example approach, so it is recommended to obtain this chapter's example content from here: `https://github.com/PacktPublishing/CMake-Best-Practices/tree/main/chapter_13`. For all the examples, assume that you will be using the container provided by this project: `https://github.com/PacktPublishing/CMake-Best-Practices`.

Let's start by learning some basics about reusability in CMake.

What is a CMake module?

A **CMake module** is a logical entity that contains CMake code, functions, and macros that are put together to serve a particular purpose. A module can provide functions and macros for other CMake code and execute CMake commands when included. CMake is shipped with many pre-made modules by default. These modules provide extra utility to consume CMake code and allow the discovery of third-party tools and dependencies (`Find*.cmake` modules). A list of the modules that CMake provides by default is available at `https://cmake.org/cmake/help/latest/manual/cmake-modules.7.html`. The official CMake documentation categorizes modules in the following two main categories:

- Utility modules
- Find modules

As their name suggests, the utility modules provide utility, whereas the find modules are designed to search for third-party software in a system. As you can recall, we have covered the find modules thoroughly in *Chapter 4, Packaging, Deploying, and Installing a CMake Project*, and *Chapter 5, Integrating Third-Party Libraries and Dependency Management*. Therefore, we will exclusively focus on utility modules in this chapter. As you can recall, we have been using some CMake-provided utility modules in previous chapters. Some of these modules we used were the `GNUInstallDirs`, `CPack`, `FetchContent`, and `ExternalProject` modules. These modules are located under the `CMake` installation folder.

To better understand the concept of a utility module, let's start by investigating a simple utility module that CMake provides. For this purpose, we will look into the `ProcessorCount` utility module. You can find the source file for this module at `https://github.com/Kitware/CMake/blob/master/Modules/ProcessorCount.cmake`. The `ProcessorCount` module is a module that allows retrieving the CPU core count of a system in CMake code. The `ProcessorCount.cmake` file defines a CMake function named `ProcessorCount`, which takes a single parameter named `var`. The implementation of the function is roughly as follows:

```
function(ProcessorCount var)
  # Unknown:
  set(count 0)
  if(WIN32)
    set(count "$ENV{NUMBER_OF_PROCESSORS}")
  endif()
  if(NOT count)
    # Mac, FreeBSD, OpenBSD (systems with sysctl):
    # … mac-specific approach … #
  endif()

  if(NOT count)
    # Linux (systems with nproc):
    # … linux-specific approach … #
  endif()
# … Other platforms, alternative fallback methods … #
# Lastly:
set(${var} ${count} PARENT_SCOPE)
endfunction()
```

The `ProcessorCount` function attempts several different approaches to retrieve the CPU core count of the host machine. The usage of the `ProcessorCount` module is simple, as follows:

```
include(ProcessorCount)
ProcessorCount(CORE_COUNT)
message(STATUS "Core count: ${CORE_COUNT}")
```

As you can see in the preceding example, using a CMake module is as easy as including the module in the required CMake file. The `include()` function is transitive, so the code after the `include` line can consume all CMake definitions contained in the module.

We now have a rough idea of what a utility module looks like. Let's continue by learning more about the fundamental building blocks of a utility module: functions and macros.

Fundamental building blocks of modules – functions and macros

It is clear that we need some basic building blocks to create utility modules. The most fundamental building blocks for utility modules are functions and macros, so it is essential to learn their working principles well. Let's start by learning about functions.

Functions

Let's remember what we have learned in *Chapter 1, Kickstarting CMake*, about functions. A **function** is the CMake language feature to define a logical code block that can be invoked to execute CMake commands. A function starts with `function(...)`, has a body that contains CMake commands, and ends with the `endfunction()` CMake command. The `function()` command needs a name as the first argument, and optional function argument names, shown as follows:

```
function(<name> [<arg1> ...])
  <commands>
endfunction()
```

A function defines a new variable scope, so the changes made on CMake variables are only visible in the function's body. Separate scoping is the most crucial property of the function. Having a new scope means we can't accidentally leak variables to the caller or modify the caller's variables *unless we want to*. Most of the time, we will want to contain the changes in the function's scope and only reflect the function's result to the caller. As CMake has no notion of return values, we will take the *defining a variable in the caller's scope* approach to return function results to the caller.

To illustrate this approach, let's define a simple function that retrieves the current Git branch name together:

```
function(git_get_branch_name result_var_name)
  execute_process(
        COMMAND git symbolic-ref -q --short HEAD
        WORKING_DIRECTORY "${CMAKE_CURRENT_SOURCE_DIR}"
        OUTPUT_VARIABLE git_current_branch_name
        OUTPUT_STRIP_TRAILING_WHITESPACE
        ERROR_QUIET
    )
    set(${result_var_name} ${git_current_branch_name}
      PARENT_SCOPE)
endfunction()
```

The `git_get_branch_name` function takes a single argument named `result_var_name`. This argument is the name of the variable that will be defined in the caller's scope to return the Git branch name to the caller. Alternatively, we can use a constant variable name, such as `GIT_CURRENT_BRANCH_NAME`, and get rid of the `result_var_name` argument, but this may cause issues if the project already uses the `GIT_CURRENT_BRANCH_NAME` name.

The rule of thumb here is to leave the naming to the caller since it allows maximum flexibility and portability. To retrieve the current Git branch name, we have invoked the `git symbolic-ref -q --short HEAD` command with `execute_process()`. The result of the command is stored in the `git_current_branch_name` variable in the function's scope. The variable being in the function's scope means the caller cannot see the `git_current_branch_name` variable. Thus, we have used `set(${result_var_name} ${git_current_branch_name} PARENT_SCOPE)` to define a variable using the value of `result_var_name` in the caller's scope with the value of the local `git_current_branch_name` variable.

The `PARENT_SCOPE` argument alters the scoping of the `set(...)` command, so it defines the variable in the caller's scope instead of the function's scope. The usage of the `git_get_branch_name` function is as follows:

```
git_get_branch_name(branch_n)
message(STATUS "Current git branch name is: ${branch_n}")
```

Let's look at macros next.

Macros

If the function's scoping is a deal-breaker for your use case, you might consider using `macro(...)` instead. Macros start with `macro(...)` and end with `endmacro()`. Functions and macros behave similarly in every aspect but one: macros do not define a new variable scope. Returning to our `git branch` example, considering `execute_process(...)` already has the `OUTPUT_VARIABLE` parameter, it is more convenient to define `git_get_branch_name` as a macro instead of a function to get rid of `set(... PARENT_SCOPE)` at the end:

```
macro(git_get_branch_name_m result_var_name)
  execute_process(
        COMMAND git symbolic-ref -q --short HEAD
        WORKING_DIRECTORY "${CMAKE_CURRENT_SOURCE_DIR}"
        OUTPUT_VARIABLE ${result_var_name}
        OUTPUT_STRIP_TRAILING_WHITESPACE
        ERROR_QUIET
    )
endmacro()
```

The usage of the `git_get_branch_name_m` macro is exactly the same as the `git_get_branch_name()` function:

```
git_get_branch_name_m(branch_nn)
message(STATUS "Current git branch name is: ${branch_nn}")
```

We have learned how we can define a function or macro when needed. Up next, we will define our first CMake module together.

Writing your first very own CMake module

In the previous section, we learned about how to use functions and macros to provide useful utility in CMake projects, Now, we can learn about how we can move these functions and macros to a separate CMake module.

Creating and using a simple CMake module file is extremely simple:

1. Create a `<module_name>.cmake` file under your project.
2. Define any macros/functions in the `<module_name>.cmake` file.
3. Include `<module_name>.cmake` in the desired file.

Alright, let's follow these steps and create a module together. As a follow-up to our previous `git` branch name example, let's extend the scope and write a CMake module that provides the ability to retrieve the branch name, head commit hash, current author name, and current author email information by using the `git` command. For this part, we will follow the `chapter_13/ex01_git_utility` example. The example folder contains a `CMakeLists.txt` file and a `git.cmake` file under the `.cmake` folder. Let's start by taking a look at the `.cmake/git.cmake` file first. The contents of the file are as follows:

```
# ...
include_guard(DIRECTORY)
macro(git_get_branch_name result_var_name)
    execute_process(
        COMMAND git symbolic-ref -q --short HEAD
        WORKING_DIRECTORY "${CMAKE_CURRENT_SOURCE_DIR}"
        OUTPUT_VARIABLE ${result_var_name}
        OUTPUT_STRIP_TRAILING_WHITESPACE
        ERROR_QUIET
    )
endmacro()
# ... git_get_head_commit_hash(), git_get_config_value()
```

The `git.cmake` file is the CMake utility module file that contains three macros named `git_get_branch_name`, `git_get_head_commit_hash`, and `git_get_config_value` respectively. Additionally, there is an `include_guard(DIRECTORY)` line at the top of the file. This is analogous to the `#pragma once` preprocessor directive in C/C++ and prevents the file from being included more than once. The `DIRECTORY` parameter denotes that `include_guard` is defined at the directory scope and this file can be included once at most within the current directory and below. Alternatively, the `GLOBAL` parameter can be specified instead of `DIRECTORY` to limit the inclusion of the file once, regardless of the scope.

To see how we can use the `git.cmake` module file, let's investigate the `CMakeLists.txt` file of `chapter_13/ex01_git_utility` together:

```
cmake_minimum_required(VERSION 3.21)
project(
  ch13_ex01_git_module
  VERSION 1.0
```

```
  DESCRIPTION "Chapter 13 Example 01, git utility module
    example"
  LANGUAGES CXX)
# Include the git.cmake module.
# Full relative path is given, since .cmake/ is not in the
  CMAKE_MODULE_PATH
include(.cmake/git.cmake)

git_get_branch_name(current_branch_name)
git_get_head_commit_hash(current_head)
git_get_config_value("user.name" current_user_name)
git_get_config_value("user.email" current_user_email)

message(STATUS "-----------------------------------------")
message(STATUS "VCS (git) info:")
message(STATUS "\tBranch: ${current_branch_name}")
message(STATUS "\tCommit hash: ${current_head}")
message(STATUS "\tAuthor name: ${current_user_name}")
message(STATUS "\tAuthor e-mail: ${current_user_email}")
message(STATUS "-----------------------------------------")
```

The `CMakeLists.txt` file includes the `git.cmake` file by specifying the full relative path to the module file. The `git_get_branch_name`, `git_get_head_commit_hash`, and `git_get_config_value` macros provided by the module are used to retrieve the branch name, commit hash, author name, and email to the `current_branch_name`, `current_head`, `current_user_name`, and `current_user_email` variables respectively. Lastly, these variables are printed on the screen by the `message(...)` command. Let's configure the example project to see whether the `git` module we've just written works as expected:

```
cd chapter_13/ex01_git_utility/
cmake -S ./ -B ./build
```

The output of the command should look similar to this:

```
-- The CXX compiler identification is GNU 9.4.0
-- Detecting CXX compiler ABI info
-- Detecting CXX compiler ABI info - done
```

```
-- Check for working CXX compiler: /usr/bin/c++ - skipped
-- Detecting CXX compile features
-- Detecting CXX compile features - done
-- ----------------------------------------------
-- VCS (git) info:
--      Branch: chapter-development/chapter-13
--      Commit hash: 1d5a32649e74e4132e7b66292ab23aae
          ed327fdc
--      Author name: Mustafa Kemal GILOR
--      Author e-mail: mustafagilor@gmail.com
-- ----------------------------------------------
-- Configuring done
-- Generating done
-- Build files have been written to:
/home/toor/workspace/CMake-Best-Practices/chapter_13
/ex01_git_utility/build
```

As we can see, we have succeeded in retrieving the information from the `git` command. Our first CMake module works as expected.

Case study – dealing with project metadata files

Let's continue with another example. Assume that we have an environment file that contains key-value pairs per line. It is not unusual to have external files in the project that contain some metadata about the project (for example, project version and dependencies). The file may be in different formats, such as JSON or new-line separated key-value pairs, as we have in this example. The task at hand is to create a utility module that reads the environment variable file and defines a CMake variable per key-value pair in the file. The contents of the file will look similar to this:

```
KEY1="Value1"
KEY2="Value2"
```

For this section, we will follow the `chapter_13/ex02_envfile_utility` example. Let's begin by examining the contents of `.cmake/envfile-utils.cmake` first:

```
include_guard(DIRECTORY)
function(read_environment_file ENVIRONMENT_FILE_NAME)
    file(STRINGS ${ENVIRONMENT_FILE_NAME} KVP_LIST ENCODING
      UTF-8)
    foreach(ENV_VAR_DECL IN LISTS KVP_LIST)
        string(STRIP ENV_VAR_DECL ${ENV_VAR_DECL})
        string(LENGTH ENV_VAR_DECL ENV_VAR_DECL_LEN)
        if(ENV_VAR_DECL_LEN EQUAL 0)
            continue()
        endif()
        string(SUBSTRING ${ENV_VAR_DECL} 0 1
          ENV_VAR_DECL_FC)
        if(ENV_VAR_DECL_FC STREQUAL "#")
            continue()
        endif()
        string(REPLACE "=" ";" ENV_VAR_SPLIT
          ${ENV_VAR_DECL})
        list(GET ENV_VAR_SPLIT 0 ENV_VAR_NAME)
        list(GET ENV_VAR_SPLIT 1 ENV_VAR_VALUE)
        string(REPLACE "\"" "" ENV_VAR_VALUE
          ${ENV_VAR_VALUE})
        set(${ENV_VAR_NAME} ${ENV_VAR_VALUE} PARENT_SCOPE)
    endforeach()
endfunction()
```

The `envfile-utils.cmake` utility module contains a single function, `read_environment_file`, that reads an environment file in the format of a list of key-value pairs. This function reads all the lines in the file to the `KVP_LIST` variable, then iterates through all lines. Each individual line is split by the (=) equals token, then the left side of the equals token is used as the variable name, whereas the right side is used as the variable value to define each key-value pair as a CMake variable. Empty lines and comment lines are skipped. As for the usage of the module, let's have a look into the `chapter_13/ex02_envfile_utility/CMakeLists.txt` file:

```
# Add .cmake folder to the module path, so subsequent
  include() calls
```

```
# can directly include modules under .cmake/ folder by
  specifying the name only.
set(CMAKE_MODULE_PATH ${CMAKE_MODULE_PATH}
  ${PROJECT_SOURCE_DIR}/.cmake/)
add_subdirectory(test-executable)
```

You may have noticed that the `.cmake` folder is added to the `CMAKE_MODULE_PATH` variable. The `CMAKE_MODULE_PATH` variable is the collection of paths that the `include(…)` directive will search in. By default, it is empty. This allows us to include the `envfile-utils` module directly by name in the current and children `CMakeLists.txt` files. Lastly, let's take a look at the `chapter_13/ex02_envfile_utility/test-executable/CMakeLists.txt` file:

```
# ....
# Include the module by name
include(envfile-utils)
read_environment_file("${PROJECT_SOURCE_DIR}/
  variables.env")
add_executable(ch13_ex02_envfile_utility_test)
target_sources(ch13_ex02_envfile_utility_test PRIVATE
  test.cpp)
target_compile_features(ch13_ex02_envfile_utility_test
  PRIVATE cxx_std_11)
target_compile_definitions(ch13_ex02_envfile_utility_test
  PRIVATE PROJECT_VERSION="${TEST_PROJECT_VERSION}"
    PROJECT_AUTHOR="${TEST_PROJECT_AUTHOR}")
```

As you can see, the `envfile-utils` environment file reader module is included by name. This is because the folder that contains the `envfile-utils.cmake` file is appended to the `CMAKE_MODULE_PATH` variable before. The `read_environment_file()` function is called to read the `variables.env` file in the same folder. The `variables.env` file contains the following key-value pairs:

```
# This file contains some metadata about the project
TEST_PROJECT_VERSION="1.0.2"
TEST_PROJECT_AUTHOR="CBP Authors"
```

So, after calling the `read_environment_file()` function, we expect the `TEST_PROJECT_VERSION` and `TEST_PROJECT_AUTHOR` variables to get defined in the current CMake scope, with their respective values specified in the file. To verify that, an executable target named `ch13_ex02_envfile_utility_test` is defined, and the `TEST_PROJECT_VERSION` and `TEST_PROJECT_AUTHOR` variables are passed into the target as macro definitions. Lastly, the target's source file, `test.cpp`, prints the `TEST_PROJECT_VERSION` and `TEST_PROJECT_AUTHOR` macro definitions to the console:

```
#include <cstdio>
int main(void) {
    std::printf("Version '%s', author '%s'\n",
      TEST_PROJECT_VERSION, TEST_PROJECT_AUTHOR);
}
```

Alright, let's compile and run the application to see whether this works or not:

```
cd chapter_13/ex02_envfile_utility
cmake -S ./ -B ./build
cmake --build build
./build/test-executable/ch13_ex02_envfile_utility_test
# Will output: Version '1.0.2', author 'CBP Authors'
```

As we can see, we have successfully read a key-value pair-formatted file from our source tree, defined each key-value pair as CMake variables, and then exposed these variables as macro definitions to our application.

Although writing CMake modules is very straightforward, there are a few extra recommendations to consider:

- Always use unique names for your functions/macros.
- Use a common prefix for all module functions/macros.
- Avoid using constant names for non-function scope variables.
- Use `include_guard()` for your module.
- If your module prints messages, provide a quiet mode for your module.
- Do not expose your module's internals.
- Use macros for simple command wrappers, and functions for everything else.

With that said, we conclude this part of the chapter. Next, we will take a look into ways of sharing CMake modules between projects.

Recommendations for sharing CMake modules between projects

The recommended way of sharing CMake modules is by maintaining a separate project for CMake modules and then incorporating the project as an external resource, either directly by Git submodules/subtree or CMake's `FetchContent/ExternalProject`. This way, all reusable CMake utilities can be maintained under a single project and can be propagated to all downstream projects. Putting CMake modules into a repository in an online Git hosting platform (such as GitHub or GitLab) will make using the module convenient for most people. Since CMake supports fetching content directly from Git, it will be straightforward to use the shared libraries.

To demonstrate how we can use an external CMake modules project, we will use an open source CMake utility module project named `Hadouken`. This project is accessible from `https://github.com/mustafakemalgilor/hadouken`. This project contains CMake utility modules for tool integrations, target creation, and feature checks.

For this part, we will follow the `chapter_13/ex03_hadouken` example. This example will fetch `Hadouken`, and then use the `Hadouken` project's **target creation helper utility**. Let's have a look at the `CMakeLists.txt` file as usual:

```
cmake_minimum_required(VERSION 3.21)
project(
  ch13_ex03_hadouken
  VERSION 1.0
  DESCRIPTION "Chapter 13 Example 03, external CMake
    modules (hadouken) example"
  LANGUAGES CXX)
include(FetchContent)
# Declare hadouken dependency details.
FetchContent_Declare(hadouken
    GIT_REPOSITORY https://github.com/mustafakemalgilor
      /hadouken.git
    GIT_TAG        7d0447fcadf8e93d25f242b9bb251ecbcf67f8cb
    SOURCE_DIR "${CMAKE_CURRENT_LIST_DIR}/.hadouken"
)
FetchContent_MakeAvailable(hadouken)
set(CMAKE_MODULE_PATH ${CMAKE_MODULE_PATH}
  ${PROJECT_SOURCE_DIR}/.hadouken/cmake/modules/)
include(misc/Log)
```

```
include(misc/Utility)
include(core/MakeCompilationUnit)
include(core/MakeTarget)
# Create an executable target by using Hadouken's
  make_target() utility function
make_target(TYPE EXECUTABLE)
```

In the preceding example, we have used `FetchContent_Declare` and `FetchContent_MakeAvailable` to retrieve `Hadouken` into our project. Then, the `Hadouken` project's module directory is appended into `CMAKE_MODULE_PATH` to use the `Hadouken` project's CMake utility modules via the `include(...)` directive. Consequently, the `Log`, `Utility`, `MakeCompilationUnit`, and `MakeTarget` modules are included. Lastly, the `make_target()` function is called to ensure that we can use an external CMake module project's functions.

The `make_target(...)` function is provided by the `Hadouken` project's `core/MakeTarget` module and is a wrapper function around the `add_executable` and `add_library` CMake commands. The `make_target(TYPE EXECUTABLE)` call is supposed to discover all source files under the `src/` folder and create an executable target by calling the `add_executable(...)` CMake command. Let's configure and build the project to see whether that is the case or not:

```
cd chapter_13/ex03_hadouken
cmake -S ./ -B build/
cmake --build build
```

The output should look similar to this:

```
 [ 50%] Building CXX object
  CMakeFiles/ch13_ex03_hadouken.dir/src/main.cpp.o
[100%] Linking CXX executable ch13_ex03_hadouken
[100%] Built target ch13_ex03_hadouken
```

The target, `ch13_ex03_hadouken`, is defined and the source file, `main.cpp`, is included as a source file in the target. This confirms that we can use an external CMake modules project in our CMake code.

We have reached another chapter's end together. Up next, we'll summarize what we have learned in this chapter and what we will learn in the upcoming chapter.

Summary

In this chapter, we have learned how to structure a CMake project to support reusability. We have learned how to implement the CMake utility modules, how to share them, and how to use utility modules written by others. Having the ability to leverage CMake modules enables us to better organize our projects and better collaborate with our team members in unison. CMake projects will be much easier to maintain with this knowledge on hand. The common, reusable code between CMake projects will grow into an extensive collection of useful modules that makes writing projects with CMake easier.

I want to remind you that CMake is a scripting language and should be treated as such. Use software design principles and patterns to make CMake code more maintainable. Organize your CMake code into functions and modules. Reuse and share the CMake code as much as possible. Please do not neglect your build system code, or you may have to write it from scratch.

In the next chapter, we will be learning about the ways of optimizing and maintaining CMake projects.

See you soon in the next chapter!

Questions

After completing this chapter, you should be able to answer the following questions:

1. What are the most fundamental building blocks of reusability in CMake?
2. What is a CMake module?
3. How can a CMake module be used?
4. What is the `CMAKE_MODULE_PATH` variable used for?
5. Name one way to share CMake modules between projects.
6. What is the principal difference between a function and a macro in CMake?

14
Optimizing and Maintaining CMake Projects

Software projects tend to live for a long time and for some, it's not unheard of to be under more or less active development for a decade or more. But even if projects do not live that long, they tend to grow over time and attract certain clutter and legacy artifacts. Often, maintaining a project does not just mean refactoring code or adding a feature once in a while, but also keeping the build information and dependencies up to date.

As projects grow in complexity, build times often increase dramatically to the point that development might get tedious because of the long wait times. Long build times are not just inconvenient, they might also encourage developers to take shortcuts because they make trying things out hard. It is hard to try out something new if each build takes hours to complete and if each push to the CI/CD pipeline takes hours to return, which does not help either.

Apart from choosing a good, modular project structure to increase the effectiveness of incremental builds, CMake has a few features to help with profiling and optimizing build times. And if CMake alone is not enough, using technologies such as **compiler cache** (**ccache**) for caching build results or precompiled headers can further help speed up incremental builds.

Optimizing build times can yield good results, improve the daily life of developers considerably, and even be a cost-saving factor because a CI/CD pipeline might need fewer resources to build projects. However, there are pitfalls that heavily optimized systems may become brittle and break down more easily and that, at one point, optimizing for build time might be a tradeoff with easy project maintenance.

In this chapter, we will cover a few general tips for maintaining projects and structuring them to keep the maintenance effort in check. Then, we will dive into analyzing build performance and see how the builds can be sped up. The following topics will be covered in this chapter:

- Keeping a CMake project maintainable
- Profiling a CMake build
- Optimizing build performance

Technical requirements

As with the previous chapters, all examples are tested with CMake 3.21 and run on either of the following compilers:

- GCC 9 or newer
- Clang 12 or newer
- MSVC 19 or newer

For viewing the profiling data, a third-party viewer for Google trace format is needed; arguably the most widely used is Google Chrome.

Examples using ccache are tested with Clang and GCC but not with MSVC. To obtain ccache, use either the package manager of your operating system or obtain it from `https://ccache.dev/`.

All examples are available at `https://github.com/PacktPublishing/CMake-Best-Practices`.

Keeping a CMake project maintainable

When maintaining a CMake project over a long time, there are often a few tasks that regularly come up. There are the usual things, such as new files being added to the project or versions of dependencies increasing, which are usually relatively trivial to handle with CMake. Then, there are things such as adding new toolchains or platforms for cross-compiling, and lastly, there are updates to CMake itself, when new features such as presets are available.

Regularly updating CMake and making use of new features can help keep projects maintainable. While it is often not practical to update every single new version, checking for new big features of CMake and using them when they are released may make projects easier to maintain. For example, the introduction of CMake presets in version 3.19 of CMake is such a feature that has the potential to make many complicated `CMakeLists.txt` files much simpler.

Keeping dependencies up to date and under control is often a task that keeps maintainers busy. Here, using a consistent concept for handling dependencies will make maintaining a project easier. In that regard, we recommend using package managers, as described in *Chapter 5, Integrating Third-Party Libraries and Dependency Management*, for any but the smallest project. As package managers are designed to shift the complexity of managing dependencies to the package manager instead of exposing it to the maintainer, they often have great potential to make the maintainers' lives much easier.

At the root of making a project maintainable is choosing an effective project structure, so things are found easily and can be improved independently from each other. The exact structure to choose depends heavily on the context and size of the project, so what works for one project might not work for another.

The biggest gain to keeping large projects maintainable is to use a project structure that fits the need. While the details of project organization depend on the actual situation a project is developed in, there are a few good practices that will help keep an overview of the project. Keeping a project maintainable starts with the `CMakeLists.txt` root of a project. For large projects, the `CMakeLists.txt` root should handle the following things:

- The basic setup of the whole project, such as handling the `project()` call, fetching toolchains, supporting programs, and helper repositories. This also includes setting language standards, search behavior, and project-wide setting of compiler flags and search paths.

- Handling cross-cutting dependencies, especially large frameworks such as Boost and Qt, should be included at the top level. Depending on the complexity of the dependencies, creating and including a subdirectory with it its own `CMakeLists.txt` to handle acquiring the dependencies might help with keeping the project maintainable. Using `add_subdirectory` is recommended over using `include` for including the dependencies because, this way, any temporary variables used for searching the dependencies are scoped to the subdirectory unless they are explicitly marked as cache variables.
- If there are more than just a few build targets, moving them to their own subdirectories and including them with `add_subdirectory()` will help to keep the individual files small and self-contained. Aiming for a design principle of loose coupling and high internal cohesion will make the libraries and executables easier to maintain independently. The file and project structure should reflect that, which might mean that each library and executable in a project gets its own `CMakeLists.txt`.
- Whether the unit tests are kept close to the units that they test against or as a subfolder of a tests folder on the root level is a matter of personal preference. Keeping the tests in their own subdirectory with their own `CMakeLists.txt` makes it easier to handle test-specific dependencies and compiler settings.
- Packaging and installation instructions for the project should be centralized and included at the top level of the project. If the installation instructions and packaging instructions are too large, they can be put in their own `CMakeLists.txt` and included from the `CMakeLists.txt` root.

Structuring a project in this way will simplify the navigation inside the project and will help to avoid unnecessary code duplication in the CMake files, especially when projects get larger over time.

A good project setup might make the difference between fighting daily with the build system and running smoothly. Using the techniques and practices from this book will help to make a CMake project maintainable. Having a clearly defined build environment by using CMake presets and build containers or sysroots, as described in *Chapter 9, Creating Reproducible Build Environments*, and *Chapter 12, Cross-Platform and Cross-Toolchain Building*, will help to make the build more portable between developers and the CI system. And last but not least, organizing your custom CMake code into macros and functions, as described in *Chapter 13, Reusable CMake Code*, will help to avoid redundancy and duplication.

Apart from the complexity of the CMake files, longer configuration and build times are often another concern when projects grow bigger. To manage those growing build and configuration times, CMake offers a few features to optimize them.

Profiling a CMake build

When CMake projects get big, configuring them might take quite a long time, especially if there is external content loaded or if there are lots of checks done for toolchain features. A first step to optimize this is to check what part of the configuration process takes up how much time. Since version 3.18, CMake includes command-line options to produce nice profiling graphs to investigate where the time is spent during configuration. By adding the `--profiling-output` and `--profiling-format` profiling flags, CMake will create profiling output. At the time of writing this book, only the Google trace format for the output format is supported. Despite this, the format and the file need to be specified to create the profiling information. A call to CMake to create a profiling graph could look like this:

```
cmake -S <sourceDir> -B <buildDir> --profiling-output
./profiling.json --profiling-format=google-trace
```

This will write the profiling output to the `profiling.json` file in the current directory. The output file can be viewed with Google Chrome by typing `about://tracing` into the address bar. A tracing output for a cached build of the GitHub project to this book could look like this:

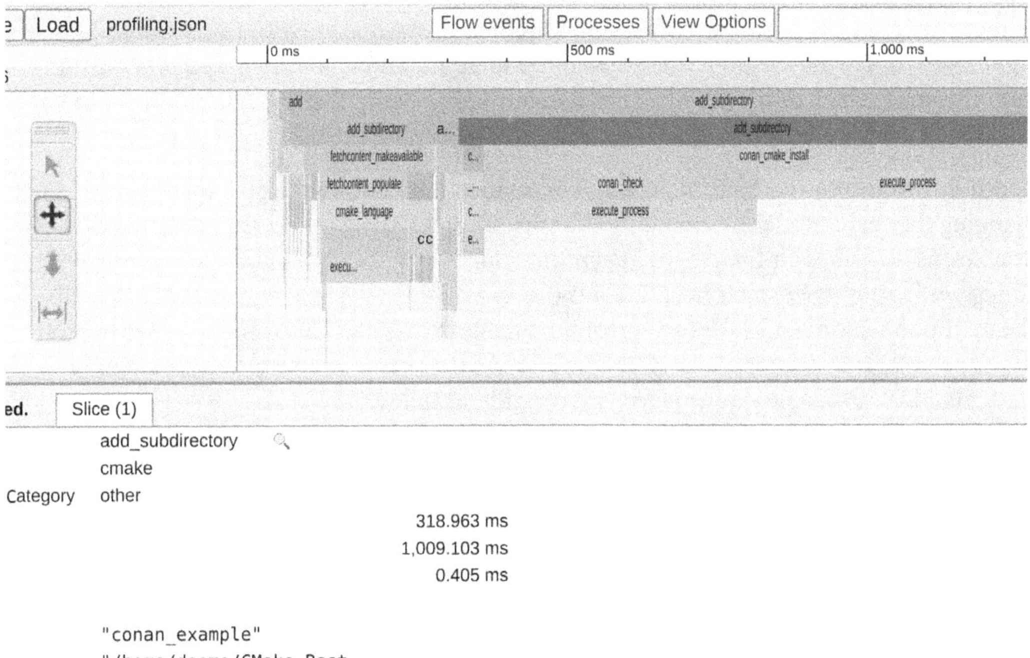

Figure 14.1 – An example profiling graph for a CMake project displayed in Google Chrome

In the preceding figure, it is pretty obvious that there is one call to `add_subdirectory` that takes up the majority of the time when configuring the project. In this case, this is the `chapter_5` subdirectory, taking a bit more than 3 seconds to complete. By drilling down a bit, it becomes apparent that these are the examples that use the Conan package manager, namely the two calls to `conan_cmake_install` that make the configuration relatively expensive. In this case, centralizing the calls to Conan in a directory further up would cut the time CMake would take for a configuration run in half.

In order to correctly interpret the profiling output, it helps to compare different runs of CMake with each other, especially comparing CMake running on a clean cache with one that makes use of cached information. If only the CMake runs on a clean cache take their time, but the incremental runs are fast enough, this might still be acceptable for the developers. However, if the incremental CMake runs take their time as well, this might be more problematic. Profiling them may help you find out if there are unnecessary steps done for each configuration run.

Fixing slow build steps will depend on the concrete situation, but a common culprit for long configuration times is files that are downloaded each time because there is no check whether the file exists in the first place. Analyzing the profiling calls might often show calls such as `execute_process` or `try_compile` consuming lots of execution time. The most obvious "fix" would be to try to get rid of these calls, but often these calls are there for a reason. More often, following up the call stack leading to the commands might reveal opportunities to reduce how often these functions will be called. Maybe the results can be cached, or maybe files created with `execute_process` do not need to be generated each time.

Especially when cross-compiling, `find_` commands might also take up a lot of time. Changing the search order by changing the various `CMAKE_FIND_ROOT_PATH_MODE_` variables, as described in *Chapter 5, Integrating Third-Party Libraries and Dependency Management*, might help a bit here. For a more thorough analysis of why the `find_` calls take up too much time, CMake can be told to enable debug output for them by setting the `CMAKE_FIND_DEBUG_MODE` variable to `true`. As this will print out a lot of information, it is a good idea to enable this only for certain calls, as follows:

```
set(CMAKE_FIND_DEBUG_MODE TRUE)
find_package(...)
set(CMAKE_FIND_DEBUG_MODE FALSE)
```

The profiling options of CMake allow profiling the configuration stage of the build process; profiling the actual compilation and time have to be done by using the respective generator. Most generators either support some profiling option or log the needed information. For Visual Studio generators, the `vcperf` tool (`https://github.com/microsoft/vcperf`) will give a lot of insights. When using Ninja, the `.ninja_log` file can be converted into Google trace format using the `ninjatracing` tool (`https://github.com/nico/ninjatracing`). While CMake does not offer support to profile the actual compiling and linking of software, it does offer ways to improve build times, which we will see in the next section.

Optimizing build performance

Apart from raw compilation time, the main driver for long build times in C++ projects is often unnecessary dependencies between the targets or files. If targets have unnecessary linking requirements between each other, the build system will be limited in executing build tasks in parallel and some of the targets will be frequently relinked. Creating a dependency graph of the targets, as described in *Chapter 6, Automatically Generating Documentation*, will help identify the dependencies. If the resulting graph looks more like a snarl of rope than a tree, optimizing and refactoring the project structure might bring a lot of performance gains. Tools such as *include what you use* and *link what you use*, as described in *Chapter 7, Seamlessly Integrating Code Quality Tools with CMake*, may further help identify unnecessary dependencies. Another common theme is C or C++ projects that expose too much private information in public headers, often causing frequent rebuilds and reducing the effectiveness of incremental builds.

A relatively safe option to bring performance improvements is to set the `CMAKE_OPTIMIZE_DEPENDENCIES` cache variable to `true`. This will cause CMake to remove some dependencies for static or object libraries at generation time if they are not needed. If working with a lot of static or object libraries and a deep dependency graph, this might already produce some gains regarding compile time.

Generally speaking, optimizing the project structure and modularizing the code will often have a greater effect on build performance than optimization of the code. On average, compiling and linking a project that consists of many small files takes longer than projects consisting of a few large files. CMake can help improve build performance with so-called unity builds, which merge several files into a larger file.

Using unity builds

The unity builds that CMake can support may help with build performance by concatenating multiple files into larger files, thus reducing the number of files to be compiled. This might bring a decrease in build time because `include` files are processed only once instead of for every smaller file. So, this will have the biggest effect if many of the files include the same header files and if the header files are heavy to digest by the compiler. Generally speaking, these are headers containing lots of macros or template metaprogramming. Creating a unity build may improve build time significantly, especially when using large header-only libraries, such as the Eigen math library. On the other hand, unity builds have the downside that incremental builds might take longer as, usually, larger chunks of the project have to be recompiled and linked when only a single file changes.

By setting the `CMAKE_UNITY_BUILD` cache variable to `true`, CMake will concatenate the sources into one or more unity sources and build them instead of the original files. The generated files use the `unity_<lang>_<Nr>.<lang>` pattern and are located in a folder called `Unity` in the build directory for the project. Unity files for C++ would be named `unity_0_cxx.cxx`, `unity_1_cxx.cxx`, and so on, while C files are named `unity_0_c.c` and so on. This variable is not intended to be set in `CMakeLists.txt` but rather to be passed over the command line or a preset, as it might depend on the context of whether a unity build is needed or not. CMake will decide on the language of the project if merging the files is needed and possible. For instance, as header files are not compiled, they will not be added to the unity sources. For C and C++, this works quite well; for other languages, this might not work.

Unity builds work best for projects that consist of many small files. If the source files are already large on their own, unity builds might run the risk of running out of memory when compiling. If only a few files are problematic in this regard, they can be excluded from the unity build by setting the `SKIP_UNITY_BUILD_INCLUSION` property on the source files, like this:

```
target_sources(ch14_unity_build PRIVATE
  src/main.cpp
  src/fibonacci.cpp
  src/eratosthenes.cpp
)

set_source_files_properties(src/eratosthenes.cpp PROPERTIES
SKIP_UNITY_BUILD_INCLUSION YES)
```

In the example, the `eratosthenes.cpp` file would be excluded from the unity build, while `main.cpp` and `fibonacci.cpp` would be included in a single compilation unit. If the preceding project is configured, the `unit_0_cxx.cxx` file would contain something like this:

```
/* generated by CMake */

#include "/chapter_14/unity_build/src/main.cpp"
#include "/chapter_14/unity_build/src/fibonacci.cpp"
```

Note that the original source files are only included in the unity file, not copied into the file.

Since CMake 3.18, unity builds support two modes that are controlled with the `CMAKE_UNITY_BUILD_MODE` variable or the `UNITY_BUILD_MODE` target property. The mode can either be `BATCH` or `GROUP`, with `BATCH` being the default if not specified. In the `BATCH` mode, CMake determines which files are grouped together by default, in the order in which they were added to the target. All files of a target will be assigned to batches unless they are explicitly excluded. In the `GROUP` mode, each target has to specify explicitly how the files are grouped together. Files not assigned to a group will be compiled individually. While group mode offers more precise control, using batch mode is often the preferred one, as it has significantly lower maintenance overhead.

By default, CMake will collect the files in batches of eight files when the `UNITY_BUILD_MODE` property is set to `BATCH`. By setting the `UNITY_BUILD_BATCH_SIZE` property of a target, this can be changed. To set the batch size globally, the `CMAKE_UNITY_BUILD_BATCH_SIZE` cache variable is used. The batch size should be selected carefully, as setting it too low will bring little gain in performance, while setting it too high might cause the compiler to use too much memory or the compilation unit to hit other size constraints. If the batch size is set to `0`, then all files of a target will be combined in a single batch, but this is discouraged because of the reasons mentioned earlier.

In group mode, no batch size is applied, but the files have to be assigned to groups by setting the `UNITY_GROUP` property of the source file, as illustrated in the following example:

```
add_executable(ch14_unity_build_group)

target_sources(ch14_unity_build_group PRIVATE
  src/main.cpp
  src/fibonacci.cpp
  src/eratosthenes.cpp
```

```
  src/pythagoras.cpp
)
set_target_properties(ch14_unity_build_group PROPERTIES
  UNITY_BUILD_MODE GROUP)
set_source_files_properties(src/main.cpp src/fibonacci.cpp
  PROPERTIES UNITY_GROUP group1)
set_source_files_properties(src/erathostenes.cpp
  src/pythagoras.cpp PROPERTIES UNITY_GROUP group2)
```

In the example, the `main.cpp` and `fibonacci.cpp` files would be grouped together, and `erathostenes.cpp` and `pythagoras.cpp` would be compiled in a different group. In group mode, the generated files are named `unity_<groupName>_<lang>.<lang>`. So, in this example, the files would be named `unity_group1_cxx.cxx` and `unity_group2_cxx.cxx`.

Depending on the anatomy of a project, using unity builds can have a significant effect on build performance. Another technique often used to improve build times is using precompiled headers.

Precompiled headers

Precompiled headers are often a significant boost for the compile time, especially in cases where processing the headers is a significant part of the compile time or when header files are included in many different compilation units. In a nutshell, precompiled headers work by compiling some headers into a binary format that is easier to process for the compilers. Since CMake 3.16, direct support precompiled headers and most of the major compilers support some form of precompiled headers.

Precompiled headers are added to a target with the `target_precompile_headers` command, which has the following signature:

```
target_precompile_headers(<target>
  <INTERFACE|PUBLIC|PRIVATE> [header1...]
  [<INTERFACE|PUBLIC|PRIVATE> [header2...] ...])
```

The `PRIVATE`, `PUBLIC`, and `INTERFACE` keywords have the usual meaning. In the majority of the cases, `PRIVATE` should be used. The headers specified in the command will be collected in a `cmake_pch.h` or `cmake_pch.hxx` file in the build folder, which will be force-included in all source files by the respective compiler flag, so there is no need for the source files to have a `#include "cmake_pch.h"` directive.

The headers may be specified either as plain filenames, with angle brackets, or with double quotes, in which case they have to be escaped with double square brackets:

```
target_precompile_headers(SomeTarget PRIVATE myHeader.h
   [["external_header.h"]]
     <unordered_map>
)
```

In this example, `myHeader.h` would be searched from the current source directory, while `external_header.h` and `unordered_map` are searched for in the `include` directories.

In large projects, precompiled headers that are used in multiple targets are relatively common. Instead of redefining them every time, the `REUSE_FROM` option of `target_precompile_headers` can be used:

```
target_precompile_headers(<target> REUSE_FROM
  <other_target>)
```

Reusing precompiled headers will introduce an automatic dependency from `target` to `other_target`. Both targets will have the same compiler options, flags, and definitions enabled. Some compilers will warn if this is not the case, but some might not.

Precompiled headers from another target may only be used if the current target does not define its own set of precompiled headers. If the target already has precompiled headers defined, CMake will halt with an error.

Precompiled headers are most effective in improving build times when the headers included rarely change. Any headers provided by the compiler, system, or external dependencies are generally good candidates to include in precompiled headers. Which headers exactly bring the most benefit is something that needs to be tried out and measured.

Together with unity builds, precompiled headers can improve compile time significantly, especially for projects with frequent header reuse. A third way to optimize build time for incremental builds is the use of compiler caches, namely ccache.

Using a compiler cache (ccache) to speed up rebuilds

Ccaches work by caching compilations and detecting when the same compilation is done again. At the time of writing this book, the most popular program for caching compile results is the **ccache**, which is open source and distributed under the LGPL 3. Ccache not only affects incremental builds but also fresh builds, as long as the cache is not deleted between the two runs. The cache created is portable between systems running the same compilers and can be stored in remote databases, so multiple developers may access the same cache. Officially, ccache supports GCC, Clang, and NVCC but people claim to have run it for MSVC and Intel compilers. When using ccache with CMake, it works best with Makefile and Ninja generators. At the time of writing this book, Visual Studio was not supported.

To use ccache with CMake, the `CMAKE_<LANG>_COMPILER_LAUNCHER` cache variable is used, where `<LANG>` is replaced with the respective programming language. The recommended way is to pass this in using a preset, but to enable ccache for C and C++ inside `CMakeLists.txt`, the following code can be used:

```
find_program(CCACHE_PROGRAM ccache)
if(CCACHE_PROGRAM)
     set(CMAKE_C_COMPILER_LAUNCHER ${CCACHE_PROGRAM})
     set(CMAKE_CXX_COMPILER_LAUNCHER ${CCACHE_PROGRAM})
endif()
```

Passing the variable from a preset or from the command line is also a good alternative, especially because the configuration of ccache is done easiest by using environment variables.

Using ccache with the default configuration might already bring a considerable improvement regarding build times, but if the build is a bit more complex, further configuration might be necessary. To configure ccache, certain environment variables starting with `CCACHE_` can be used; for full documentation of all the configuration options refer to the ccache documentation. Common scenarios that need special attention are combining ccache with precompiled headers, managing dependencies that are included using `FetchContent`, and combining ccache with other compiler wrappers, such as *distcc* or *icecc* for distributed builds. For these scenarios, the following environment variables are used:

- To work efficiently with precompiled headers, set `CCACHE_SLOPPINESS` to `pch_defines,time_macros`. The reason for this is that ccache cannot detect changes in `#defines` in the precompiled header and it cannot tell whether `__TIME__`, `__DATE__`, or `__TIMESTAMP__` is used when creating precompiled headers. Optionally, `include_file_mtime` to `CCACHE_SLOPPINESS` might further increase the cache hit performance, but it carries a very small risk of a race condition.

- When including big dependencies that are built from source, for instance, by using `FetchContent`, setting `CCACHE_BASEDIR` to `CMAKE_BINARY_DIR` might increase the cache hit rate; this might bring a performance boost especially if there are many (sub)projects fetching the same dependency. On the other hand, if the sources in the project itself are the ones that take more time to compile, setting this to `CMAKE_SOURCE_DIR` might bring better results. It needs to be tried out to learn which one brings the better result.
- To work with other compiler wrappers, the `CCACHE_PREFIX` environment variable is used to add the commands for the other compiler wrapper. It is recommended to use ccache first when chaining multiple wrappers so the result of the other wrappers may also be cached.

Passing the environment variables to CMake using a configure preset, as described in *Chapter 9, Creating Reproducible Build Environments*, is the recommended way; this can either be combined with detecting ccache inside the `CMakeLists.txt`, or the `ccache` command may also be passed using the following preset:

```
{
"name" : "ccache-env",
...
  "environment": {
    "CCACHE_BASEDIR" : "${sourceDir}",
    "CCACHE_SLOPPINESS" : "pch_defines,time_macros"
  }
},
```

With these configurations, using ccache can yield very large benefits to the compile time, but caching compiler results is a complicated matter, so to get the full benefit, the ccache documentation should be consulted. In most cases, using ccache will probably bring the most performance benefit with a relatively trivial setup. Other tools, such as *distcc* for distributed builds, work very similarly from the CMake perspective, but require a bit more setup work.

Distributed builds

Distributed builds work by shoveling off some part of the compilation to different machines on a network. This requires setting up the servers that can accept connections and then configuring the clients to be able to connect to these servers. Setting up a server for *distcc* happens with the following command:

```
distccd --daemon --allow client1 client2
```

Here, `client1` and `client2` are the hostnames or IP addresses of the respective build servers. On the client side, configuring CMake to use *distcc* will work similarly to using ccache by setting `CMAKE_<LANG>_COMPILER_LAUNCHER` to the `distcc` command. The list of potential servers is either configured over a configuration file or by the `DISTCC_HOSTS` environment variable. Unlike the ccache configuration, this is very host-specific, so the configuration should be put into a user preset, not the project-specific preset. The respective preset might look like this:

```
{
"name" : "distcc-env",
...
  "environment": {
    "DISTCC_HOSTS" : "localhost buildsrvr1,cpp,lzo
      host123,cpp,lzo"
  }
},
```

Note the `cpp` postfix after the `buildsrvr1` host. This puts *distcc* into so-called *pump mode*, which further increases compilation speed by also distributing the preprocessing to the servers. The `lzo` postfix tells *distcc* to compress the communication.

The downside to distributed builds is that in order to gain a speed benefit, the network has to be fast enough or else the cost of transferring the information for compiling might be higher than the reduced build time. However, in most local networks, this is easily the case. Distributed builds work well if the machines are similar regarding processor architecture, compilers, and operating systems. While cross-compiling using *distcc* is possible, it can be quite a bit of work to set up. By combining good coding practices, precompiled headers and compiler caches working on large projects still work without waiting minutes for every single build.

Summary

In this chapter, we discussed some general tips on structuring and maintaining CMake projects, especially larger projects. With increased project size, configuring and building times usually increase, which could be a hindrance in the developer workflow. We looked at how the CMake profiling feature may be a useful tool to find performance hogs in the configuration process, even though it cannot be used to profile the compilation itself.

To help with long compilation times, we showed how to use unity build and precompiled headers from CMake to improve the compile times themselves. And if all this does not yet bring the desired effect, using a compiler cache such as ccache, or a distributed compiler such as *distcc*, may be used by prefixing the compiler command.

Optimizing build performance is a very satisfying affair, even if finding the right combination of tools and methods to get the most out of CMake might be a bit tedious. The downside of heavily optimized builds is, however, that the builds might be more prone to failure and the added complexity in the build process might need a deeper understanding and more expert knowledge to maintain in the long run.

In the next chapter, we will outline some high-level strategies for migrating from any build system to a CMake project.

Questions

Answer the following questions to test your knowledge of this chapter:

1. What command-line flags are used to generate profiling information from CMake?
2. On a very high level, how do unity builds optimize compile time?
3. What is the difference between the `BATCH` and `GROUP` modes for unity builds?
4. How are precompiled headers added to a target?
5. How does CMake handle compiler caches?

15
Migrating to CMake

While CMake is evolving into a de facto industry standard for C++ and C projects, there are still projects—sometimes large ones—that use different build systems. Of course, there is nothing wrong with that as long as it fits your needs. However, at some point, and for whatever reason, you might wish to switch to CMake. For instance, maybe the software should be buildable by different IDEs or on different platforms, or the dependency management has become cumbersome. Another common situation is when the repository structure changes from a big mono-repo, where all libraries are checked in, to distributed repositories for each library project. Whatever the reason, migrating to CMake can be a challenge, especially for large projects, but the results could be worth it.

While converting a project in one go would be the preferred way, often, there are non-technical requirements that might not make this possible. For instance, development might still need to go on in some parts during the migration, or some parts of the project cannot be migrated right from the start because of various requirements that are outside a team's control.

So, a step-by-step approach is often needed. Changing the build system will most likely influence any CI/CD process, so this should also be considered. In this chapter, we will take a look at a few high-level strategies regarding how projects could be migrated, step by step, to CMake. However, note that the concrete migration paths are very dependent on the individual situation. For example, migrating from a project based on Makefiles in a single repository will work differently from moving from a Gradle-based build that spans multiple repositories.

Changing the build system and, possibly, the project structure too can be very disruptive for all those involved, as they will have become accustomed to working with the existing structure and build system. Therefore, the decision to switch to a build system should not be taken lightly and should only be done if the benefits are significant.

While this chapter focuses on the CMake aspect of migrating projects, often, migrations are not done with the goal of switching the build system but have other primary goals instead, such as simplifying the project structure or reducing the coupling between parts of the project to make them easier to maintain independently. When talking about the benefits, remember that they do not necessarily have to be purely technical benefits, such as having faster build speeds as a result of being able to better parallelize the build. The benefits could also be more from a "social" side, for instance, having a standardized, well-known way to build software will reduce the ramp-up time for new developers.

In this chapter, we will cover the following topics:

- High-level migration strategies
- Migrating small projects
- Migrating large projects to CMake

In this chapter, we will introduce some higher-level concepts for migrating from any build system to CMake. As you will see, migrating small projects might be quite straightforward, whereas large, complex projects require more planning upfront. By the end of the chapter, you will have a good idea of the different strategies for migrating projects of various sizes to CMake. Additionally, we will provide a few hints regarding what to check for when migrating, along with a rough step-by-step guide for migrations and how to interact with legacy build systems.

Technical requirements

This chapter does not have specific technical requirements, as it shows concepts rather than concrete examples. However, it is recommended that when migrating to CMake, the newest version of CMake is used. The examples in this chapter assume that CMake 3.21, or newer, is being used.

High-level migration strategies

Before migrating a software project to CMake, first, it pays to answer a few questions about the existing project and define what the endpoint should look like. At a very high level of abstraction, usually, software projects define how the following things are handled:

- How the individual parts of the software, that is, the libraries and executables, are compiled and how they are linked together
- Which external dependencies are used, how they are found, and how they are used in the project
- Which tests to build and how to run them
- How the software is to be installed or packaged
- Providing additional information such as license information, documentation, changelogs, and more

Some projects might only define a subset of the preceding points. But typically, these are the tasks that we, as developers, want to be handled in a project setup. Often, these tasks are defined in a structured way such as with Makefiles or IDE-specific project definitions. There are countless ways regarding how projects are organized and structured, and what works for one setting might not work for another. So, in any case, an individual assessment of the situation is necessary.

A few tools exist that can automatically convert some build systems, such as qmake, Autotools, or Visual Studio, into CMake, but the quality of the resulting CMake files is doubtful at best, and they tend to assume certain conventions. Because of this, using them is not recommended.

Additionally, a project might define how it is built, tested, and deployed in a CI/CD pipeline, and while this is closely related, the definition of the CI/CD pipeline is not often seen as *part of* the project description, but rather as something that *uses* the project definition. Changing from one build system to another will invariably affect the CI/CD pipeline, and often, the desire to modernize or change the CI/CD infrastructure might be the trigger to change the build system.

It is important to realize that migrating is only complete when the old way of building is no longer used. So, we recommend removing any old build instructions once the projects are migrated to CMake, to eliminate the need to maintain backward compatibility with the old way of building.

In an ideal world, all parts of the project will be migrated to CMake. However, there are situations where this is not possible, or it is economically questionable if a part of a project actually should be migrated. For instance, a project could rely on a library that is no longer actively maintained and is destined to be phased out soon. The best case is that the migration effort can be used as a trigger to actually remove the dependency; however, more often than not, this is not feasible. In the cases where the legacy dependency cannot be completely removed, it might be a good idea to remove it from the project. This is so that it is no longer considered an internal dependency but an external dependency with its own release cycle. Also, if this is not possible or the effort is too big, making an exception for this particular library and using the legacy build system with `ExternalProject` might be the solution for a limited time. For the migration strategies discussed in this chapter, we differentiate between internal and external dependencies. Internal dependencies are those that are actively developed by the same organization or person of the project that is to be migrated, so the developers can potentially change the build process. External dependencies are those where developers have limited or no control over the build process or the code.

One thing to consider when migrating projects is how many people will be blocked from working on the project during its migration and how long the old way of building software and CMake have to be maintained side by side. Changing the build system is very intrusive for the developer's workflow. There will likely be times when some part of a project cannot be worked on until it has been fully migrated. The easiest way to work around this is to stop feature development for the moment and get everybody to help with the migration. However, if this is not possible, good communication and good partitioning of the work are often what is needed. Having said that, avoid the trap of stopping the migration halfway through: having some parts of a large project migrated while some are still using the old way to build is very likely to bring in the disadvantages of both ways to build the software while bringing the benefit of neither.

So, how do you go forward when migrating a project from one build system to another? For small projects that mainly have external dependencies, this might be quite straightforward.

Migrating small projects

We define *small projects* as any project that contains only a few targets and which are usually all deployed together. Small projects are self-contained inside a single repository, and usually, you can get a relatively quick overview of them. These might be projects that build a single library or an executable with a few external dependencies. In these cases, migrating to CMake is often relatively trivial. For small projects, in the first iteration, putting everything inside a single file will probably be the easiest way to go for a relatively quick and early result. Rearranging the files and splitting up the `CMakeLists.txt` file into multiple parts to be used with `add_subdirectory()` is much easier if the project has already been built correctly.

A general approach for migrating to CMake could be the following:

1. Create an empty `CMakeLists.txt` file inside the root of the projects.
2. Identify the targets and associated files in the project, and create the appropriate targets inside the `CMakeLists.txt` file.
3. Find all of the external dependencies and include paths, and add them to the CMake targets where necessary.
4. Identify the necessary compiler features, flags, and compiler definitions if any, and make them available to CMake.
5. Migrate any tests to CTest by creating the necessary targets and calling `add_test()`.
6. Identify any installation or packaging instructions to CMake, including the need to install any resource files and more.
7. Clean up and make the project nice. Create presets, rearrange files and folders if necessary, and split up the `CMakeLists.txt` file if needed.

Naturally, what exactly has to be done for each step depends very much on how the original project is organized and which technology is being used. Often, a migration will require several iterations of the `CMakeLists.txt` file until everything works, and if the first implementation of a CMake project does not yet look particularly nice, that is often quite normal.

For small projects, handling dependencies is one of the more difficult tasks to do, as there are some implicit assumptions about where to find the dependencies and how they are internally structured or hidden inside the project. Using a package manager, as described in *Chapter 5, Integrating Third-Party Libraries and Dependency Management*, might reduce the complexity of handling dependencies significantly.

Usually, the process of migrating small, mostly self-contained projects is relatively straightforward, although depending on how messy the original setup was, it might be quite a bit of work to get everything organized and working again. In larger organizations, several such smaller projects might be used together in a software portfolio, which, again, might be described as a project. Their migration needs a bit more planning to go forward.

Migrating large projects to CMake

Migrating large projects that contain a number of libraries and several executables can be quite a challenge. On a closer look, those projects might, in fact, be multiple hierarchically nested projects, with one or more root projects that pull together multiple subprojects, which, in turn, contain or require multiple subprojects themselves. Depending on the size and complexity of the software portfolio of an organization, many root projects that share common subprojects might exist side by side, which might complicate migration. Creating a dependency graph, such as the one in the following diagram, of the projects and subprojects often helps us to figure out the migration order. Each project might, in itself, contain multiple projects or targets that have their own dependencies:

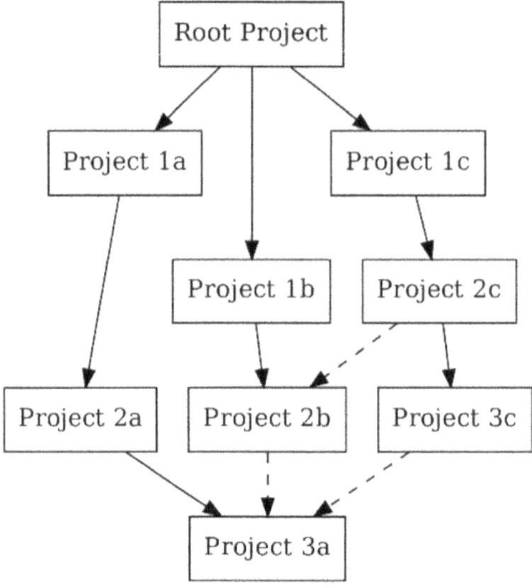

Figure 15.1 – An example project hierarchy showing the various dependencies

Before migrating, the first thing to do is a thorough analysis of how the projects depend on each other and in what order they need to be built. Depending on the state of the project, the resulting graph might be quite large, so figuring out where to start might be a bit of a challenge. In reality, dependency graphs are often not as neat as the ones shown in this book. Whether it is easier to first disentangle the project and then move to CMake or move to CMake first and then disentangle the project depends on the actual situation. If projects are very large and complex, start by finding islands in the graph that are as self-contained as possible and start from there.

For complex, hierarchical projects, there are two general migration strategies to consider. One is a top-down approach, where the root project(s) are migrated and replaced first, and then the child projects are ordered by which one project has the *least incoming* dependencies. The second is a bottom-up approach, where the individual projects are migrated one by one, starting with the one that has the *most incoming dependencies*.

A top-down approach will have the benefit of ensuring the full project can be built, tested, and packaged with CMake early on, but it requires the existing build system to be integrated into CMake with `ExternalProject`. The downside of a top-down approach might be that, in the early stages, the resulting CMake project contains a lot of custom code to work with packages built by the old system. In practice, using a few temporary workarounds to include existing projects in the build is often the most pragmatic way of getting good results relatively quickly, and it partially mitigates the effort of maintaining two build systems for the same subprojects.

A bottom-up approach has the benefit that each library that is migrated to CMake can use dependencies that have already been migrated. The downside is that the root project can only be replaced once all the child projects are made buildable with CMake. Even though the projects are migrated from the bottom up, a good practice is for the root CMake project to be created early on. It lives side by side with the root project in the original build system. This allows you to put in external dependencies and install configuration and packaging instructions inside the new CMake project early on.

The following diagram illustrates how the top-down and bottom-up strategies look side by side. The numbers beside the boxes represent the order of migration:

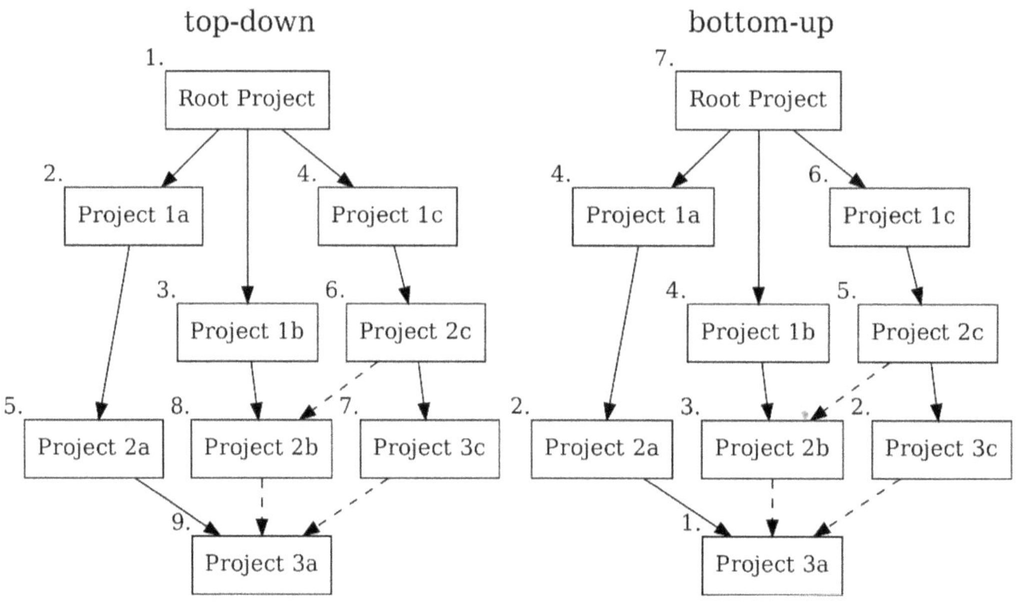

Figure 15.2 – An example of migration orders

Apart from the overall migration strategy, another thing to consider is whether the project is going to be set up as a superbuild or as a regular project. When working with a top-down approach, a superbuild structure might be easier to migrate, as one of its benefits is that it is easier to integrate non-CMake projects by design. More information about superbuild structures can be found in *Chapter 10, Handling Big Projects and Distributed Repositories in a Superbuild.*

No matter whether you choose a top-approach down or a bottom-up approach for migrating the individual projects, the general strategy for migrating large projects will look similar to the following:

1. Analyze what the dependencies, project hierarchy, and deployment units are.
2. Decide on a migration strategy and whether a regular project structure or a superbuild is intended.
3. Create or migrate the root project and pull in all not-yet-converted projects using `ExternalProject`, `FetchContent`, or intermediate find modules if working with binary packages.
4. Handle project-wide dependencies using CMake.

5. Convert the child projects, one by one, into CMake, as described in the last section of this chapter. If using intermediate find modules, replace them one by one:
    i. If desired, at this point, change the dependency handling to a package manager.
    ii. Find common options and propagate them toward the root and create presets.
6. When migrating the children, organize the packaging in CMake if not already done.
7. Clean up, reorganize the files and projects, improve performance, and more.

Starting the migration by analyzing the existing project hierarchy and dependencies will help you to set up a migration plan to communicate with all of the people involved. Creating a visualization such as the one from earlier is often a good tool, although, for very large projects, this can become quite a challenge in itself. Another important point to bear in mind when creating the migration plan is to identify what is commonly deployed together and which subproject has which frequency for releases. Projects that are rarely changed and released might not be as critical to migrate as those that are frequently touched and released. Identifying the deployment units is closely related to how projects are packaged. Depending on how this is organized, it might not be possible to migrate the packaging to CMake until all the projects have been migrated.

So far, we have mostly talked about subprojects, but while analyzing the existing structure, it is important to recognize which of the subprojects are actually projects that should be buildable as standalone and inside the full project context and which can be handled as regular CMake targets that are rarely built outside the project context.

Creating the root `CMakeLists.txt` file will cover the basic project setup and include the necessary modules such as `FetchContent`, `CTest`, `CPack`, and more. While not directly in the `CMakeLists.txt` file, setting up toolchain files, build containers, or sysroots for cross-compiling will also be done here. For large projects, often, the root `CMakeLists.txt` file does not contain the targets directly. Rather, it includes them either with `add_subdirectory` or `FetchContent`, or in the case of a superbuild, with `ExternalProject`. The root `CMakeLists.txt` file should have the following structure:

1. The **project definition** and the minimum required version of CMake.
2. **Global properties and default variables** such as the minimum language standard, the custom build types, and the search and module paths.
3. Any **modules and helper functions** that are used project-wide.

4. Project-wide **external dependencies** that are included via `find_package()`.
5. The **build targets and subprojects**, possibly added with `add_subdirectory`, `FetchContent`, or in the case of a superbuild, `ExternalProject`.
6. **Tests** for the whole project. Typically, each subproject will have its own unit tests, but integration or system tests might be at the top level.
7. **Packaging** instructions for CPack.

Depending on the complexity of the definitions, it might help to put the handling of external dependencies, tests, and packaging into their own subdirectories to keep the CMake file short and concise.

The external dependencies that are used project-wide might be large software frameworks, such as Qt or Boost, or small but common utility libraries that are used frequently.

For a top-down approach, the subprojects will be imported at the beginning and then migrated one by one. When using a bottom-up strategy, the build targets and subprojects will most likely be empty at the beginning and then become filled as the projects are migrated. When migrating the subprojects, keep your eyes open for common dependencies or build options that can be propagated to the root project or moved to presets.

Once all of the child projects have been migrated, typically, there are some maintenance tasks still open, such as organizing the packaging and orchestrating and grouping tests together. Also, it is not unusual for there to still be some clutter left in the CMake files after migrating everything, so having an extra round of cleaning up the centralizing functions will make sure that the migrated project is ready to use.

Often, migrating large projects is a challenge, especially if the build process is complicated and – unfortunately, as is frequently the case – lacks proper documentation. Software is built in many different ways, and the strategy described in this section tries to give a general approach. However, in the end, each migration will be unique in its own way. There are cases where the build system is so complex that the migration strategies described earlier are more of a hindrance than a help; for instance, including non-migrated projects into CMake is so difficult that a step-by-step migration might be more effort than just starting the build from scratch. Let's take a closer look at how the subprojects that use the original build system could be included when starting with a top-down approach.

Integrating legacy projects when migrating top down

For the top-down migration strategy, the existing projects are made available to CMake at the beginning. The easiest way is by using `ExternalProject`, regardless of whether a superbuild is intended or not. The imported targets can either be defined directly or with `find` modules. For regular projects, this is just an intermediate step to be able to build the full project relatively quickly and hand over control of the configuration and build order to CMake. The resulting CMake code might not look particularly nice, but the first goal is to get the root project building with CMake. However, be sure to clean it up, step by step, when migrating the subprojects. For regular projects that consist of a mono-repo or that pull in dependencies with Git submodules or similar, `ExternalProject_Add` might omit downloading by specifying the `SOURCE_DIR` property. The resulting CMake code for including the Autotools project might look like this:

```
include(ExternalProject)
set(ExtInstallDir ${CMAKE_CURRENT_BINARY_DIR}/install)
ExternalProject_Add(SubProjectXYZ_ext
    SOURCE_DIR ${CMAKE_CURRENT_LIST_DIR}/SubProjectXYZ/
    INSTALL_DIR ${ExtInstallDir}
    CONFIGURE_COMMAND <SOURCE_DIR>/configure --prefix <INSTALL_
      DIR>
    INSTALL_COMMAND make install
    BUILD_COMMAND make
)
add_library(SubProjectXYZ::SubProjectXYZ IMPORTED SHARED)
set_target_properties(SubProjectXYZ::SubProjectXYZ
    PROPERTIES IMPORTED_LOCATION " ${CMAKE_CURRENT_LIST_DIR}
      /SubProjectXYZ/lib/libSubProjectXYZ.so"
    INTERFACE_INCLUDE_DIRECTORIES "${CMAKE_CURRENT_LIST_DIR}
      /SubProjectXYZ/include"

   IMPORTED_LINK_INTERFACE_LANGUAGES "CXX"
)
...
add_dependencies(SomeTarget SubProjectXZY_ext)
target_link_libraries(SomeTarget SubProjectXYZ:: SubProjectXYZ)
```

As `ExternalProject` only makes the content available during build time, this approach only works for sub-projects that are already available in a local folder. Since they include directories of an imported target that have to exist at configuration time when using them in `target_link_libraries`, the exported location should point to the source directory rather than to the installation location of the external project.

> **These Practices Are for Temporary Workarounds**
>
> The practices using `ExternalProject` and `FetchContent` described here are meant for temporary workarounds to be able to include legacy projects in the CMake build while migrating. These are not good practices to use in a production environment. This pattern allows the use of the original build system and will provide an imported target to link against projects that have already been migrated. Whether the effort of creating such an intermediate project structure is justified by being able to build the full project with CMake early on, this has to be considered for each case separately.
>
> If migrating from Microsoft Visual Studio instead of using `ExternalProject`, the `include_external_msproject()` function might be used to directly include the project files.

With this, you should have all the concepts to migrate to CMake from another build system.

Summary

In this chapter, you learned about some concepts and strategies for migrating projects of various sizes to CMake. The effort and actual work to be done when migrating projects to CMake will very much depend on the individual setup of a project. However, with the approaches described here, choosing the right strategy will hopefully be easier. Changing the build processes and developer workflow is often disruptive, so you have to carefully consider whether the effort is worth it. Nevertheless, switching a project to CMake will open up the possibilities of all the features and practices for building quality software, as described in this book. Additionally, having a clean and well-maintained build system to work with will allow developers to focus on their main task, which is writing code and shipping software.

This brings us to the final chapter of this book, which is about getting access to the CMake community, finding further reading material, and contributing to CMake itself.

Questions

Answer the following questions to test your knowledge of this chapter:

1. What are the two main strategies for migrating large projects?
2. When choosing a bottom-up approach for migrating projects, which subprojects or targets are migrated first?
3. When choosing a top-down approach, which projects should be migrated first?
4. What are the advantage and disadvantages of a top-down approach?
5. What are the benefits and drawbacks of using a bottom-up approach for migrating?

16 Contributing to CMake and Further Reading Material

It has been a long journey since we started, and we have learned a great deal about CMake. But as you may have figured by now, CMake is a huge ecosystem, and one book is not enough to cover the vast number of topics that can be discussed. Thus, in this chapter, we are going to take a look at the resources that may help you to get to know CMake even better and also the ways of getting involved in the CMake project itself.

CMake is a flexible tool and used in many projects in the software industry. As a result, CMake has an ever-growing community supporting it. There are plenty of resources available online for learning and troubleshooting the CMake problems that you may experience.

To understand the skills shared in this chapter, we'll cover the following main topics:

- Where to find the CMake community
- Contributing to CMake
- Recommended books and blogs

Let's begin!

Prerequisites

This is a read-through chapter that has no practices or examples. So, the only requirement is a compatible device, a quiet place, and, of course, your time.

Where to find the CMake community

After getting involved with CMake, you may feel the need to exchange ideas and find a platform for asking questions of people that will likely know the answers. For that purpose, I have a few online platform recommendations for you.

Stack Overflow

Stack Overflow is a popular Q&A platform and the go-to place for most developers. If you are having an issue with CMake or have any questions, you can search for answers to your questions on Stack Overflow first. It is highly likely that someone has experienced that issue or asked the same or similar questions before. You can also take a look at the popular questions list to discover new ways of doing things with CMake.

When asking questions, ensure that you are labeling your questions with the `cmake` tag. This will allow individuals who are interested in answering CMake-related questions to find your question easier. You can access the Stack Overflow home page at `https://stackoverflow.com/`.

Reddit (r/cmake)

Reddit is a popular place that has dedicated, separate bulletin-like areas named subreddits to exchange ideas around topics. Reddit also has an `r/cmake` subreddit that contains CMake-specific questions, announcements, and shares. You can discover many useful GitHub repositories, get notified about recent CMake releases, and discover blog posts and materials that can help you. You can access the `r/cmake` subreddit at `https://www.reddit.com/r/cmake/`.

The CMake Discourse forum

The CMake Discourse forum is the main place for CMake developers and users to meet. It is completely dedicated to CMake-specific matters only. The forum contains announcements, guides on how to use CMake, a community space, a CMake development space, and much more content that you may be interested in. You can access the discourse forum at `https://discourse.cmake.org/`.

The Kitware CMake GitLab repository

Kitware's CMake repository is also a good resource for finding solutions for issues you may experience. Try searching for the issue you have in the issue list available at `https://gitlab.kitware.com/cmake/cmake/-/issues`. There is a good chance that somebody else may already have filed an issue about the topic. If that's not the case, you can create a new issue by adhering to CMake's contributing rules.

The preceding list is non-exhaustive and many more forums are available online. These four platforms will be sufficient to get you started. Next, we will talk about ways of contributing to the CMake project itself.

Contributing to CMake

As you already know, CMake is open source software developed by Kitware. Kitware maintains development activity for CMake in a dedicated GitLab instance at `https://gitlab.kitware.com/cmake`. Having everything available as open source and transparent means that getting involved in CMake is pretty easy. You can view the issues, merge requests, and get involved in the development of CMake. If you believe you have discovered a bug in CMake or want to make a feature request, you can create a new issue at `https://gitlab.kitware.com/cmake/cmake/-/issues`. If you have an idea about improving CMake, discuss that idea by creating an issue first. You can also take a look at open merge requests at `https://gitlab.kitware.com/cmake/cmake/-/merge_requests` to assist in reviewing the code being developed.

Contributing to open source software is crucial for the sustainability of the open source world. Please don't hesitate to help the open source community in whatever ways are convenient. The help you provide may only be small, but small contributions quickly add up to greater achievements. Next, we will look at reading material that you may find useful.

Recommended books and blogs

There are a lot of books, blogs, and resources regarding CMake. The following is a curated list of hand-selected resources that you may find useful. This list will allow you to discover more about CMake and expand your horizons:

- *CMake official documentation*: `https://cmake.org/documentation/`

  The official documentation for CMake. It is very broad and up to date.
- `awesome-cmake`: `https://github.com/onqtam/awesome-cmake`

  A vast curation of CMake-related resources. It is very extensive and regularly updated.

- *Getting Started with CMake: Helpful Resources*: `https://embeddedartistry.com/blog/2017/10/04/getting-started-with-cmake-helpful-resources/`

  A curation of helpful CMake resources gathered together by Embedded Artistry.

- *An Introduction to Modern CMake*: `https://cliutils.gitlab.io/modern-cmake/`

  An online book and a good resource about learning modern CMake in detail. It is written by Henry Schreiner and many other contributors.

- *More Modern CMake*: `https://hsf-training.github.io/hsf-training-cmake-webpage/01-intro/index.html`

  This is a follow-up book to *An Introduction to Modern CMake*, written by the HEP Software Foundation.

- *More Modern CMake*: `https://www.youtube.com/watch?v=y7ndUhdQuU8`

  This YouTube video is a presentation performed by Deniz Bahadir at Meeting C++ 2018. Its main aim is to give tips about using CMake correctly.

- *Deep CMake for Library Authors*: `https://www.youtube.com/watch?v=m0DwB4OvDXk`

  This YouTube video is a CppCon talk given by Craig Scott, co-maintainer of the CMake project. It covers CMake topics oriented toward library development.

- Daniel Pfeifer's *Effective CMake*: `https://www.youtube.com/watch?v=bsXLMQ6WgIk`

  This YouTube video is a talk given by Daniel Pfeifer about using CMake effectively. It covers overall CMake usage.

- *Professional CMake: A Practical Guide*: `https://crascit.com/professional-cmake/`

  A comprehensive book written by CMake's co-maintainer, Craig Scott. It is very extensive and contains many details that you can't find elsewhere.

- `learning-cmake`: `https://github.com/Akagi201/learning-cmake`

  This repository has a collection of examples for the purpose of learning different use cases in CMake.

- `cmake-examples`: `https://github.com/ttroy50/cmake-examples`

  Another good collection of examples for the purpose of learning different use cases in CMake.

With that said, we have reached another chapter's end. Next, we'll summarize what we have learned in this chapter.

Summary

In this chapter, we briefly discussed the CMake communities that you can find online, contributing to CMake, and also good recommendations for reading and watching. There are tons of material and talks about CMake, and the content is growing day by day. Always keep an eye out for updates about CMake and regularly visit your forums of choice to keep yourself in the loop.

With that said, if you are here and reading this paragraph, then congratulations! You have reached the end of the topics we wanted to cover in this book. This was the last chapter to cover. Don't forget to apply and practice what have you learned from this book in your everyday workflow. We have enjoyed the journey we were on together and hope that the knowledge you have gained from this book proves itself useful.

Assessments

This section is for answers to questions from all chapters.

Chapter 1, Kickstarting CMake

1. `cmake -S /path/to/source -B /path/to/build`.
2. `cmake -build /path/to/build`.
3. `ctest`.
4. `cpack`.
5. Targets are logical units around which CMake organizes a build. They can be executables, libraries, or contain custom commands.
6. Unlike variables, properties are attached to a specific object or scope.
7. CMake presets are used to share working configurations for a build.

Chapter 2, Accessing CMake in the Best Ways

1. Here are the answers:

   A. `cmake -S . -B ./build -DCMAKE_CXX_COMPILER:STRING= "/usr/bin/clang++ "`

   B. `cmake -S . -B ./build -G "Ninja"`

   C. `cmake -S . -B ./build -DCMAKE_BUILD_FLAGS_DEBUG:STRING= "-Wall"`

2. The project previously configured in question 1 can be built using CMake via the command line, as follows:

   A. `cmake --build ./build --parallel 8`

   B. `cmake --build ./build -- VERBOSE=1`

3. `cmake --install ./build --prefix=/opt/project.`
4. `cmake --install ./build --component ch2.libraries.`
5. It is a CMake cache variable that is marked as *advanced* to hide it in GUIs via the `mark_as_advanced()` function.

Chapter 3, Creating a CMake Project

1. `add_executable.`
2. `add_library.`
3. By adding the `SHARED` or `STATIC` keyword or by setting the `BUILD_SHARED_LIBS` global variable.
4. Object libraries are libraries that are compiled but not linked. They are useful for internally separating code and reducing compile time.
5. By globally setting the `<LANG>_VISIBILITY_PRESET` property.

Chapter 4, Packaging, Deploying, and Installing a CMake Project

1. It can be achieved via the `install(TARGETS <target_name>)` command.
2. The output artifacts of the specified targets.
3. No, because header files are not classified as output artifacts of a target. They must be installed separately via the `install(DIRECTORY)` command.
4. The `GNUInstallDirs` CMake module provides system-specific default paths for installation, such as `bin`, `lib`, and `include`.
5. With the help of the `PATTERN` and `FILES_MATCHING` parameters of the `install(DIRECTORY)` command.

Chapter 5, Integrating Third-Party Libraries and Dependency Management

1. The answer is `find_file`, `find_path`, `find_library`, `find_program`, and `find_package`.
2. The `IMPORTED_LOCATION` and `INTERFACE_INCLUDE_DIRECTORIES` properties.
3. `HINTS` takes precedence over `PATHS`.

4. `ExternalProject` downloads external content at build time.
5. `FetchContent` downloads external content at configuration time.

Chapter 6, Automatically Generating Documentation

1. Doxygen is the de facto standard of documentation generation tools for C and C++ projects.
2. Since CMake already provides a `find` module for Doxygen, this can be done by using the `find_package(...)` CMake command.
3. Yes – Doxygen can draw graphs, given that graphing software such as DOT, Graphviz, and PlantUML are available in the environment. To enable DOT graphing, setting `HAVE_DOT` to `TRUE` is sufficient. For PlantUML, `PLANTUML_JAR_PATH` needs to be set to the path that contains the `plantuml.jar` file.
4. `@startuml` and `@enduml`.
5. `PLANTUML_JAR_PATH` needs to be set to the path that contains the `plantuml.jar` file.
6. With the help of the `install(DIRECTORY)` command.

Chapter 7, Seamlessly Integrating Code-Quality Tools with CMake

1. Tests are defined by using the `add_test` function.
2. Either by using a regular expression on the test's name with `ctest -R` or by using the test number using `ctest -I`.
3. By calling `ctest --repeat:until-pass:n` or `ctest --repeat:until-fail:n`.
4. By running `ctest -j <num_of_jobs> --schedule-random`.
5. By setting the `RESOURCE_LOCK` or `RESOURCE_GROUP` property for the respective tests.
6. Static code analyzers are enabled by passing the command line including any arguments to the respective target properties.
7. By either adding them to the `CMAKE_CONFIGURATION_TYPES` property for multi-configuration type generators or by adding them to the `CMAKE_BUILD_TYPE` property.

Chapter 8, Executing Custom Tasks with CMake

1. Commands that are added with `add_custom_command` are executed at build time, while commands that are added with `execute_process` are executed at configuration time.
2. One signature is used to create custom build steps, while the other is used to generate files.
3. Only `POST_BUILD` is reliably supported across all generators.
4. Variables can be defined either as `${VAR}` or as `@VAR@`.
5. Variable substitution can be controlled by either passing `@ONLY`, which only replaces variables defined as `@VAR@`, or by specifying the `COPYONLY` option, which does not perform any substitution at all.
6. With `cmake -E`, common tasks can be executed directly. With `cmake -P`, `.cmake` files can be executed as scripts.

Chapter 9, Creating Reproducible Build Environments

1. `CMakePresets.json` is usually maintained and delivered together with the project, while `CMakeUserPresets.json` is maintained by the user. Regarding the syntax and the contents, there is no difference.
2. By calling `cmake --preset=presetName`, `cmake --build --preset=presetName`, or `ctest --preset=presetName`.
3. There are configure, build, and test presets. Build and test presets depend on a configure preset to determine the `build` directory.
4. A configure preset should define a name, the generator, and the build directory to be used.
5. The first preset to set a value takes precedence.
6. Either using the native support of an editor for build containers, running the editor from within the container, or starting the container each time for invoking single commands inside.
7. The system name, the location of sysroot, the compilers to use, and how the `find_` commands behave.

Chapter 10, Handling Big Projects and Distributed Repositories in a Superbuild

1. A super-build is a method of building a software project that spans multiple repositories.
2. Where we don't have a package manager and want to make the project able to satisfy its own dependencies.
3. Use anchor points such as branches, tags, or commit hashes.

Chapter 11, Automated Fuzzing with CMake

1. Fuzzing is a testing technique that is based on feeding computer-generated data to a system or function to check whether the target behaves as expected.
2. Corpus data is the set of all interesting inputs that persisted between fuzzing runs. The corpus can grow over time.
3. Via passing `-fsanitize=fuzzer` to both compiler and linker flags of the target (`target_compile_options` and `target_link_options`)
4. OpenSSL – Heartbleed, and Bash – Shellshock

Chapter 12, Cross-Platform-Compiling Custom Toolchains

1. Toolchain files are passed either by the `--toolchain` command-line flag, the `CMAKE_TOOLCHAIN_FILE` variable, or with the `toolchainFile` option in a CMake preset.
2. Usually, the following things are done in a toolchain file for cross-compiling:

   A. Defining the target system and architecture
   B. Providing paths to any tools needed to build the software
   C. Setting default flags for the compiler and linkers
   D. Pointing to the sysroot and possibly any staging directory if cross-compiling
   E. Setting hints for the search order for any `find_` commands of CMake

3. The staging directory is set with the `CMAKE_STAGING_PREFIX` variable and is used as a place to install any built artifacts if the sysroot should not be modified.

4. The emulator command is passed as a semicolon-separated list in the `CMAKE_CROSSCOMPILING_EMULATOR` variable.
5. Any call to `project()` or `enable_language()` in a project will trigger detection of the features.
6. The configuration context for compiler checks can be stored with `cmake_push_check_state()` and restored to a previous state with `cmake_pop_check_state()`.
7. If `CMAKE_CROSSCOMPILING` is set, any call to `try_run()` will only compile the test but not run it unless an `emulator` command is set.
8. Build directories should be fully cleared because the temporary artifacts for compiler checks might not be rebuilt properly when just deleting the cache.

Chapter 13, Reusing CMake Code

1. Functions and macros.
2. A CMake module is a logical entity that contains CMake code, functions, and macros to serve a particular purpose.
3. By including it in the required scope.
4. To add additional paths to the `include(...)` directive's search path.
5. By using `git` submodules/subtrees or CMake's `FetchContent/ExternalProject`.
6. Functions define a new variable scope; macros do not.

Chapter 14, Optimizing and Maintaining CMake Projects

1. The --profiling-output <filename> --profiling-format=google-trace flags are used.
2. By grouping together various compilation units into a single one, the need for relinking happens less often.
3. In `BATCH` mode, CMake automatically groups the sources together; in `GROUP` mode, the grouping has to be specified by the user. By default, `BATCH` mode groups all sources for a unity build, while `GROUP` only adds the explicitly marked files to the unity build.

4. By using the `target_precompile_headers` function. Precompiled headers are automatically included, without the need for a `#include` directive in the files.
5. By prefixing the compiler command with the command specified in `CMAKE_<LANG>_COMPILER_LAUNCHER`.

Chapter 15, Migrating to CMake

1. Large projects can be migrated from the top down or the bottom up.
2. When working with a bottom-up approach, those projects or targets with the most incoming dependencies should be migrated first.
3. When choosing a top-down approach, the projects with the least incoming dependencies should be migrated first.
4. Top-down approaches quickly allow you to build the whole project using CMake as an entry point. Additionally, for each migrated project, the old build system can be discarded when the project is done. The downside is that a top-down approach will require some intermediate code.
5. A bottom-up approach will require less intermediate code than a top-down approach and allow for clean CMake code right from the start. The downside is that the full project can only be built when all of the subprojects have been migrated.

Index

A

B

C

D

E

F

G

H

I

J

K

L

M

N

O

P

Q

R

S

W

Y

Packt.com

Subscribe to our online digital library for full access to over 7,000 books and videos, as well as industry leading tools to help you plan your personal development and advance your career. For more information, please visit our website.

Why subscribe?

- Spend less time learning and more time coding with practical eBooks and Videos from over 4,000 industry professionals
- Improve your learning with Skill Plans built especially for you
- Get a free eBook or video every month
- Fully searchable for easy access to vital information
- Copy and paste, print, and bookmark content

Did you know that Packt offers eBook versions of every book published, with PDF and ePub files available? You can upgrade to the eBook version at `packt.com` and as a print book customer, you are entitled to a discount on the eBook copy. Get in touch with us at `customercare@packtpub.com` for more details.

At `www.packt.com`, you can also read a collection of free technical articles, sign up for a range of free newsletters, and receive exclusive discounts and offers on Packt books and eBooks.

Other Books You May Enjoy

If you enjoyed this book, you may be interested in these other books by Packt:

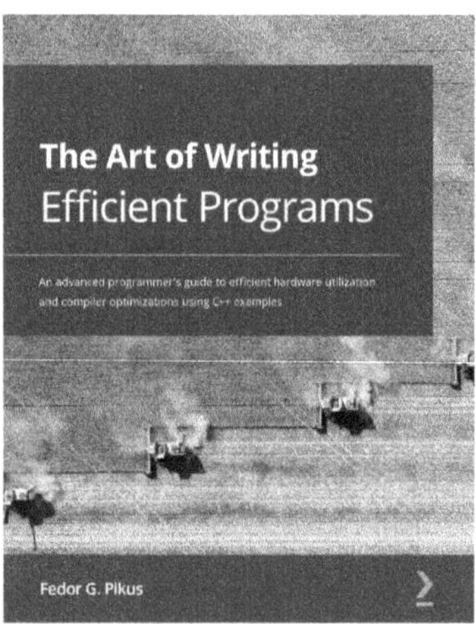

The Art of Writing Efficient Programs

Fedor G. Pikus

ISBN: 978-1-80020-811-7

- Discover how to use the hardware computing resources in your programs effectively
- Understand the relationship between memory order and memory barriers
- Familiarize yourself with the performance implications of different data structures and organizations
- Assess the performance impact of concurrent memory accessed and how to minimize it
- Discover when to use and when not to use lock-free programming techniques
- Explore different ways to improve the effectiveness of compiler optimizations
- Design APIs for concurrent data structures and high-performance data structures to avoid inefficiencies

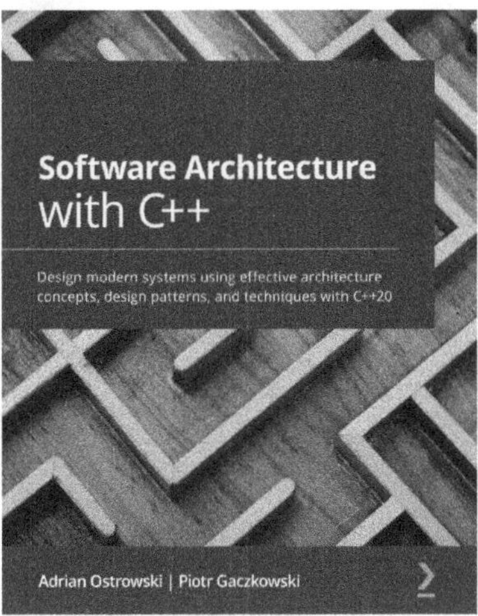

Software Architecture with C++

Adrian Ostrowski, Piotr Gaczkowski

ISBN: 978-1-83855-459-0

- Understand how to apply the principles of software architecture
- Apply design patterns and best practices to meet your architectural goals
- Write elegant, safe, and performant code using the latest C++ features
- Build applications that are easy to maintain and deploy
- Explore the different architectural approaches and learn to apply them as per your requirement
- Simplify development and operations using application containers
- Discover various techniques to solve common problems in software design and development

Packt is searching for authors like you

If you're interested in becoming an author for Packt, please visit `authors.packtpub.com` and apply today. We have worked with thousands of developers and tech professionals, just like you, to help them share their insight with the global tech community. You can make a general application, apply for a specific hot topic that we are recruiting an author for, or submit your own idea.

Share Your Thoughts

Now you've finished *CMake Best Practices*, we'd love to hear your thoughts! Scan the QR code below to go straight to the Amazon review page for this book and share your feedback or leave a review on the site that you purchased it from.

https://packt.link/r/1-803-23972-7

Your review is important to us and the tech community and will help us make sure we're delivering excellent quality content.

www.ingramcontent.com/pod-product-compliance
Lightning Source LLC
Chambersburg PA
CBHW062034050326
40690CB00016B/2940

* 9 7 8 1 8 0 3 2 3 9 7 2 9 *